Sustainable Manufacturing

Handbooks in Advanced Manufacturing

Sustainable Manufacturing

Edited by

KAPIL GUPTA
Mechanical and Industrial Engineering Technology,
University of Johannesburg, Johannesburg,
Republic of South Africa

KONSTANTINOS SALONITIS
Sustainable Manufacturing Systems Centre, School of
Aerospace, Transport and Manufacturing, Cranfield
University, Bedfordshire, United Kingdom

Series Editors-in-Chief

J. PAULO DAVIM

KAPIL GUPTA

ELSEVIER

Elsevier
Radarweg 29, PO Box 211, 1000 AE Amsterdam, Netherlands
125 London Wall, London, EC2Y 5AS, United Kingdom
50 Hampshire Street, 5th Floor, Cambridge, MA 02139, United States

British Library Cataloguing-in-Publication Data
A catalogue record for this book is available from the British Library

Library of Congress Cataloging-in-Publication Data
A catalog record for this book is available from the Library of Congress

ISBN: 978-0-12-818115-7

For Information on all Elsevier publications
visit our website at https://www.elsevier.com/books-and-journals

Publisher: Matthew Deans
Acquisitions Editor: Brian Guerin
Editorial Project Manager: John Leonard
Production Project Manager: Sojan P. Pazhayattil
Cover Designer: Victoria Pearson Esser

Typeset by MPS Limited, Chennai, India

Contents

List of contributors

Rajeev Agrawal
Department of Mechanical Engineering, Malaviya National Institute of Technology, J.L.N. Marg, Jaipur, India

Mahmood Anwar
School of Engineering and Science, CDT250, Curtin University, Miri, Malaysia

Peter Ball
University of York, York, United Kingdom

Animesh Kumar Basak
Adelaide Microscopy, University of Adelaide, Adelaide, SA, Australia

Senol Bayraktar
Department of Mechanical Engineering, Faculty of Engineering and Architecture, Recep Tayyip Erdogan University, Rize, Turkey

Berat Baris Buldum
Engineering Faculty, Department of Mechanical Engineering, Mersin University, Mersin, Turkey

Fiona Charnley
Exeter Center for Circular Economy, Science, Innovation, Technology and Entrepreneurship Department, The Business School, University of Exeter, Exeter, United Kingdom

Suleyman Cinar Cagan
Engineering Faculty, Department of Mechanical Engineering, Mersin University, Mersin, Turkey

Sujan Debnath
School of Engineering and Science, CDT250, Curtin University, Miri, Malaysia

Amir Dehghanghadikolaei
School of Mechanical, Industrial and Manufacturing Engineering, Oregon State University, Corvallis, OR, United States

Steve Evans
Institution for Manufacturing, University of Cambridge, United Kingdom

Behzad Fotovvati
Department of Mechanical Engineering, The University of Memphis, Memphis, TN, United States

Neelesh Kumar Jain
Department of Mechanical Engineering, Indian Institute of Technology Indore, Simrol, India

Anbesh Jamwal
Department of Mechanical Engineering, Malaviya National Institute of Technology, J.L.N. Marg, Jaipur, India

Jayakrishna Kandasamy
School of Mechanical Engineering, VIT University, Vellore, India

Vishal Kharka
Department of Mechanical Engineering, Indian Institute of Technology Indore, Simrol, India

Xinyi Kuang
Institution for Manufacturing, University of Cambridge, United Kingdom

Manish Kumar
Centre for Product Design and Manufacturing, Indian Institute of Science, Bangalore, India

E. Manavalan
School of Mechanical Engineering, VIT University, Vellore, India

Monto Mani
Centre for Sustainable Technologies and Centre for Product Design and Manufacturing, Indian Institute of Science, Bangalore, India

Mehrshad Mehrpouya
Faculty of Engineering Technology, University of Twente, Enschede, the Netherlands

Okechukwu Okorie
Exeter Center for Circular Economy, Science, Innovation, Technology and Entrepreneurship Department, The Business School, University of Exeter, Exeter, United Kingdom

Michail Papanikolaou
Sustainable Manufacturing Systems Centre, Manufacturing Theme, Cranfield University, Cranfield, United Kingdom

Lijo Paul
Department of Mechanical Engineering, St. Joseph's College of Engineering and Technology Palai, Choondacherry, India

M.E. Peralta
Department of Design Engineering, University of Seville, Seville, Spain

Handson Pimenta
Federal Institute of Education, Science and Technology Rio Grande do Norte, Brazil

Alokesh Pramanik
School of Civil and Mechanical Engineering, Curtin University, Bentley, WA, Australia

Konstantinos Salonitis
Sustainable Manufacturing Systems Centre, School of Aerospace, Transport and Manufacturing, Cranfield University, Bedfordshire, United Kingdom

Prateek Saxena
Sustainable Manufacturing Systems Centre, Manufacturing Theme, Cranfield University, Cranfield, United Kingdom

Jilse Sebastian
Department of Mechanical Engineering, St. Joseph's College of Engineering and Technology Palai, Choondacherry, India

Monica Sharma
Department of Mechanical Engineering, Malaviya National Institute of Technology, J.L.N. Marg, Jaipur, India

V. Soltero
Department of Design Engineering, University of Seville, Seville, Spain

M. Thanigai Arasu
School of Mechanical Engineering, VIT University, Vellore, India

Janak Valaki
Mechanical Engineering Department, Government Engineering College, Bhavnagar, India

Alireza Vosooghnia
Department of Civil and Environmental Engineering, Sapienza University of Rome, Rome, Italy

Miying Yang
College of Engineering, Mathematics and Physical Sciences, University of Exeter, United Kingdom

Series Foreword

Dear Readers,

This series of handbooks on advanced manufacturing covers four major areas, namely, advanced machining and finishing, advanced welding and deforming, additive manufacturing, and sustainable manufacturing. The series aims to not only present the advancements in various manufacturing technologies but also provide a fundamental and detailed understanding about them. It encompasses a wide range of manufacturing technologies with their mechanisms, working principles, salient features, applications, and research, development, and innovations in there. Fundamental research, latest developments, and case studies conducted by international experienced researchers, engineers, managers, and professors are mainly presented. *Handbook 1* on advanced machining and finishing majorly covers advanced machining of difficult-to-machine materials; hybrid, high speed, and micromachining; and burnishing, laser surface texturing, and advanced thermal energy–based finishing processes. *Handbook 2* on advanced welding and deforming covers ultrasonic welding, laser welding, and hybrid welding type advanced joining processes and also describes advanced forming techniques such as microwave processing, equal channel angular pressing, energy-assisted forming, and so on. *Handbook 3* additive manufacturing sheds light on 3D and 4D printing, rapid prototyping, laser-based additive manufacturing, advanced materials and postprocessing in additive manufacturing. *Handbook 4* on sustainable manufacturing presents advancements, results of experimental research, and case studies on sustainability interventions in production and industrial technologies.

We hope that this series of handbooks would be a good source of knowledge and encourage researchers and scientists to conduct research, developments, and innovations to establish these fields further.

J. Paulo Davim and Kapil Gupta

Preface

Sustainability is a prime concern these days. To comply with strict environment regulations and considering a social responsibility, manufacturers are busy adopting various sustainable manufacturing practices and techniques. Whether its sustainable manufacturing of products or manufacturing of sustainable products to attain a clean, green, and safe environment is the major objective. This handbook details some of the important sustainable manufacturing practices and techniques. Basic knowledge, case studies, literature review, technological developments, latest research, and future directions are covered.

The handbook consists of a total of 15 selected chapters on sustainable manufacturing. It starts with Chapter 1 that introduces sustainable manufacturing and suggests some future directions. Chapter 2 highlights sustainability interventions in additive manufacturing. A case study on sustainable grinding is detailed in Chapter 3. Roles of remanufacturing and refurbishment in this era of industry 4.0 are discussed in Chapter 4. Chapter 5 presents effectiveness of nanofluid-based minimum quantity lubrication system for sustainable machining operations. Chapter 6 discusses how circular economy can be a promising solution to achieve sustainability. Simulation-driven optimization of sustainable casting processes are described in Chapter 7. Chapter 8 is on diffusion of sustainability practices through supplier performance assessment. Various aspects of dry cutting as a sustainable machining technology are highlighted in Chapter 9. Chapter 10 presents cryogenic cooling–based sustainable machining. Chapter 11 details assessment of sustainability in manufacturing. A comprehensive detail on life cycle engineering is given in Chapter 12. Recycling and regeneration of plastics are discussed in Chapter 13. Chapter 14 discusses various aspects of sustainable supply chain management. Finally, Chapter 15 presents a case study on achieving sustainability in cylindrical gear machining using minimum quantity lubrication.

We hope that this handbook would be a good source of knowledge and information for researchers, engineers, scholars, and technical experts and specialists working in the area of sustainable manufacturing.

We sincerely acknowledge Elsevier for this opportunity and their professional support. Finally, we would like to thank all contributors for their time and efforts.

Kapil Gupta and Konstantinos Salonitis

January 2021

CHAPTER 1

Sustainable manufacturing: needs for future quality development

M.E. Peralta and V. Soltero
Department of Design Engineering, University of Seville, Seville, Spain

1.1 Manufacturing and sustainable future

The Horizon 2020 programs (Filos, 2013) and the set of lines of research related to Sustainable Manufacturing Engineering reflect the interest and need to manage the life cycle of products and processes from the point of view of sustainable development. The proposal of new forms of manufacturing fabrication (Rahimifard, Sheldrick, Woolley, Colwill, & Sachidananda, 2013) through clean technologies and the continuous improvement and optimization of existing ones (including sustainability criteria), has oriented research in recent years to the search for new methods of design, modeling, simulation and optimization; these should allow the integration of the ecological and social dimensions at operational level within the manufacturing processes, maintaining the economic and technical viability achieved so far.

This growing interest in new forms of sustainable action is motivated by the problems derived from the manufacturing industry and its installations; the set of impacts and damages generated is due to its configuration as an open system in terms of linear flow. Moreover, the way of approaching manufacturing processes (Clark & Foster, 2009) without considering the nature and biosphere recovery cycles has generated a complex problem with different consequences: *high environmental, social, and economic impact (characterized by the expenditure of energy resources and materials), environmental degradation, destruction of ecosystems, loss relative to biodiversity, deterioration of the population's health, minimization of well-being and quality of life.*

Sustainable Manufacturing
DOI: https://doi.org/10.1016/B978-0-12-818115-7.00011-0

This situation is called by Marxist Ecology as metabolic rift (Peralta Álvarez, Marcos Bárcena, & Aguayo González, 2016); or the metric that describes the system where the biosphere (nature) and the Technosphere (industry) coexist with different rhythms of evolution, although conserving their dependence.

The metabolic rift is a consequence of the operating mode of the productive systems in the successive industrial revolutions. Seeking an improvement in needs with the aim of speeding up production times and maximizing demand, human beings separate themselves from the ecosystem by changing their "nature": **society and its activities are now an isolated and modifying element of the environment** (Foster, 2000; Marx, 1894). This implies, as Fig. 1.1 shows, that the Biosphere and the Technosphere coexist in an uncoupled way, although maintaining their dependence. Specifically, the Technosphere inefficiently subtracts materials, resources and energy from the nature sphere; moreover, this causes environmental and social impacts resulting from poor management and design of productive activities.

Gradually, Marx's concept of metabolic rift has evolved into a more operational strategy called Triple Bottom Line (3E-TBL) (Elkington, 1998) that allows to design and execute concrete solutions in sustainable manufacturing (as shows Fig. 1.2).

In order to understand the evolution of manufacturing systems and the origin of sustainable manufacturing, there is a very interesting reading that analyses production systems throughout history, identifying in detail how *technology, work, and their conception* transform the environment and place it at the service of human needs and economic growth. Lewis Mumford

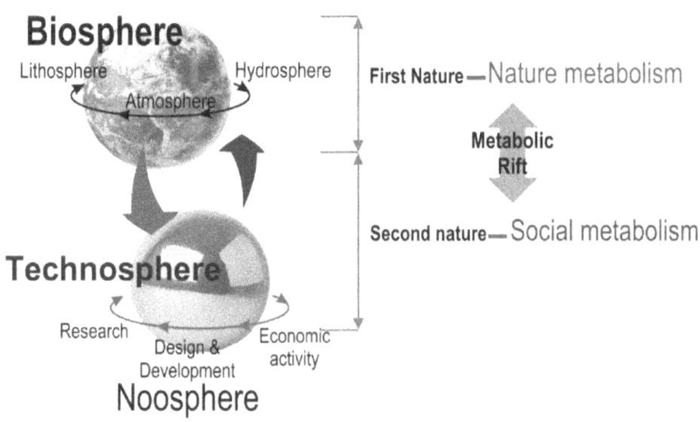

Figure 1.1 Separation of technical, social and natural metabolisms.

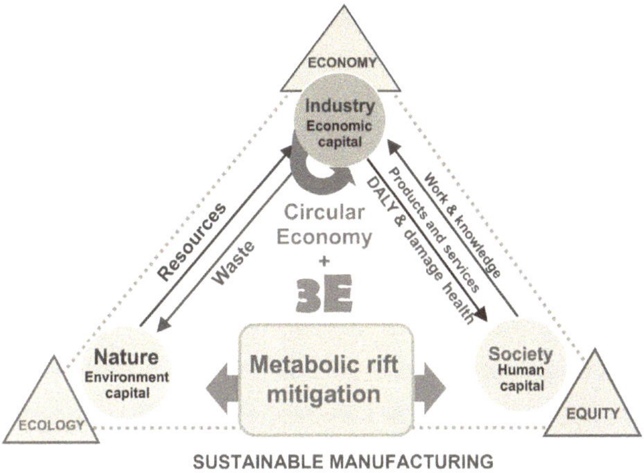

Figure 1.2 Evolution towards the triple bottom line (3E-TBL) framework.

(1934) in his book "Technic and Civilization" poses a set of questions whose answers define the evolution from "manufacturing" to "sustainable manufacturing":

> *.../... During the last thousand years the material basis and the cultural forms of Western civilization have been profoundly modified by the development of the machine. How did this come about? Where did it take place? What were the chief motives that encouraged this radical transformation of the environment and the routine of life: what were the ends in view: what were the means and methods: what unexpected values have arisen in the process? .../...*
>
> ***(Mumford, 1934)***

The answers to the preceding questions have been identified and some have been solved in the last two decades with the implementation of new sustainable values, although one of the major questions remains unresolved:

> *.../... By now, it is plain, a new world has come into existence; but it exists only in fragments. New forms of living have for long been in process; but so far they have likewise been divided and unfocussed: indeed, our vast gains in energy and in the production of goods have manifested themselves in part in a loss of form and an impoverishment of life. What has limited the beneficence of the machine? Under what conditions may the machine be directed toward a fuller use and accomplishment? .../... Can one distinguish and define the specific properties of a technics directed toward the service of life: properties that distinguish it morally, socially, politically, aesthetically from the cruder forms that preceded it? .../...*
>
> ***(Mumford, 1934)***

This question defines the fundamental principle of Sustainable Manufacturing:

BOX 1.1. Fundamental principle of Sustainable Manufacturing
Identify the necessary changes in the manufacturing processes, in the type and quantity of resources used and in the management (prevention, reduction and control) of wastes and emissions generated; all this taking into account social, environmental, and economic factors with the aim of achieving a cyclical, innocuous and efficient exchange of goods and services both in the contexts of exploitation (zero direct impact) and in the contexts affected (zero indirect impacts or externalities).

1.2 From manufacturing to sustainable manufacturing

Mumford divides the evolution of society around the technique into three stages: Eotechnic, Paleotechnic and Neotechnic. The first phase, Eotechnic, places the origin of the machine in the 11th century and the ideological preparation of the population; this stage guides the change towards the new system of values and forms of work characteristic of later phases, that is: *putting the environment and the planet at the disposal of human needs*. The following phases, Paleotechnic (in which the First and Second Industrial Revolutions take place) and Neotechnic (with the Third and Fourth Industrial Revolutions) correspond to the consolidation, universalization and improvement of *manufacturing and technology* as the central axis of the economy. Evolution and characteristics are represented in Fig. 1.3.

In the Neotechnic phase, the concept of sustainability was born, extending and reaching all social scales, becoming an indispensable requirement in any governmental, industrial, business, economic, or social activity. In addition, with the proposal of the Brundtland Report (United Nations, 1987), *manufacturing engineering* begins an exhaustive search for the most efficient modes of production that can satisfy the needs of the present without compromising the needs of future generations.

Trying to solve the groups of impacts identified in Fig. 1.3, the manufacturing systems have undergone a set of changes that can be classified according to the orientation and approach of the solutions in four phases: reactive, constructive, proactive eco-efficient and proactive eco-effective manufacturing; these define the evolution towards "SUSTAINABLE MANUFACTURING".

PHASE I: Reactive manufacturing. Because of the pressure of some environmental and ecological institutions and organizations, the

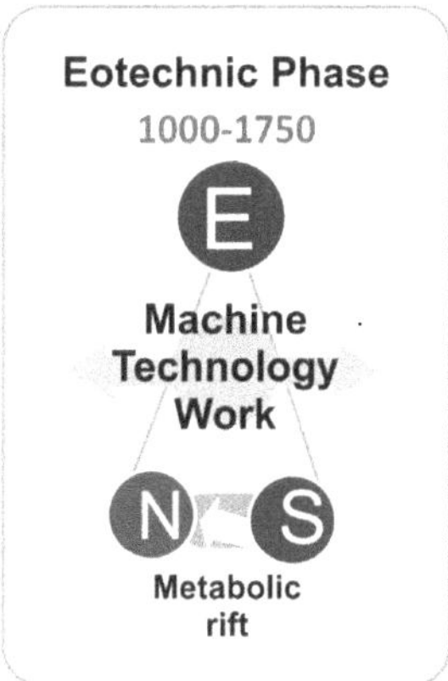

Technological complex:
{mechanical clock, windmills, transport, machine tool}
Materials: wood
Energy: water and wind
Change engine: natural power

Significant damage and impacts:
Energy exploitation
Dismissal of renewable energies
Change in land use
Deforestation and destruction of ecosystems
Social inequalities
Division of labour

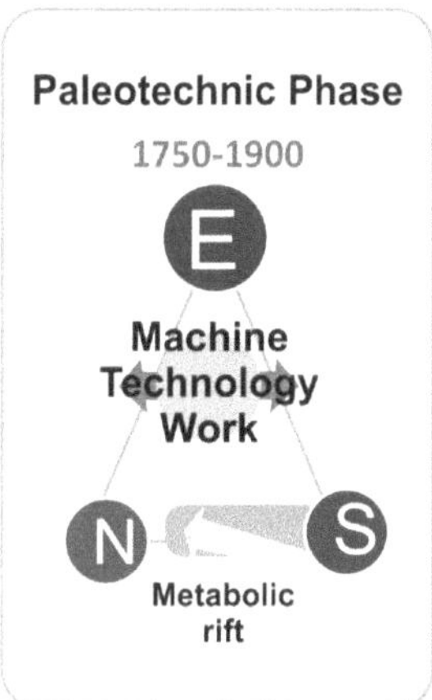

Significant events:
First and Second Industrial Revolution (IR)
Technological complex:
{steam engine, mass production}
Materials: steel
Energy: carbon
Change engine: **(I-IR) Mechanic and fluids**
(II-IR) Electricity
Significant damage and impacts:
Economic inequalities
Atmospheric pollution
Aquatic pollution
Terrestrial pollution
Social inequalities
Division of labour

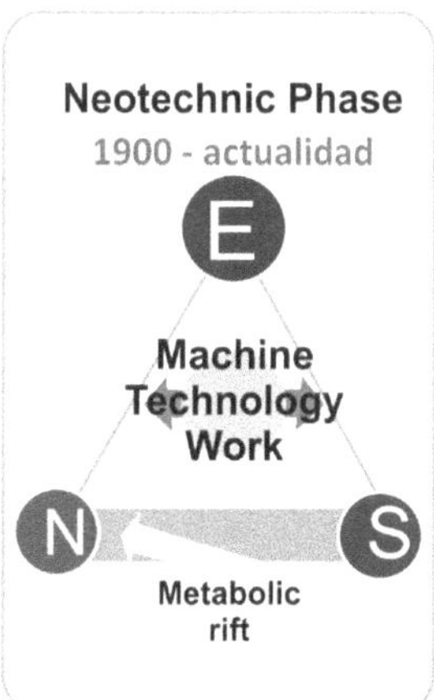

Significant events:
Third and Fourth Industrial Revolution (IR)
Technological complex:
{economization of production, combustion engine, synthesized materials, biomimicry and bionic design}
Materials: alloy
Energy: electricity; non renewable and renewable energy
Change engine: **(III-IR): electronic**
(IV-IR): computing
Significant damage and impacts:
Economic inequalities
Decentralization of factories
Standard impacts: midpoint and endpoint

Figure 1.3 Evolution of manufacturing systems with respect to sustainable development.

interest for the care of the planet begins; also the demand of modification of those more polluting manufacturing operations. Faced with social pressure, the industry's first reactions were *to dilute the pollution and hide the impacts generated*.

PHASE II: Constructive manufacturing, with end-of-pipe technologies. Government institutions began to demand greater pollution control. The first measures came with the imposition of sanctions on industry for the environmental impact generated; as a solution, "end-of-pipe" technologies (that capture pollution and prevent its dilution in the environment) became the Best Available Techniques for impact management. With them, and for some sectors, the damage was minimized. Although with clear benefits, these technologies are now considered transitional solutions to treat waste and emissions at the end of the production line, because they require a demand for energy, materials and specialized equipment, with an additional cost for investment and maintenance, and do not solve the cause of the impact. These drawbacks increase interest in Clean Technologies.

PHASE III: Proactive eco-efficiency manufacturing with clean technologies. Proving that the optimization of activities and processes is the next best option, proactive eco-efficiency strategies are developed, proposing a vision of the future based on the profitability of preventing pollution and not combating it.

Eco-efficiency (Avlonas & Nassos, 2013) is part of the current industrial strategy. Determines actions to reduce, minimize, sustain, limit (including prohibitions and penalties) to stop the environmental impact by increasing the efficiency of results or minimize the negative effects; involves a comprehensive assessment of manufacturing processes to minimize resource consumption and the generation of toxic substances, as shown in Fig. 1.4; in addition, priority actions are established to control outputs (substances and wastes) with two objectives: (1) reduction of midpoint impacts, or intermediate mechanisms of the biosphere such as global warming, toxicity, acidification or ozone depletion; and (2) reduction of endpoints or damage effects on protected elements: human health, ecosystem and resources. For this purpose, eco-effective models use management tools and impact analysis; they are described in detail in further sections.

This model currently has great operational value but limited scope: it offers short-term solutions, since it minimizes impacts but does not eliminate them, slowing down the environmental and social problems.

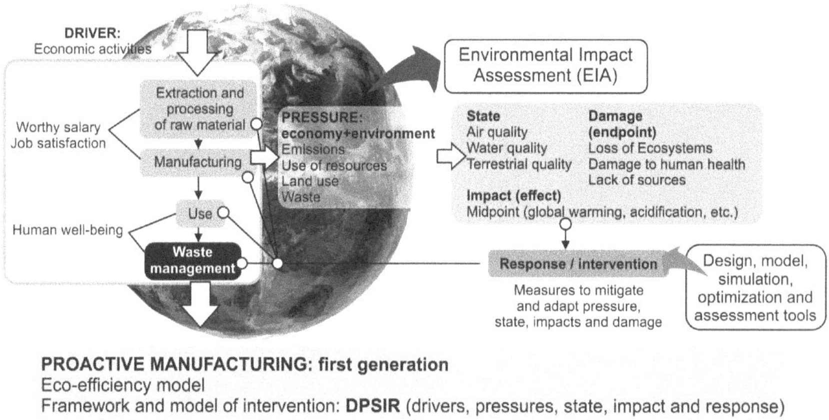

Figure 1.4 Sustainability from eco-efficiency.

PHASE IV: Proactive eco-effectiveness manufacturing with Clean technologies. New perspectives such as Industrial Ecology (Graedel & Allenby, 2003) arise to provide solutions beyond process efficiency. Eco-effectiveness proposes the identification of activities, operations and products that cause environmental and social impacts, to replace them with new solutions; in other words, it intends to solve the impacts eradicating the problem completely. Its strategies are based on eco-innovation (Fig. 1.5): *production in closed cycles; closure of material cycles; dematerialization; search for a balance between social, economic and environmental benefits; and new clean technologies that do not require waste management processes, with zero environmental impacts, and that allow the recovery of the damage generated and the revitalization of ecosystems.*

Linked to the above and as a solution to the linear model of production from cradle to grave, the perspective of Circular Economy is developed; it is currently the main objective of the projects promoted by the European Union in terms of sustainability and regenerative economy. (European Commission, 2015; European Commission, 2015). Circular Economy redefines the concept of economic growth to solve environmental and social problems. The aim is to conceive a social metabolism that gradually decouples "consumption" from "use of (finite) resources" (Schandl et al., 2016). To this end, it proposes the creation of a closed-loop system for the exchange of goods and services (Korhonen, Honkasalo, & Seppälä, 2018), which must be self-optimized (eco-efficient) and regenerative (eco-effective, with zero impact and with the

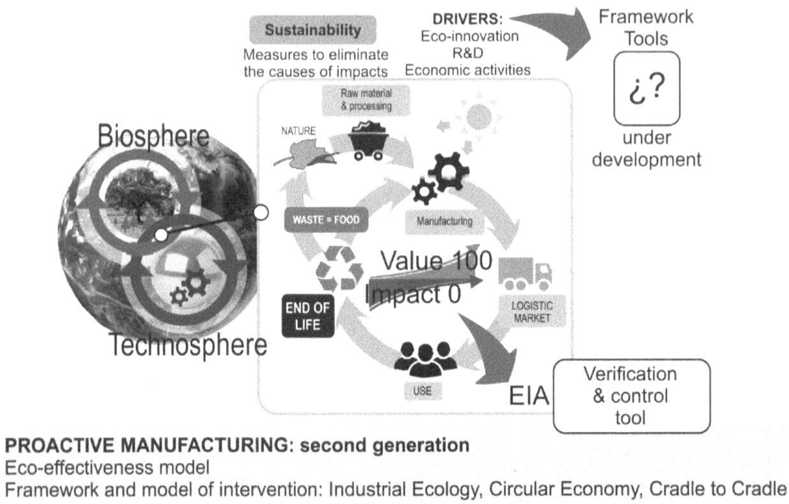

Figure 1.5 Sustainability from eco-effectiveness and circular economy.

capacity to reverse the damage caused). For this, the global challenge is to convert the Technosphere into an artificial ecosystem that functions with the same synergy as natural ecosystems; this implies configuring productive systems as a second nature (Peralta-Alvarez, 2016). **Industrial Ecology** (Graedel & Allenby, 2003), **Biomimicry** (Benyus, 2009), **Blue Economy** (Pauli, 2010), **Cradle to Cradle** (Braungart, McDonough, & Bollinger, 2007), **Performance Economy** (Stahel, 2010), **Regenerative design** (Lyle, 1994), or **Natural Capitalism** (Hawken, Lovins, & Lovins, 1999) have the aim of care of natural capital (lithosphere, hydrosphere, atmosphere and biotic elements) and develop an economy centered on increasing the productivity of natural resources, the change to bioinspired production models, the transition to business models based on services, and the regeneration of damaged natural capital.

These four phases have contributed to the development of the field of knowledge of Sustainable Manufacturing with different approaches: Cleaner Production, Green Manufacturing, Clean Technologies, Industrial Ecology, Environmentally Benign Manufacturing or Social Manufacturing, among others; they all work by combining eco-effective solutions (phase III) with eco-efficient solutions (phase IV) simultaneously.

Although an eco-effective approach must be a priority (Huesemann, 2004), integrating eco-effective solutions involves a great challenge for organizations and introduces a high level of complexity into activities. In addition, the materialization and exploitation of these solutions implies difficulties from the point of view of technical and economic viability (CSUTORA, 2011). In many occasions, it is not possible to replace the conventional solution with an eco-efficient one: optimized processes and technologies cannot be eliminated radically, so the only way to control the impact is eco-efficiency. At other occasions, the eco-effective solution is either unknown or simply a conceptual idea or viable in the long term (Abukhader, 2008). So, for now, an integration of the two approaches is the most appropriate solution; this situation defines *two generations* in the processes (Fig. 1.6), which currently guide the principles and objectives of Sustainable Manufacturing.

Over the past twenty years, these two generations have consolidated the knowledge base of sustainable manufacturing, more developed for eco-efficiency and still reduced for eco-effectiveness; in addition, they have oriented the lines of research and defined the drivers of innovation for technological development, consolidating the current manufacturing paradigms.

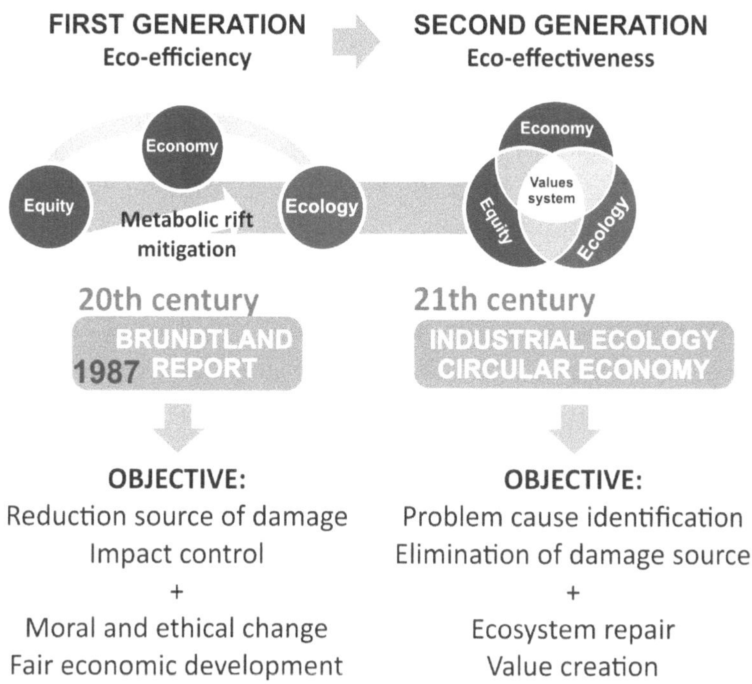

Figure 1.6 Evolution of sustainable approaches towards 3E-TBL.

1.3 Sustainable manufacturing paradigms

Industrial production is, together with the final consumer products and the expenditure of materials, one of the most damaging activities for the planet (UNEP, 2010, 2014). Specifically, industry is responsible for one-third of the world's final energy consumption and almost 40% of total CO_2 emissions. (OECD/IEA, 2009). With the development of commitment and corporate accountability, manufacturing engineering has progressively incorporated new standards, operating criteria, indicators, and good practices through the optimization of cost, quality, variety, product specialization, responsiveness and integral sustainability (3E), making the knowledge base evolve through the so-called "manufacturing paradigms", represented in Fig. 1.7 (Bi, Bi, & Zhuming, 2011; Feigenbaum, 1991; Stamm, Neitzert, & Singh, 2009).

The most recent, ***Sustainable Manufacturing*** includes Clean Production and Green Manufacturing as the main and most developed paradigms available for their materialization; it also relies on Industrial Ecology (Graedel & Allenby, 2003), Lean Manufacturing (Miller, Pawloski, & Standridge, 2010), Environmentally Benign Manufacturing (Gutowski et al., 2005) and Circular Economy (Kalmykova, Sadagopan, & Rosado, 2018). It aims to create quality products from processes that (1) minimize or eliminate negative environmental impacts, enhance efficiency (energy, water, raw materials, transport, etc.) and conservation of natural resources; (2) are safe for workers (occupational safety), communities and consumers; (3) are economically viable (Mani, Madan, Lee, Lyons, & Gupta, 2013). In addition, it proposes new production strategies that, thanks to the coordination between the different stakeholders of the supply chain, allow the closure of material cycles.

The lines of research in Sustainable Manufacturing are divided into seven fields (Gunasekaran & Spalanzani, 2012): (1) sources of sustainable challenges and problems, (2) business development in manufacturing and

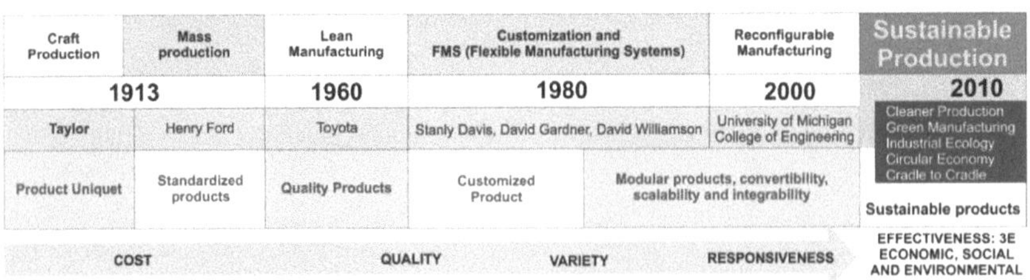

Figure 1.7 Evolution of manufacturing towards sustainable development.

services, (3) product and process design and development, (4) supply operations, (5) production operations, (6) distribution chain operations and (7) remanufacturing, recycling and reverse logistics. To facilitate the implementation of the results, there are procedures for integrating clean technologies and conventional manufacturing processes, including strategies to be followed (Rashid, Evans, & Longhurst, 2008), the appropriate project structure (Westkämper, Alting, & Arndt, 2001), methodologies, tools and techniques for implementation (Westkämper, 2008), and benefits in sustainable business development, as well as case studies and successful applications (Despeisse, Mbaye, Ball, & Levers, 2012).

Covering the lines of Sustainable Manufacturing but with a more operational focus, **Cleaner Production and Green Manufacturing** develop specific applications from different levels, usually coinciding with macro level (supply chain), mesolevel (organization and factory) and micro level (process and machine).

On one hand, Cleaner Production is a specific environmental protection initiative for preventive processes and for the implementation of sustainable manufacturing projects and organizational environmental management covering the three levels mentioned (micro, meso and macro-level); was defined by UNEP in 1991 as *The continuous application of an integrated environmental strategy to processes, products and services to increase efficiency and reduce risks to humans and the environment* (Huhtala & Ciccozzi, 2003). Being a framework that has experienced progress since 1991 (UNIDO, & UNEP, 2010) it is now possible to see the benefits of its application at a global level and in a large number of companies.

On the other hand, **Green Manufacturing** develops solutions linked to the micro-level (process); among its main objectives is the minimization of the environmental impact during the whole life cycle of the "green processes". The areas of work are classified in the following lines: (1) minimization of the consumption of resources, which implies maximizing the rate of use, reuse and remanufacture of products, (2) minimization of the discharge of pollutants into the environment and (3) proposal of available optimization solutions to achieve the two previous objectives, harmonizing the economic-social relationship with the maximum benefit. (Ahn, 2014; Liu, Yin, Cao, & He, 2005).

The change in the last few years in the lines of research related to sustainable manufacturing is significant. Most of the existing publications on manufacturing include some criteria of study and sustainable application, which has allowed the consolidation of the knowledge base. Table 1.1 compiles representative and reference publications.

Table 1.1 Representative publications on sustainable manufacturing.

Paradigm	References
Sustainable Manufacturing	Despeisse et al. (2012), Fratila (2014), Gunasekaran and Spalanzani (2012), Mani et al. (2013), Rashid et al. (2008), Westkämper et al. (2001), Westkämper, 2008
Cleaner Production	Huhtala and Ciccozzi (2003), Nowak and Cichy (2008), Rennings and Zwick (2003), Schaltegger, Bennett, Burritt, and Jasch (2008), Staniskis (2005), Wu and Low (2013)
Green Manufacturing	Ahn (2014), Davim and Dornfeld (2013), Deif (2011), Liu et al. (2005), Lopes Silva, Delai, Soares de Castro, and Ometto (2013), Mittal and Sangwan (2014), Nowak and Cichy (2008), Pawlewski and Dossou (2011), Pirraglia and Saloni (2011), Rennings and Zwick (2003), Schaltegger et al. (2008), Staniskis (2005), Tomovic, Anderson, Anglin, and Al (2009), Wu and Low (2013)
3E-TBL Manufacturing	Azevedo (2013), Despeisse et al. (2012), Duflou et al. (2012), Fratila (2014), Gunasekaran and Spalanzani (2012), Marxer, Keferstein, and Weckenmann (2011), Pawlewski and Greenwood (2014), Rashid et al. (2008), Seliger (2012), Smith and Ball (2012), Takata and Umeda (2007), Tao and Xun (2012), Westkämper et al. (2001), Westkämper, 2008

1.4 Operational framework 3E-TBL in sustainable manufacturing

From the point of view of the materialization of results, sustainability is an essentially dynamic and multidimensional concept; this makes the implementation of solutions complex when it is intended to achieve objectives associated with the three fundamental dimensions: *economic* (profitability, technical feasibility and future business opportunities), *ecological* (compatibility of activities with the environment and its cycles) and *equity* (attention to quality of life and individual or collective human well-being).

In many occasions, the satisfaction of several simultaneous objectives can be incompatible: these can be totally or partially opposed, that is to say, improvements in one dimension, can harm the remaining ones. So balancing the benefits is the main requirement for sustainable manufacturing 3E.

Fig. 1.8 summarizes the 3E-TBL framework, which divides the objectives into the phases of (1) management and (2) operation where the

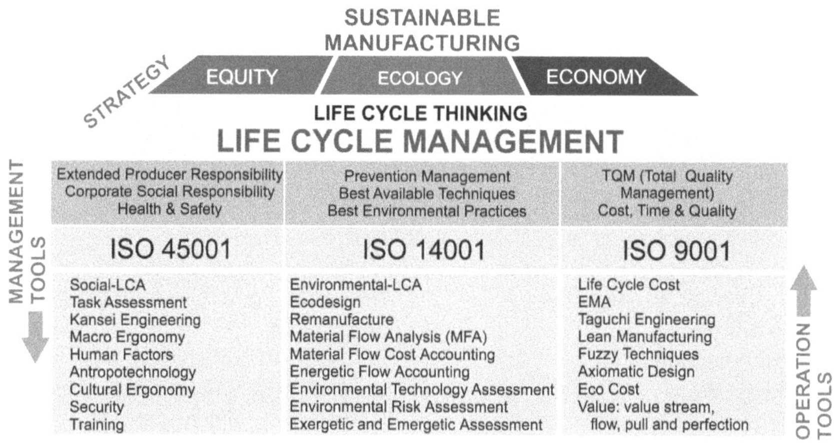

Figure 1.8 Working domains of sustainable manufacturing with 3E-TBL.

integration of eco-effective and eco-efficient improvements will be sought according to the technical and economic feasibility of the results.

At the management level, the *life cycle thinking* strategy (LCT) is used (Thabrew, Wiek, & Ries, 2009; UNEP, 2012) for integrating the following systems:

1. Occupational environment with Occupational Health and Safety Assessment Series, ISO: 45001; it includes the social dimension of sustainability with care for worker health and safety, Corporate Social Responsibility of the Organization and Extended Producer Responsibility (care of the user and the environment);
2. Environmental context with the ISO 14001 standard and supported by the ISO 14000 series
3. Economic context and quality management with ISO 9000 series

The results of the three levels of management are shown in Table 1.2; in an integrated way, they configure a sustainable manufacturing project.

Secondly, and at the operational level, 3E-TBL seeks a balance between the *economic, environmental and social operating parameters* of the manufacturing process. In particular:

- The economic dimension improves processes and operations by taking into account economic performance, maintenance of technical feasibility and quality assurance of results. To this end, the "conventional factors" are optimized: *technical feasibility, time, process performance, product variety, customer responsibility, investment, cost, economic benefits and quality.*

Table 1.2 Generic parameters and variables for Sustainable Manufacturing Management.

Sustainable manufacturing management	
Social dimension	
External: creating value for customers/suppliers/community	
Community (philanthropy). Socially responsible investment. Development cooperation: externalities • Territory exploitation compensations • Donations and other Community investments **Suppliers**. Supply chain control. • Priority to premises • Promotion of clusters and relations from the circular economy	**Clients** • Safe and quality products and services, customization and on-demand production. • Development and continuous improvement, solution of new needs. Resolution of expectations • Transparent relationships • Extended producer responsibility. End-of-life control of products
Internal: corporate governance and transparency	
• Communication and consultation. Participation and relationships • Occupational well-being: safety and health (safety, hygiene, ergonomics) • Education, training and coaching. Recycling (knowledge update) • Diversity (special populations) and equality.	• Benefits and compensations. • Labor integration. • Continuous improvement of conditions. Human rights. • Encouraging local commitment. • Worker Empowerment. • Reconciliation of family life.
Economic dimension	
Profitability	Indirect impact control: economic externalities.
Job offers	Deontology: professional, personal, legal.
Consumption patterns	New markets and sustainability
Production patterns	Social economy
Corporate communication policies: transparency and social audits.	Legal and regulatory compliance (national, international)
ECOLOGICAL DIMENSION	
Control and reduction of pollution	Materials
Waste management	Water

(*Continued*)

Table 1.2 (Continued)

Sustainable manufacturing management	
Emissions Effluents Solid waste Efficient use of resources	Energy Continuous improvement of cleaner operations. Care for biodiversity Priority management policies: reuse, remanufacturing, recycling, recovery and disposal

- The **environmental dimension** is aimed at reducing or eliminating the environmental burden generated by processes and operations, managing all inputs (raw materials, energy, water, fluids, other resources and equipment) and outputs (product, by-products or reusable flows, and non-reusable flows). It is supported by Best Available Techniques (BAT) and Best Environmental Practices (BEP) that promote cleaner production, waste and emission management and pollution prevention for the care of abiotic (water, atmosphere and soil) and biotic (human, flora and fauna) elements. As parameters to be optimized, it takes into *account energy consumption, consumption of raw materials (virgin and recycled), consumption of fluids (water, chemical products, refrigerants, lubricants, etc.), generation of emissions and waste, process toxicity and optimum use of machinery, tools and equipment.*
- Finally, the social dimension is in charge of the study of individual or collective human well-being based on a healthy environment. It considers the relationship between employee and company, safety, health and comfort of workers, also includes the study of the functionality of the results, customer relationship interference with any stakeholder in the process (consumer and society). As parameters to be optimized in the process, it takes into account all those related to the design of the workplace and the context-environment, focusing on the human factor measured through: *performance, reduction of the physical and cognitive workload, reduction of the toxic load of the process (human toxicity), improvement of productivity, motivation, and training, and information in the workplace.*

All sustainable manufacturing approaches and solutions require a design procedure and results assessment. To this end, there are a variety of tools adapted to each management and operational activity. In the literature, there are review works that collect available techniques and tools (European Commission, 2006; Hörisch, Ortas, Schaltegger, & Álvarez, 2015; Ness, Urbel-Piirsalu, Anderberg, & Olsson, 2007), analysis and assessment indicators (Siew, 2015) and specifically, tools and indicators for sustainable manufacturing planning (Amrina & Yusof, 2011; Joung, Carrell, Sarkar, & Feng, 2013; Kim, Leong, & Chen, 2012). Table 1.3 summarizes a set of representative application tools.

Despite great progress, there is much to be developed. Manufacturing faces the following challenges to achieve 3E-TBL sustainability (Gunasekaran & Spalanzani, 2012; Jovane, Westkämper, & Williams, 2009; Peralta, Marcos, & Aguayo, 2017; Pusavec, Krajnik, & Kopac, 2010; Pusavec, Stoic, & Kopac, 2010):

1. In general, there is no connection between theoretical models (paradigms and frameworks) and their implementation. Publications that have a 3E-TBL sustainability scope are very general; semi-quantitative or involve long-term technical–economic feasibility. So far few proposals cover sustainability in a comprehensive way, being the objective stated by scientific community *the development of a methodology, as well as a set of assessment indicators, for the design, development and sustainable optimization of manufacturing taking into account the three dimensions in a concrete and systematic way.*

Table 1.3 Tools available for the design and evaluation of sustainable manufacturing.

Management activities and procedures	Activities and operating procedures
Strategic Environmental Assessment (SEA)	Life Cycle Assessment
Territorial Impact Assessment (TIA)	Eco-cost
Cumulative Environmental Impacts	Material flow analysis (MFA) and substance flow analysis (SFA)
Cumulative Effects Assessment	Material flow cost accounting
Appropriate assessment Cost-benefit analysis	Energetic flow accounting
Sustainable supply chain management	Environmental Technology Assessment (EnTA)
Input-output Life Cycle Assessment	Environmental risk assessment
Environmental Accounting	Exergetic and emergetic assessment

2. The existing methods and tools that integrate the 3E are of theoretical and qualitative development. This is due to the fact that, in many cases, integration is complex, resources are not available, or knowledge is insufficient; in other cases, the set of 3E-TBL optimization variables are unknown.
3. Quantitative methods have a reduced scope: they focus on the optimization of an operating factor or variable of one of the three dimensions, checking its influence on any of the others; that is, the methods study the equilibrium of the dimensions two to two. The most usual relationships are environmental—economic and social—economic.
4. In general, the research effort has focused on the economic and environmental dimension, in addition to studying more exhaustively those parameters and indicators that may affect process performance: energy, quality of results, costs and consumption of resources (as fluids, lubricants, raw materials, etc.)
5. The social dimension has been the least considered, being included in studies in the last decade of the twentieth century. In addition, it is worth highlighting the lack of development of models, methodologies and optimization techniques; most of the studies are qualitative or do not deal in detail with social characteristics, or correspond to analysis work where tools are transferred from other areas of knowledge such as ergonomics or industrial hygiene.

1.5 New challenges and future opportunities: sustainable smart manufacturing

The Fourth Industrial Revolution establishes the principles of design, planning and management of "self-regulated", "flexible", "autonomous", "intelligent", and "sustainable" production. (Kagermann, Wahlster, & Helbig, 2013); They are intended to *optimize the use of resources, increase productivity, reduce costs, and maximize yield, taking into account well-being, quality of life, and the promotion of ecosystem diversity.*

The different innovation drivers that define the European roadmaps in Horizon 2020 (Filos, 2013) and proposals for international manufacturing research associations such as Factories of the Future 2020 (EFFRA, & European Commission, 2013) or Industry 4.0 (Bauernhansl, Ten Hompel, & Vogel-Heuser, 2014), modify the concept of sustainability in manufacturing to the next level: *Sustainable Smart Manufacturing*. The new challenges are the design and development of intelligent, adaptive and self-optimized clean technologies which, through the control of operating parameters in real time, adapt their behavior to the needs of the environment. The Cyber Physical Systems (CPS) and the Human-centered manufacturing are the new challenges for working on eco-effectiveness and eco-efficiency in manufacturing processes. The CPS will have the ability to self-assess their behavior in real time through data management (recovery, analysis and prediction) to self-regulate and optimize it. This will allow to strike a balance between the social, environmental, and economic benefits according to the circumstances of the context, eliminating the disadvantages that a pre-determined design of the process currently implies. That is to say, the Sustainable Smart Manufacturing Systems (Kamble, Gunasekaran, & Gawankar, 2018) will allow, in the short, medium and long term, the development of commitment and integrated social, economic and environmental responsibility.

The two research and working frameworks that are establishing the basis for the following paradigm "*Sustainable Smart Manufacturing*" are described below.

1.5.1 Sustainability in EFFRA: factories of the future

The "Factories of the Future 2020" project (EFFRA, & European Commission, 2013) sets out a long-term manufacturing strategy that includes transforming factories into intelligent, adaptive, digital, virtual and fully resource-efficient factories, with mobile enterprise management and customer-centered, human-centered manufacturing. In the proposals, *Sustainable Smart Manufacturing* is characterized by the following principles:

- **Factory and nature: green and sustainable**. Promote the similarity between artificial and natural industrial ecosystems, encouraging their integration through the use of sources of consumption of less energy (slim, clean, green), minimizing the consumption of resources and closing product cycles.
- **Factory as "good neighbor": close to the worker and the customer**. Encourages a manufacturing system close to people (in cities and metropolitan areas), where the factory is integrated and accepted by the consumer, ecosystems and the natural environment.
- **Factory and humans: human-centered manufacturing**. The factories of the future will be human-oriented interfaces, where the simulation and visualization of processes and forms of work are carried out from the skill, education and training of the worker; that is, it is technology that adapts to the human being.

The *Factories of the Future model* establishes different domains of innovation and research for the development of *Eco-Factory and Green Products Manufacturing*. Proposals should always be guided by the triple bottom line (3E-TBL) (European Commission, 2010) taking into account:

- **Environmental sustainability in manufacturing: Environmental friendliness**. Reduce resource consumption and the generation of waste and impacts at an optimized cost.
- **Economic sustainability in manufacturing: Economic growth**. Combine high performance and quality with the cost of effective productivity, creating reconfigurable, adaptive and evolutionary factories, capable of producing on a small scale in an economically viable way and with optimized resource management and efficient production processes.
- **Social sustainability in manufacturing: Social well-being**. Optimize human–machine interaction through safety, hygiene and ergonomics, but redefining the human role in the manufacturing environment: integrate human capabilities into technology.

1.5.2 Sustainability in industry 4.0: cyber physical systems

The concept **Industry 4.0** (Bauernhansl et al., 2014), also called intelligent industry or cyber-industry, is the approach that promotes the transformation of 21st century organizations into intelligent manufacturing environments. This German initiative project was born out of the need to conceive manufacturing systems that can adapt to new forms of market, characterized by increased competition and unpredictable dynamics of demand, achieving results of quality and variety in short periods of time. The objectives of the project are under development and proposes a set of lines of research to be addressed in the coming decades. In them, the concept of sustainable manufacturing is transformed into intelligent manufacturing adapted to the human and the environment, which includes the properties of agility, continuous improvement and control in real time of the process to prevent and avoid environmental and social impacts. This will be done by Cyber Physical Systems and Internet of Things (IoT) (Thiede, 2018). Thanks to the study of the following technological areas and key lines (Kagermann et al., 2013): *reference architectures and standardization, management of complex systems, broadband infrastructure, safety and health, design and organization of work, education, training, and professional development, regulated framework, and resource efficiency*.

1.6 Global principles for sustainable quality manufacturing

To close this chapter, the following is a set of principles that must be taken into account in order to achieve quality in Sustainable Manufacturing. They are a derivative of the knowledge base developed so far and future challenges (Anastas & Warner, 1998; Benyus, 2009; Blount, Clarimón, Cortés, Riechmann, & Romano, 2003; EFFRA, & European Commission, 2013; Kalmykova et al., 2018; Lee, Bagheri, & Kao, 2015; Lee, Lapira, Bagheri, & Kao, 2013; Matías Hernández & Vizán Idoipe, 2013; McDonough & Braungart, 2002; Peralta, 2016; Rolstadas, 2006; Staniskis, 2005; Yuan, 2013).

PRINCIPLES DERIVED FROM BIOMIMICRY

- Cyclicality: closing material cycles
- Respect the biological (natural) and technical (industrial) cycles
- Respect the planet's regeneration cycles
- Promote the use of renewable energy sources and resources
- Respect diversity

PRINCIPLES DERIVED FROM THE DAMAGE CAUSED AND LESSONS LEARNED FROM INDUSTRY

- Not filling it all, leaving enough environmental space for others
- Systematically increase eco-efficiency by obtaining greater human well-being with less energy and material inputs
- Respect the limits: social metabolism must be based on the limits of absorption and regeneration of ecosystems
- Thinking about tomorrow
- Social equality: equal environmental space for all human beings
- Caution and prudence: do not underestimate long-term ecological risks

DERIVATIVES OF THE CIRCULAR ECONOMY

- Proactive approach (eco-effectiveness versus eco-efficiency)
- Closed-loop product life cycle and process management
- Use of renewable energies
- Scope of balanced economic, environmental and social benefits
- Consideration of *waste = food*
- Bioinspired eco-innovation, manufacturing as second nature (biomimicry)
- Cyclical use: internal cyclicality (closing and narrowing the cycles of matter and energy); intensify (lengthen the cycle of use of products and resources); cascading (slow down the use of resources by

respecting the sequence of correct use of resources > reuse, remanufacturing > recycling > valorization > waste); prioritize pure inputs (dematerialize) and non-toxic resources

CLEAN PRODUCTION DERIVATIVES

- Process health and safety
- Management of chemicals and reduction of production waste
- Energy Efficiency
- Efficient processes and resources. End-of-Life Strategies: Recovery and Recycling

GREEN MANUFACTURING DERIVATIVES

- Green innovation: technology, materials and design.
- Green production: zero impact, reduction of energy and material consumption, and reduction of waste.
- Green recycling: of products and waste.
- Supply chain: alliances in the value chain. Cooperation and coordination.
- Green certification: infrastructure, product y process.
- Green logistics: logistics, storage and eco-design of packaging

GREEN CHEMISTRY DERIVATIVES

- Prevention: prevent waste before treating waste
- Atomic economy: synthesis methods should be designed to maximize the incorporation of all materials used in the process into the final product.
- Safe synthesis: synthesis methods must be designed to use and generate substances that are non-toxic to human health and the environment.
- Chemically safe products: fulfill their function minimizing their toxicity
- Safe solvents: the use of auxiliary substances must be minimal and innocuous
- Energy efficiency: energy requirements should be considered for their economic and environmental impact; methods should be prioritized at ambient temperature and pressure.
- Use of renewable raw materials, as long as it is technically and economically possible
- Reduce derivatives, as they require additional reactive and generate waste
- Catalysis: use of catalytic reactive instead of stoichiometrics
- Design for degradability: products should be designed so that at the end of their useful life they can be safely degraded and do not persist in the environment.
- Real-time analysis to prevent pollution and hazardous substances.

- Prevention of accidents: substances should be chosen to minimize the potential risk of accidents (emissions, explosions and fires)

DERIVATIVES FROM LEAN MANUFACTURING

- Minimize waste by generating value for the customer.
- Just-in-time production: production with pull system, flexibility and regular continuous flow; adjustment on demand (takt time)
- Jidoka/ISCP: quality and perfection (no defects thanks to continuous improvement)
- Identify value chain, standardization and 5 S management.

INDUSTRY 4.0 DERIVATIVES AND THE NEW "SMART SUSTAINABILITY" CHALLENGES

- Factory as second nature
- Interoperability: real-time monitoring, connection (Internet of Things), communication, and real-time intelligent control
- Modularity and service-orientation (Internet of Service).
- Virtualization of performance, temporary activity logs, memory, and intelligent data
- Cognition and intelligence, embedded and ubiquitous computing
- Real-time capabilities and resilience: auto-configuration, auto-adjustment, and auto-optimization
- Human-centered factory

1.7 Conclusions

This chapter has introduced the interest of manufacturing engineering as a body of knowledge and operation for sustainability; acting on the causes to solve the problems caused by industry and the set of sectors that have contributed to the set of environmental and social impacts generated is the main objective of the new perspectives of sustainability in manufacturing. In addition, it has made it possible to highlight the effort made by manufacturing engineering in relation to the study of sustainability. It is clear that there is a consolidated knowledge base, although many of the objectives, strategies and good practices proposed by the different paradigms and frameworks have yet to be developed: Smart Sustainability, Circular Economy, Cleaner Production, Green Manufacturing, Industrial Ecology, etc.

The next challenge for sustainable manufacturing is structuring, simplifying and reducing complexity around sustainability 3E-TBL. It will be necessary, therefore, to provide manufacturing engineering, firstly, with a system of values which, from the root (ethics, professional deontology, morals and type of thinking), encourage the mitigation of impacts; secondly, an operational framework that allows all the variables required for the elimination of impacts generated with efficient support from the human, environmental, and technical systems to be addressed with minimum complexity. This framework should have an eco-effective approach (zero impact, value creation and recovery of the damage generated); it will also incorporate the perspective of support, verification and analysis from eco-efficiency (control and minimization of impact). It must also be oriented to achieving sustainable objectives, but it must clearly structure the actions and provide a set of facilitators and tools that allow the incorporation of clean measures and proposals to the reality of the manufacturing processes without compromising their technical and economic viability; in other words, it must incorporate sustainability in a natural way, creating the optimum conditions for its integration in the tasks of design, modeling, simulation and optimization of the manufacturing process.

References

Abukhader, S. M. (2008). Eco-efficiency in the era of electronic commerce – should 'Eco-Effectiveness' approach be adopted? *Journal of Cleaner Production, 16*(7), 801–808. Available from https://doi.org/10.1016/J.JCLEPRO.2007.04.001.

Ahn, S.-H. (2014). An evaluation of green manufacturing technologies based on research databases. *International Journal of Precision Engineering and Manufacturing-Green Technology, 1*(1), 5–9. Available from https://doi.org/10.1007/s40684-014-0001-8.

Amrina, E., & Yusof, S.M. (2011). Key performance indicators for sustainable manufacturing evaluation in automotive companies. In *2011 IEEE International Conference on Industrial Engineering and Engineering Management* (pp. 1093–1097). Ieee. https://doi.org/10.1109/IEEM.2011.6118084.

Anastas, A., & Warner, J. (1998). *Green Chemistry: Theory and Practice*. Oxford: Oxford University Press.

Avlonas, N., & Nassos, G. P. (2013). *Practical Sustainability Strategies*. Hoboken, NJ: John Wiley & Sons, Inc. Available from https://doi.org/10.1002/9781118787472.

Azevedo, A. (2013). *Advances in Sustainable and Competitive Manufacturing Systems: 23rd International Conference on Flexible Automation & Intelligent Manufacturing*. Berlín, Heidelberg: Springer Science & Business.

Bauernhansl, T., Ten Hompel, M., & Vogel-Heuser, B. (2014). *Industrie 4.0 in Produktion, Automatisierung und Logistik*. (B. Bauernhansl, Thomas, ten Hompel, Michael, Vogel-Heuser, Ed.). https://doi.org/10.1007/978-3-658-04682-8.

Benyus, J. M. (2009). *Biomimicry*. New York City: William Morrow & Co.

Bi, Z., Bi., & Zhuming. (2011). Revisiting system paradigms from the viewpoint of manufacturing sustainability. *Sustainability, 3*(9), 1323–1340. Available from https://doi.org/10.3390/su3091323.

Blount, E., Clarimón, L., Cortés, A., Riechmann, J., & Romano, D. (2003). *Industria como naturaleza: hacia la producción limpia*. Madrid: Los Libros de la Catarata.

Braungart, M., McDonough, W., & Bollinger, A. (2007). Cradle-to-cradle design: creating healthy emissions? a strategy for eco-effective product and system design. *Journal of Cleaner Production, 15*(13–14), 1337–1348. Available from https://doi.org/10.1016/j.jclepro.2006.08.003.

Clark, B., & Foster, J. B. (2009). Ecological imperialism and the global metabolic rift unequal exchange and the guano/nitrates trade. *International Journal of Comparative Sociology, 50* (3–4), 311–334. Available from https://doi.org/10.1177/0020715209105144.

CSUTORA, M. (2011). From eco-efficiency to eco-effectiveness? the policy-performance paradox. *Society and Economy, 33*(1), 161–181. Available from https://doi.org/10.1556/SocEc.33.2011.1.12.

Davim, P., & Dornfeld, D. A. (2013). *Green manufacturing: fundamentals and applications*. Argonne, IL: Springer. Available from https://doi.org/10.2172/1072353.

Deif, A. M. (2011). A system model for green manufacturing. *Journal of Cleaner Production, 19*(14), 1553–1559. Available from https://doi.org/10.1016/j.jclepro.2011.05.022.

Despeisse, M., Mbaye, F., Ball, P. D., & Levers, A. (2012). The emergence of sustainable manufacturing practices. *Production Planning & Control, 23*(5), 354–376. Available from https://doi.org/10.1080/09537287.2011.555425.

Duflou, J. R., Sutherland, J. W., Dornfeld, D., Herrmann, C., Jeswiet, J., Kara, S., & Kellens, K. (2012). Towards energy and resource efficient manufacturing: a processes and systems approach. *CIRP Annals - Manufacturing Technology, 61*(2), 587–609. Available from https://doi.org/10.1016/j.cirp.2012.05.002.

EFFRA, & European Commission. (2013). *Factories of the future: multi-annual roadmap for the contractual PPP under Horizon 2020.* https://doi.org/10.2777/29815.

Elkington, J. (1998). Cannibals with forks: the triple bottom line of 21st century business. Conscientious commerce. Chichester: Capstone Publishing Ltd, https://doi.org/0865713928.

European Commission. Factories of the Future. Strategic multi-annual roadmap (2010). Retrieved from http://ec.europa.eu/research/research-eu.

European Commission. (2015). Closing the loop on the ground: 10 EU projects working towards a circular economy. Retrieved from https://ec.europa.eu/easme/en/news/closing-loop-ground-10-eu-projects-working-towards-circular-economy.

European Commission. (2006). *Handbook for Trade Sustainability Impact Assessment. Assessment.*

European Commission. (2015). *Closing the loop - An EU action plan for the Circular Economy. COM(2015) 614 final.* Brussels. Retrieved from https://ec.europa.eu/transparency/regdoc/rep/1/2015/EN/1-2015-614-EN-F1-1.PDF.

Feigenbaum, A. V. (1991). *Total quality control: achieving productivity, market penetration and advantage in the global economy*. McGraw-Hill Higher Education.

Filos, E. (2013). *Manufacturing innovation and horizon 2020. Digital product and process development systems* (pp. 1–10). Berlin Heidelberg: Springer Berlin Heidelberg. Available from https://doi.org/10.1007/978-3-642-41329-2_1.

Foster, J. B. (2000). *Marx's ecology: materialism and nature*. New York, USA: Monthly Review Press.

Fratila, D. (2014). Environmentally friendly manufacturing processes in the context of transition to sustainable production. *Comprehensive Materials Processing, 8*, 163–175, Health, Safety and Environmental. Available from https://doi.org/10.1016/B978-0-08-096532-1.00815-3.

Graedel, T. E., & Allenby, B. R. (2003). *Industrial Ecology*. Upper Saddle River: Prentice Hall.

Gunasekaran, A., & Spalanzani, A. (2012). Sustainability of manufacturing and services: investigations for research and applications. *International Journal of Production Economics, 140*(1), 35–47. Available from https://doi.org/10.1016/j.ijpe.2011.05.011.

Gutowski, T., Murphy, C., Allen, D., Bauer, D., Bras, B., Piwonka, T., ... Wolff, E. (2005). Environmentally benign manufacturing: observations from Japan, Europe and the United States. *Journal of Cleaner Production, 13*(1), 1–17. Available from https://doi.org/10.1016/j.jclepro.2003.10.004.

Hawken, P., Lovins, A. B., & Lovins, L. H. (1999). *Natural capitalism: creating the next industrial revolution*. Earthscan.

Hörisch, J., Ortas, E., Schaltegger, S., & Álvarez, I. (2015). Environmental effects of sustainability management tools: an empirical analysis of large companies. *Ecological Economics, 120*, 241–249. Available from https://doi.org/10.1016/j.ecolecon.2015.11.002.

Huesemann, M. H. (2004). The failure of eco-efficiency to guarantee sustainability: Future challenges for industrial ecology. *Environmental Progress, 23*(4), 264–270. Available from https://doi.org/10.1002/ep.10044.

Huhtala, A., & Ciccozzi, E. (2003). Financing cleaner production investments—UNEP experience. *Clean Technologies and Environmental Policy, 5*(2), 87–91. Available from https://doi.org/10.1007/s10098-003-0181-3.

Joung, C. B., Carrell, J., Sarkar, P., & Feng, S. C. (2013). Categorization of indicators for sustainable manufacturing. *Ecological Indicators, 24*, 148–157. Available from https://doi.org/10.1016/j.ecolind.2012.05.030.

Jovane, F., Westkämper, E., & Williams, D. (2009). *The manufuture road towards competitive and sustainable high-adding value manufacturing*. Berlin: Springer Berlin Heidelberg, *Vasa*.

Kagermann, H., Wahlster, W., & Helbig, J. (2013). *Recommendations for implementing the strategic initiative INDUSTRIE 4.0. Securing the future of German manufacturing industry*.

Kalmykova, Y., Sadagopan, M., & Rosado, L. (2018). Circular economy—from review of theories and practices to development of implementation tools. *Resources, Conservation and Recycling, 135*, 190–201. Available from https://doi.org/10.1016/J.RESCONREC.2017.10.034.

Kamble, S. S., Gunasekaran, A., & Gawankar, S. A. (2018). Sustainable Industry 4.0 framework: a systematic literature review identifying the current trends and future perspectives. *Process Safety and Environmental Protection, 117*, 408–425. Available from https://doi.org/10.1016/J.PSEP.2018.05.009.

Kim, D.B., Leong, S., & Chen, C.-S. (2012). An Overview of Sustainability Indicators and Metrics for Discrete Part Manufacturing. *Proceedings of the ASME 2012 International Design Engineering Technical Conference & Computers and Information in Engineering Conference*, 1–9. https://doi.org/10.1115/DETC2012-71020.

Korhonen, J., Honkasalo, A., & Seppälä, J. (2018). Circular Economy: the concept and its limitations. *Ecological Economics, 143*, 37–46. Available from https://doi.org/10.1016/J.ECOLECON.2017.06.041.

Lee, J., Bagheri, B., & Kao, H.-A. (2015). A cyber-physical systems architecture for industry 4.0-based manufacturing systems. *Manufacturing Letters, 3*, 18–23. Available from https://doi.org/10.1016/j.mfglet.2014.12.001.

Lee, J., Lapira, E., Bagheri, B., & Kao, H. (2013). Recent advances and trends in predictive manufacturing systems in big data environment. *Manufacturing Letters, 1*(1), 38–41. Available from https://doi.org/10.1016/j.mfglet.2013.09.005.

Liu, F., Yin, J., Cao, H., & He, Y. (2005). Investigations and practices on green manufacturing in machining systems. *Journal of Central South University of Technology, 12*(2), 18–24. Available from https://doi.org/10.1007/s11771-005-0004-2.

Lopes Silva, D. A., Delai, I., Soares de Castro, M. A., & Ometto, A. R. (2013). Quality tools applied to Cleaner Production programs: a first approach toward a new methodology. *Journal of Cleaner Production, 47*, 174–187. Available from https://doi.org/10.1016/j.jclepro.2012.10.026.

Lyle, J. T. (1994). *Regenerative design for sustainable development*. John Wiley.

Mani, M., Madan, J., Lee, J.H., Lyons, K.W., & Gupta, S.K. (2013). Review on Sustainability Characterization for Manufacturing Processes. *National Institute of Standars and Technology, February*. https://doi.org/10.6028/NIST.IR.7913.

Marx, K. (1894). In F. Engels (Ed.), *Capital volume III. The process of capitalist production as a whole*. Hamburgo: International Publishers, NY.

Marxer, M., Keferstein, C. P., & Weckenmann, A. (2011). Advances in sustainable manufacturing. In G. Seliger, M, Khraisheh, & I. Jawahir (Eds.), *Engineering education*. Berlin, Heidelberg: Springer Berlin Heidelberg, https://doi.org/10.1007/978 3 642 20183-7

Matías Hernández, J. C., & Vizán Idoipe, A. (2013). *Lean manufacturing conceptos, técnicas e implantación*. Madrid: Fundación eoi.

McDonough, W., & Braungart, M. (2002). *Cradle to cradle: remaking the way we make things*. New York: Farrar, Straus and Giroux.

Miller, G., Pawloski, J., & Standridge, C. R. (2010). A case study of lean, sustainable manufacturing. *Journal of Industrial Engineering and Management*, *3*(1), 11–32. Available from https://doi.org/10.3926/jiem.2010.v3n1.p11-32.

Mittal, V. K., & Sangwan, K. S. (2014). Prioritizing drivers for green manufacturing: environmental, social and economic perspectives. *Procedia CIRP*, *15*, 135–140. Available from https://doi.org/10.1016/j.procir.2014.06.038.

Mumford, L. (1934). *Technics and civilization*. Harcourt, Brace & Company, Inc.

Ness, B., Urbel-Piirsalu, E., Anderberg, S., & Olsson, L. (2007). Categorising tools for sustainability assessment. *Ecological Economics*, *60*(3), 498–508. Available from https://doi.org/10.1016/j.ecolecon.2006.07.023.

Nowak, Z. A., & Cichy, M. J. (2008). *Phenomenological model of cleaner production. Environmental management accounting for cleaner production* (pp. 123–139). Ámsterdam: Springer Netherlands, https://doi.org/978-1-4020-8913-8_7.

OECD/IEA. (2009). *Energy technology transitions for industry the Next Industrial*. https://doi.org/987-92-64-06858-2-2.

Pauli, G. A. (2010). *The blue economy: 10 years, 100 innovations, 100 million jobs*. Paradigm Publications.

Pawlewski, P., & Dossou, P. E. (2011). *Comparison of enterprise integration concepts (PLM and EA) from the point of view green manufacturing. Information technologies in environmental engineering environmental science and engineering* (pp. 659–673). Heidelberg: Springer Berlin Heidelberg, https://doi.org/978-3-642-19536-5_51.

Pawlewski, P., & Greenwood, A. (2014). *Process simulation and optimization in sustainable logistics and manufacturing*. Cham: Springer International Publishing, https://doi.org/10.1007/978-3-319-07347-7.

Peralta, M.E. (2016). *Modeling, simulation and optimization of the sustainability of manufacturing processes for its multiscale and multilevel integration. Application to machining cells*. University of Cadiz. Retrieved from https://www.educacion.gob.es/teseo/mostrarRef.do?ref = 1281186.

Peralta, M. E., Marcos, M., & Aguayo, F. (2017). On the sustainability of machining processes. Proposal for a unified framework through the triple bottom-line from an understanding review. *Journal of Cleaner Production*.

Peralta Álvarez, M. E., Marcos Bárcena, M., & Aguayo González, F. (2016). A review of sustainable machining engineering: optimization process through triple bottom line. *Journal of Manufacturing Science and Engineering*, *138*(10), 100801. Available from https://doi.org/10.1115/1.4034277.

Peralta-Alvarez, M.E. (2016). *Modelado, simulación y optimización de la sostenibilidad de procesos de fabricación para su integración multiescala y multinivel. Aplicación a células de mecanizado*. PhD thesis. Univesidad de Cádiz.

Pirraglia, A., & Saloni, D. E. (2011). Measuring environmental improvements image in companies implementing green manufacturing, by means of a fuzzy logic model for decision-making purposes. *The International Journal of Advanced Manufacturing Technology*, *61*(5–8), 703–711. Available from https://doi.org/10.1007/s00170-011-3748-x.

Pusavec, F., Krajnik, P., & Kopac, J. (2010). Transitioning to sustainable production—part I: application on machining technologies. *Journal of Cleaner Production*, *18*(2), 174–184. Available from https://doi.org/10.1016/j.jclepro.2009.08.010.

Pusavec, F., Stoic, A., & Kopac, J. (2010). Sustainable machining process—myth or reality. *Strojarstvo: Journal for Theory and Application in Mechanical Engineering*, *52*(2), 197–204.

Rahimifard, S., Sheldrick, L., Woolley, E., Colwill, J., & Sachidananda, M. (2013). How to manufacture a sustainable future for 9 billion people in 2050. *Re-Engineering Manufacturing for Sustainability*, 1–8. Available from https://doi.org/10.1007/978-981-4451-48-2_1.

Rashid, S. H. A., Evans, S., & Longhurst, P. (2008). A comparison of four sustainable manufacturing strategies. *International Journal of Sustainable Engineering*, *1*(3), 2008. Available from https://doi.org/10.1080/19397030802513836.

Rennings, K., & Zwick, T. (2003). *Employment impacts of cleaner production* (Vol. 21). Heidelberg: Physica-Verlag HD. Available from https://doi.org/10.1007/978-3-642-57418-4.

Rolstadas, A. (2006). IMS 2020 Roadmap for Sustainable Manufacturing Research. In *Proceedings from the IMS2020 Summer School on Sustainable Manufacturing* (pp. 1–6). Zurich, Switzerland.

Schaltegger, S., Bennett, M., Burritt, R. L., & Jasch, C. (2008). In S. Schaltegger, M. Bennett, R. L. Burritt, & C. Jasch (Eds.), *Environmental management accounting for cleaner production* (Vol. 24). Dordrecht: Springer Netherlands, https://doi.org/10.1007/978-1-4020-8913-8.

Schandl, H., Hatfield-Dodds, S., Wiedmann, T., Geschke, A., Cai, Y., West, J., … Owen, A. (2016). Decoupling global environmental pressure and economic growth: scenarios for energy use, materials use and carbon emissions. *Journal of Cleaner Production*, *132*, 45–56. Available from https://doi.org/10.1016/J.JCLEPRO.2015.06.100.

Seliger, G. (2012). *Sustainable manufacturing: shaping global value creation*. Berlin, Heidelberg: Springer Berlin Heidelberg. Available from https://doi.org/10.1007/978-3-642-27290-5.

Siew, R. Y. J. (2015). A review of corporate sustainability reporting tools (SRTs). *Journal of Environmental Management*, *164*, 180–195. Available from https://doi.org/10.1016/j.jenvman.2015.09.010.

Smith, L., & Ball, P. (2012). Steps towards sustainable manufacturing through modelling material, energy and waste flows. *International Journal of Production Economics*, *140*(1), 227–238. Available from https://doi.org/10.1016/j.ijpe.2012.01.036.

Stahel, W. R. (2010). *The performance economy*. Palgrave Macmillan.

Stamm, M.L., Neitzert, T.R., & Singh, D.P.K. (2009). *TQM, TPM, TOC, Lean and Six Sigma-Evolution of manufacturing methodologies under the paradigm shift from Taylorism/Fordism to Toyotism?* Retrieved from https://core.ac.uk/download/pdf/56363032.pdf.

Staniskis, J. (2005). Cleaner production in the developing world. *Clean Technologies and Environmental Policy*, 7(3), 145–147. Available from https://doi.org/10.1007/s10098-005-0001-z.

Takata, S., & Umeda, Y. (2007). Advances in Life Cycle Engineering for Sustainable Manufacturing businesses. In S. Takata, & Y. Umeda (Eds.), *LCE, Proceedings of the 14th CIRP Conference on Life Cycle Engineering* (p. 480). London: Springer. Available from https://doi.org/10.1007/978-1-84628-935-4.

Tao, P., & Xun, X. (2012). The State of the Art in Energy Consumption Model – The Key to Sustainable Machining. In *Applied Mechanics and Materials* (Vol. 232, pp. 592–599). Retrieved from http://www.scientific.net/AMM.232.592.

Thabrew, L., Wiek, A., & Ries, R. (2009). Environmental decision making in multi-stakeholder contexts: applicability of life cycle thinking in development planning and implementation. *Journal of Cleaner Production*, *17*(1), 67–76. Available from https://doi.org/10.1016/j.jclepro.2008.03.008.

Thiede, S. (2018). Environmental Sustainability of Cyber Physical Production Systems. *Procedia CIRP*, *69*, 644–649. Available from https://doi.org/10.1016/J.PROCIR.2017.11.124

Tomovic, C., Anderson, H., Anglin, A., & Al, E. (2009). *Social issues of product lifecycle management: developing cross cultural virtual teams; supporting today's green manufacturing imperative; educating and preparing tomorrow's workforce; and impacting inter-organizational relationships in supply chain manag. Product Realization* (pp. 1–33). New York City: Springer US, https://doi.org/978-0-387-09482-3_8.

UNEP. (2010). *Assessing the Environmental Impacts of Consumption and Production. Priority Products and Materials.* United Nations Environment Programme; International Panel for Sustainable Resource Management.

UNEP. (2012). *Greening the Economy - Through Life Cycle Thinking.*

UNEP. (2014). *Assessing Global Land Use: Balancing consumption With sustainaBle supply.*

UNIDO, & UNEP. (2010). *Taking Stock and Moving Forward. The UNIDO–UNEP National Cleaner Production Centres.*

United Nations. (1987). *Report of the World Commission on Environment and Development: Report of the World Commission on Environment and Development: Our Common Futu.* Retrieved from http://www.exteriores.gob.es/Portal/es/PoliticaExteriorCooperacion/Desarrollosostenible/Documents/InformeBrundtland(Eninglés).pdf.

Westkämper, E., Alting., & Arndt. (2001). Life Cycle Management and Assessment: Aproaches and Visions Towards Sustainable Manufacturing. *Journal of Engineering Manufacture*, 599–626.

Westkämper, E. (2008). Manufuture and Sustainable Manufacturing. In Manufacturing Systems and Technologies for the New Frontier. The 41st CIRP Conference on Manufacturing Systems (pp. 11–14). https://doi.org/10.1007/978-1-84800-267-8_3

Wu, P., & Low, S. P. (2013). *Lean and Cleaner Production.* Berlin, Heidelberg: Springer Berlin Heidelberg. Available from https://doi.org/10.1007/978-3-642-42062-7.

Yuan, C. (2013). Green Manufacturing, 203–221. https://doi.org/10.1007/978-1-4419-6016-0.

CHAPTER 2

The benefits of additive manufacturing for sustainable design and production

Mehrshad Mehrpouya[1], Alireza Vosooghnia[2], Amir Dehghanghadikolaei[3] and Behzad Fotovvati[4]
[1]Faculty of Engineering Technology, University of Twente, Enschede, the Netherlands
[2]Department of Civil and Environmental Engineering, Sapienza University of Rome, Rome, Italy
[3]School of Mechanical, Industrial and Manufacturing Engineering, Oregon State University, Corvallis, OR, United States
[4]Department of Mechanical Engineering, The University of Memphis, Memphis, TN, United States

2.1 Introduction

Sustainability was introduced in the industry around a century after the industrial revolution (Kiely, 2005). Economic, social, and environmental are the three principal dimensions that can define sustainable development (Hyvarinen, 2006). These three principal dimensions are addressed with concepts such as reducing, recovering, recycling, reusing, redesigning, and remanufacturing which are also known as 6R concepts (Peng et al., 2018). The term sustainable manufacturing is discussed within the sustainable development framework and the engineering contexts (Malshe et al., 2015). Sustainability is defined as ensuring that the quality of human life does not reduce due to diminishing of resources and the future industrial and financial opportunities are still available with positive effects on social, economic, and environmental aspects of humankind (Mihelcic et al., 2003). Therefore the purpose of sustainable manufacturing lies in reducing the environmental impacts and enhancing the economic and social impacts through the entire life cycle of a specific product (Esmaeilian, Behdad, & Wang, 2016). Fig. 2.1 represents a timeline of key issues in the development of sustainable manufacturing.

Additive manufacturing (AM) has been recognized as an efficient and sustainable technology in modern manufacturing having the potential of providing a series of sustainability features (Ford & Despeisse, 2016). AM provides different fields to substitute the conventional manufacturing methods as a more sustainable manufacturing approach with minimal

Sustainable Manufacturing
DOI: https://doi.org/10.1016/B978-0-12-818115-7.00009-2

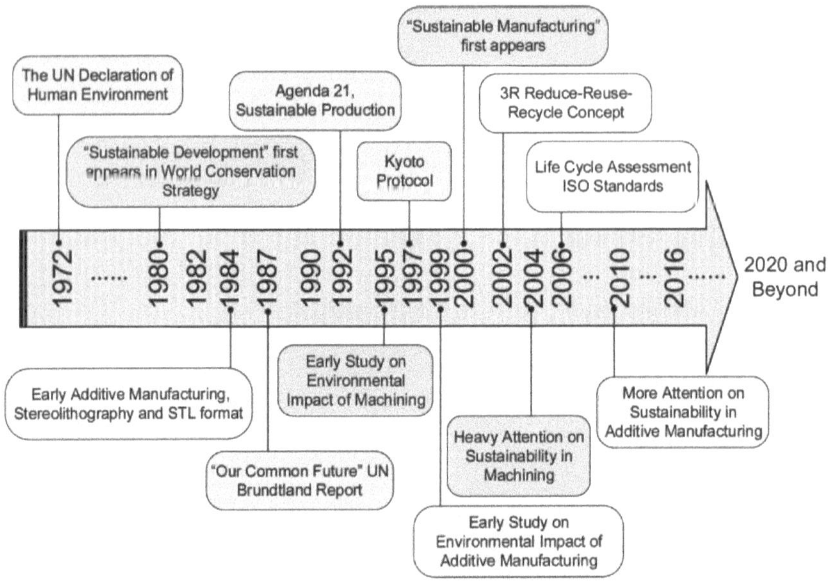

Figure 2.1 A timeline of sustainable manufacturing. *Peng et al. (2018), Sustainability of additive manufacturing: an overview on its energy demand and environmental impact. Additive Manufacturing, with kind permission of Elsevier.*

carbon footprint in different stages of product development and production. In addition, the ability of AM processes in fabricating smart materials, superalloys and strengthened metallic compounds such as Inconel 718, and reinforced stainless steels, reduces the volume of waste material to a significant extent (Mehrpouya et al., 2020a,b; Sadooghi et al., 2019). The capability of repairing, upgrading, and remanufacturing of products, provides an opportunity to reduce energy consumption and costs (Morrow et al., 2007). All these features come to action while complex geometries and even assemblies could be made in less manufacturing steps compared to conventional processes (Ibrahim et al., 2018).

2.2 Sustainability of additive manufacturing processes

AM processes introduce many significant modifications in materials processing, product design, manufacturing processes, and supply chains (Fotovvati, Etesami, & Asadi, 2019). AM technology provides more sustainable alternatives for the conventional manufacturing processes such as molding for polymer parts, and machining and casting for metallic components (Ghoreishi, Roohi, & Ghadikolaei, 2019). Overall, the advantages of sustainable manufacturing provided by AM processes are as following (Chen et al., 2015; Ford & Despeisse, 2016; Mani, Lyons, & Gupta, 2014; Peng et al., 2018; Reeves, 2008):

1. The need for raw materials in the supply chain is reduced.
2. Production/Use phase improvement of efficiency in resource consumption.
3. The need for wasteful and polluting manufacturing processes is decreased.
4. Product design is improved and has become more flexible and efficient.
5. The transport-related byproducts are decreased, and carbon footprint is enhanced.
6. Production is decentralized and closer to consumers.
7. Value chains are redesigned, supply chains are shorter and simpler, production is more localized with more innovative distribution schemes.
8. Product life is extended using sustainability concepts such as reusing, repairing, refurbishing, and remanufacturing while considering more sustainable social, economic, and environmental approaches.

AM processes provide shorter, smaller, and localized value chains. Since these value chains offer many sustainability benefits and considering the product/material life cycles, reconfigurations of the value chains for sustainability need to better understand the interactions between

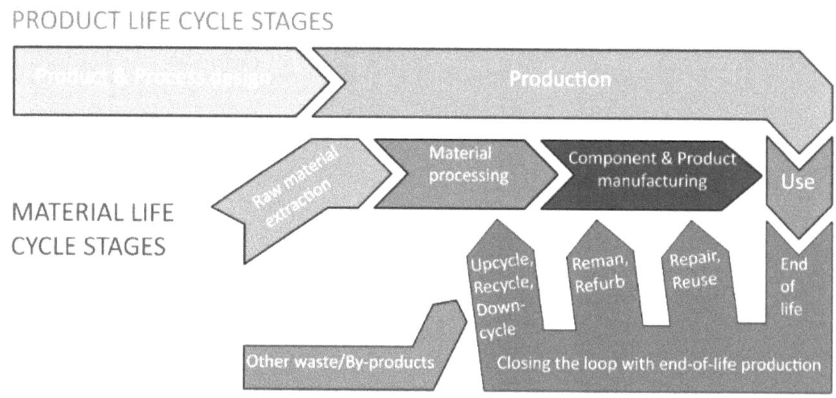

Figure 2.2 Product and material life cycle stages. *Ford and Despeisse (2016), Additive manufacturing and sustainability: an exploratory study of the advantages and challenges. Journal of Cleaner Production 137: p. 1573-1587, with kind permission of Elsevier.*

stakeholders. As illustrated in Fig. 2.2, in the life cycle of materials there are four main fields, including product/process design; material processing; component manufacturing; and product's end-of-life. In these stages, the adoption of AM is carried out to enhance resource efficiency (Despeisse & Ford, 2015). Table 2.1 represents a summary of the advantages and challenges of having a more sustainable production process.

2.3 Product design

The applications of AM are more prominent in the design process. In conventional manufacturing, prototyping is relatively costly to produce because it requires tools, molds and specialized operators, while AM significantly influences prototyping in terms of cost and time. Due to its significantly more design freedom, AM is capable of designing more complex but optimized parts. Furthermore, since AM makes multiple components in simpler configurations, it provides simpler assemblies. The lightweight and free-form structures are often nature-inspired, and biomimicry concepts are drawn based on those. All these benefits can occur over the entire life cycle of the product and the used feedstock. Other advantages of AM processes include ease of manufacturing and maintenance, higher operational efficiency, enhanced functionality, and improved durability (Ford & Despeisse, 2016; Karunakaran et al., 2012). Furthermore, AM provides multi-material fabrication of complex components which results in achieving functions that were not available before or were costly (Mehrpouya et al., 2019a). This feature, along with the digital thread of production, makes the system capable of the realization of topologically optimized structures to decrease material use (Gao et al., 2015). AM processes also provide a high degree of freedom in the design stage of a product due to its ability to fabricate complex parts and even assemblies. Elimination of various tools (i.e., cutting tools, dies, molds, etc.) combined with a highly automated fabrication process results in a more complex design with the less overall cost, compared to conventional processes. (Khorram Niaki & Nonino, 2017; Mellor, Hao, & Zhang, 2014; Tuck et al., 2008). Moreover, AM facilitates the manufacturing of geometrically complex parts which reduces the material consumption up to 40%. This reduction comes from the fact that the feedstock could be easily be recycled and used for the next parts (Achillas et al., 2015; Khorram Niaki & Nonino, 2017).

Table 2.1 Summary of advantages and difficulties of utilizing AM processes.

Advantages	Difficulties
Product design	
• Design freedom • Optimized geometries and performance • Reduced cost and time • Improved product functionality and durability • Simple assemblies • Upgradable design	• Integrating sustainability using the design for environment or eco-design • Training of designers & engineers concerning the potential utilization & benefits of AM • Certifying new components
Material processing	
• Improved resource efficiency • Decreased by-products from the waste flow • Recycling of waste materials • Localized material recycling	• Improving resource efficiency and specific material recycling • Increasing the use of recycled materials • Unknowns in the environmental performance of processing techniques
Component manufacturing	
• Less energy consumption • Less waste production • Higher access to digital designs • Fewer material inputs • Simple assemblies/supply chains • Higher efficiency in production, cost, and resource usage	• Restricted speed and reliability of AM processes • Quality of products • Increased instrumentation cost • Enhanced cost/energy-effectiveness • High volume manufacturing • Lack of environmental data • Performance of AM technologies and supply chains • Protocols for standards and regulations
Product use	
• Lightweight components • Improved operational • Improved functionality/durability	Risk of malfunction of AM-produced components and assemblies • Certifying repair and remanufacturing processes • Performing maintenance
End-of-life	
• Less waste in repairing • Faster repair • Improved utilization of the product during repair • Improved resource efficiency in the recycling process • Higher recycled material quality	• Restricted recyclability based on materials • Non-recyclability of AM-produced multi-material parts

2.3.1 Eco-design

In eco-design, the environment should be considered in all the design and manufacturing stages to make sure that a minimum environmental impact throughout the life cycle is obtained (Glavič & Lukman, 2007). The goal of eco-design tools can be broken into two stages of environmental design and environmental assessment. Life cycle analysis/assessment (LCA) is one of the most commonly used tools in the former category, and its role is to assess the environmental impact of a process and a product (see Section 2.4.2). The focus of this approach is on the technical issues of the product design from which the optimization of material consumption could be named. There are also other assessment tools, which allow us to conduct a simplified LCA analysis. Eco-indicator 99 is one of the simplified LCA approaches that allow an earlier and more precise analysis (Diegel et al., 2016). Manuals, checklists, guidelines (such as a work by Telenko), which compiled standards and guidelines of design for the environment (Telenko, Seepersad, & Webber, 2008), descriptions of current green technologies, such as the *Eco-Design Handbook* (Fuad-Luke, 2008) are some of the environmental design tools. Using them, the environmental assessment would be possible earlier in the design/development processes (Diegel et al., 2016). A review of the AM design methods is presented along with a framework, where the input is the functional design description rather than a particular part design (Tang & Zhao, 2016). A binder-jetting process was employed in a case study to produce an engine component and the result was a topologically optimized part, for which less energy is required to manufacture in comparison with the same part made by a computer numerical controlled (CNC)-milled component.

2.3.2 Sustainable design

By further elaborating on the eco-design, sustainable product design could be introduced which surpasses the simple eco-design principles. Sustainable design is defined as "design with minimized environmental impacts" while, at the same time, it achieves a considerable amount of economic benefits, especially to the society around and related to the manufacturing activities (Diegel et al., 2010; Morrow et al., 2007). Furthermore, economic gains, ethics, along with socio-economic aspects of sustainability are incorporated with sustainable designs, which use ecological principles as design methods and aims for triple bottom line (TBL) solutions. TBL is a framework that includes financial, social, and environmental aspects of a business. A product that can maximize all these areas is considered an ideal product, which is useful for the environment, the company, and society. An advantage of economic sustainability is a simple quantification of a concept since it is relatively easier to measure compared to environmental and social aspects. Social sustainability is more difficult to quantify due to the fact that its nature is relatively intangible, and it is subjected to many of the factors that are assumed to be beneficial for society. Likewise, environmental sustainability is difficult to quantify as the entire life cycle of the product has to be considered that can become quite a complicated effort (Diegel et al., 2016).

Five common models were developed to help sustainable product design management and diminishing the conflict between these three categories of impact (Charter & Tischner, 2017; Diegel et al., 2016). However, four of the five models concentrate on environmental and economic issues rather than solving social/ethical issues. In addition, none of them addresses design quality as a factor, which influences the durability and reliability of a product (Charter & Tischner, 2017; Diegel et al., 2016). Obtaining integrated functions and part consolidation is an emerging category of design philosophy for AM that has a big impact on sustainable design. It can be considered as a general design flowchart, which supports sustainable AM designs. A design that has fewer part counts without compromising its functionality is presented by this general design framework, as well. According to a case study, redesigning the existing product suggested by the proposed framework has a lower part count and decreases the environmental effect of the manufacturing process of a product. In this case study, it can be observed that design freedom, which is achieved by AM is an indispensable factor that needs to be taken into account when analyzing the environmental impacts of AM-fabricated components (Tang, Yang, & Zhao, 2016).

The potential of AM processes to benefit from a sustainable design comes from the ability to manufacture a wide range of complex components without having conventional manufacturing constraints. Design for AM enables the generation of ideas such as coupling additive and subtractive processes, called hybrid manufacturing, that are more efficient in terms of cost and resources such that they reduce environmental burdens compared to subtractive manufacturing. Moreover, extensive customer-customized products could be obtained by using AM processes. This maximizes customer satisfaction by increasing the design desirability and market acceptance. Eventually, this affects longevity and thus the sustainability of products. In this context, durability is defined as improving the in-use time of a product that can reduce its environmental effects (Malshe et al., 2015).

2.3.3 Design quality and sustainability

From a sustainability point of view, quality has a direct effect on product longevity. The term design quality includes both technical quality and desirability of a product which reflects in the satisfaction of the customers and their willingness to use that product more. AM processes let designers design products that not only function more pleasingly but are also customized for the individual needs of the customer in order to increase the loyalty of the customers to that specific product (Diegel et al., 2016). Numerous research projects have been done on the enhancement of quality in AM, focusing on esthetic aspects such as improving surface finishes or functional perspectives such as enhancing the durability/quality of parts. Considering other aspects of identity and access management services (IAMS) plays a key role in improving the final product quality, which is reached as a culmination of all activities done on the part. This requires a connection or synchronization between design improvement and preparation practices and the fabrication and post-processing steps (Eyers & Potter, 2017).

In general, in layered manufacturing processes, the layer thickness determines the product surface quality. As a result, a curved or inclined surface would be poorly approximated when using a larger layer thickness. This is called the "stair-stepping" effect, which also depends on the AM technology that is employed for building the part. For instance, extrusion AM systems utilize a larger deposition volume for fabrication and as a result, they have the largest layer thickness and lowest (200 μm) geometrical accuracy. On the other hand, laser-assisted and vat photopolymerization processes provide layer thicknesses about half of the ones in extrusion AM systems. In material jetting systems, small jetted droplets create a fine layer thickness in the range of 20 μm. Other than the fabrication process, surface quality is also dependent on the feedstock material and form. For instance, parts manufactured using powder bed AM systems have poor geometrical accuracy due to the fact that the powder particles might agglomerate and partially melt and form rough surfaces out of the tolerance in the design stage (Fotovvati, Namdari, & Dehghanghadikolaei, 2018; Gao et al., 2015).

2.3.4 Design optimization

In this section, a general method of sustainable design optimization in AM processes is presented. This design optimization approach considers the environmental effects as feedback and modifies the product design process resulting in decreased environmental effects of AM processes (Tang, Mak, & Zhao, 2016). One of the important features of AM processes for offering relevant physical properties (i.e., droplet size, geometrical accuracy, curing temperature, melting temperature, etc.) is the ability to simulate, plan, and optimize the process. The processes and process parameters that could be simulated include but not limited to droplet size in jetting, melt pool temperature in high-energy beam processes (i.e., laser beams, ion beams, electron beams), etc. However, the simulation process of AM methods could be challenging due to the fact that many different parameters from the processes, the equipment, the fabrication conditions, and the material properties are directly affecting the quality of the final product (Brenken et al., 2019; Dehghanghadikolaei et al., 2018b).

Traditional manufacturing techniques generally constrain the optimal design of products, whereas in AM it can be employed to increase product performance and functionality (Dehghanghadikolaei et al., 2018a; Malshe et al., 2015). A database of different AM processes is needed in order to fully investigate the effect of different parameters, optimize the fabrication conditions, and obtain a high-accuracy control of the process. However, all the parameters might not result in a comprehensive evaluation of the process since a number of them have a negligible effect on the quality of the products. Thus, one should maintain a filtration process of data in order to use the most important ones, known as key characteristics (KC) (Al-Meslemi, Anwer, & Mathieu, 2018; Thornton, 1999). Al-Meslemi et al. (2018) proposed a connection between the identification of KCs, which are used to decrease production's inconsistencies, and the sustainability goals. Pinkerton (2016) declared that AM processes are likely to possess fewer environmental impacts while the deposition source is an energy-efficient laser system. Some of the first-generation AM processes have high energy consumption and low production rate which could be replaced with some energy-friendly modifications that consume less energy, enhance the controllability of the process, and increase the production rate. Finding the optimum laser parameters can help to improve the microstructure and mechanical properties of samples as well (Mehrpouya & Emamian, 2017; Mehrpouya, Lavvafi, & Darafsheh, 2018;

Mehrpouya et al., 2018; Mehrpouya et al., 2019b). Computational methods could be useful approaches to enhance the design stage, simulating the lightening process of products by making hollow parts or the use of new materials as advanced replacements for conventional materials. Fulfilling all these results in a more sustainable product which is easier to design, analyze, control, and optimize (Hao et al., 2010)

2.4 Sustainability assessment

Due to the crucial role of environmental protection and preventing precious resource depletion, the environmental aspects of manufacturing processes have been carefully investigated. AM processes are no exception and one of the essential aspects of research in this field was covering their sustainability and environmental issues before introducing them as commercial tools (Liu et al., 2016; Mehrpouya et al., 2019c). In order to have the grounds to start with the sustainability analysis of AM processes, one should consider the material waste management and the product/raw material transportation as the two major impacting factors of AM sustainability. Villamil et al. (2018), have reported that investigating these two elements, that is, reduction in fuel consumption and material waste, could be the most significant challenges which reflect in overall cost reduction and a more sustainable manufacturing process. In addition to that, it is important to maintain sustainability issues as efficient tools for policy and public communication which carry information regarding the country/region of manufacturing and their performance (Mani et al., 2014). Sustainability, in general, is categorized into three different sections of environmental, social, and economic impacts which are analyzed via LCA and applied to different manufacturing processes such as AM processes. Some important aspects or dimensions of sustainability in additive manufacturing processes are presented in Fig. 2.3.

2.4.1 Costs of additive manufacturing and production

Different forms of economic impacts and sustainability measures exist which could be integrated in order to provide a powerful tool for cost reduction (Gao et al., 2015). Efficiency improvement of design, reduction in idle time of manufacturing steps, and the order of different manufacturing processes come prior to tooling design and cost reduction in the next steps of an analysis process (Fotovvati & Asadi, 2019; Götze, Peças, & Richter, 2019). A combination of all these factors results in a considerable

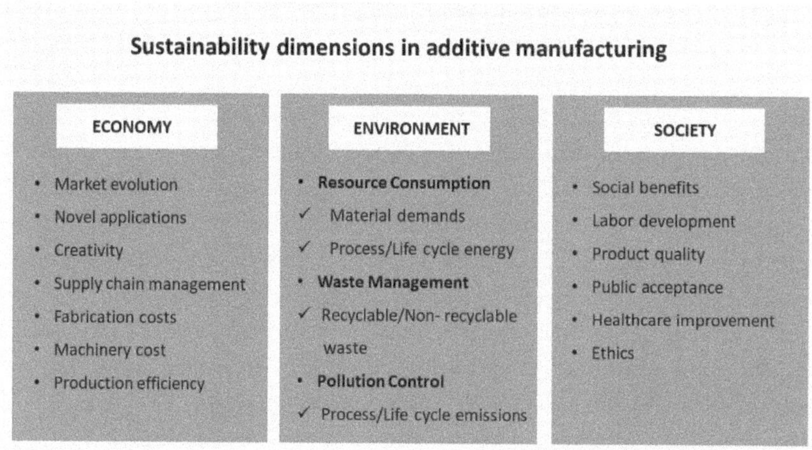

Figure 2.3 Sustainability dimensions of additive manufacturing. *Reproduced from (Peng et al., 2018) Sustainability of additive manufacturing: an overview on its energy demand and environmental impact. Additive Manufacturing, with kind permission of Elsevier.*

cost reduction in the most efficient way. Ford and Despeisse (2016), concluded that with the production of lightweight materials and components via AM processes, that is, porous structures for bone implants (Dehghanghadikolaei et al., 2019b), vehicle parts in automotive industries (Böckin & Tillman, 2019), AM of lightweight materials such as Mg-based alloys with enhanced durability and performance, etc. (Ibrahim et al., 2019). It leads to a predicted cost reduction from an energy consumption point of view in transportation, which is predicted to be around \$56–219 billion by the year 2025. Since AM processes only use the needed amount of feedstock material for fabrication, the remaining materials can be reused with a low cost of refining and preparation, the overall material waste is considered to be the minimal amount, compared to conventional processes (Attaran, 2017). Despeisse and Ford (2015), stated that a combination of lower material waste and lower feedstock material with shorter supply chain leads to a modified manufacturing process which is able to save a significant amount of expenses, that is, \$113–370 billion by 2025. Caffrey (2013), predicted the AM industry will grow to \$10.8 billion in 2021 which shows more than 300% growth compared to the growth value of \$3.07 billion in 2013. Kellens et al. (2017), estimated that in the industries with the highest potential to utilize AM processes, for example, tooling, medical, and aerospace, etc., the cost reduction can reach a peak of \$593 billion with the lowest prediction of \$170 billion. Gebler, Uiterkamp, and Visser (2014) categorized the AM-related markets as (1) a predicted potential of \$230–550 billion for AM-related processes, and (2) a predicted medical part fabrication of \$100–200 billion, \$30–50 billion in tooling and mold production for conventional processes, and a general market of \$100–300 billion for consumer products. AM processes have the potential of real part fabrication while there are numerous applications in the prototyping of 3D components for industry, research, education, etc. (Chekurov et al., 2018; Kellens et al., 2017; Wohlers 2016). As illustrated in Fig. 2.3, market preparation, handling the supply chain, manufacturing, machine, and tooling costs are the main aspects of the economic investigation of AM sustainability analysis. Thomas (2016), introduced three different stages of the economic evaluation of AM processes which are discussed as:

1. **Manufactured products and their value:** added value includes all the sections from raw materials and manufacturing processes to the cost of all the purchased accessories and instruments for fabrication and quality assurance. The main effect of added value is to raise the value of a product by assessing all the expenses during the manufacturing process such as salary of the workers, taxes, and fees paid for that particular product, and the gross operating costs after considering all the segments of a production process. In an estimation by Wohlers and Caffrey (2014), the USA portion in AM-fabricated products is around \$498 million which nearly half of the whole market that is around \$1.307 billion in 2014. Thomas (2016), evaluated the value added to the AM-fabricated products to be \$667 million in the year 2014. This amount is 0.01% of the total value added to the manufacturing products in the world.
2. **An estimate on technology distribution/adoption:** Kim and Park (2013), clarifies that different types of resources, that is, natural resources for raw materials, workers, etc., a comprehensively defined process schedule, and sufficient potentials, that is, abilities to control and integration of different processes and their flexibility in different applications, are crucial to be available for creation of specific services/products purposes. Since AM processes provide high degrees of flexibility, controllability, and integration, they are well-prepared candidates to be considered for different firms in order to provide different services. The potential of AM-related services is predicted to surpass \$4.4, 16, and 196.8 billion in 2020, 2025, and 2035, respectively

(Thomas, 2016). Due to the facts and accurate predictions, there is a crucial need to change the paths that AM processes are adopting in the current market.

3. **Cost/benefit measurement of applicable technology:** cost models are the most important stages of a production process since the overall cost, the proper manufacturing processes, and the benefits/financial outcome of them dictate their suitability for a specific market or application. Cost-effectiveness is the most important part of and effective extension of AM technology for providing different services in the current and the future market. As a result, it is important to combine the information on process properties/capabilities and a well-prepared and accurate cost model to reach the maximum precision in evaluating the potential of AM processes and increase the cost-effectiveness of the products (Cunningham et al., 2017). To this purpose, the AM processes should be compared to traditional manufacturing processes to assess the efficiency in cost reduction for a similar service/product and it should be considered that the resources must be selected wisely in order to have the least raw material consumption while maintaining the waste management and recycling/reusing the resources.

   In recent years, the advancement of cost models had amazing progress for different AM processes. Cost models, which are utilized for conventional manufacturing approaches concentrate on costs related to labor and materials, whereas new production methods require cost models capable of considering the high influence of overheads and investments on production (Cunningham et al., 2017). Recent studies have discussed the need to investigate the economic benefits of AM regarding the life cycle cost. The main existing cost models for cost estimation of AM processes are mentioned below (Schröder, Falk, & Schmitt, 2015; Thomas, 2016; Thompson et al., 2016):

   1. *Model by Hopkins and Dickens*

This model proposed the earliest generated AM cost models (Thompson et al., 2016). The model may be utilized for various elements to recognize the particular unit costs to precisely justify break-even points (Hopkinson & Dicknes, 2003; Schröder et al., 2015). The cost of AM processes is calculated according to the following conditions:

- For a single product in a year
- Applicable for high volumes
- For AM equipment with a 90% operation time

The analysis includes labor-related costs (i.e., installation, operation, and post-processing), material-related costs for components and support structures, and machine-related costs (i.e., purchase, consumables, and maintenance) (Thompson et al., 2016). Other factors including energy consumption and space-related costs, such as rental, etc., are taken into account, still, they do not exceed 1% of the total costs and are neglected in this model (Hopkinson & Dicknes, 2003; Thomas, 2016). *Hopkins and Dickens* proposed a model to compare AM processes, such as Stereolithography (SLA), Fused Deposition Modeling (FDM), Selective Laser Sintering (SLS), and Selective Laser Melting (SLM) (Schröder et al., 2015). They reported that SLS was the most desired option (economically) for about 14000 pieces for small parts and about 700 pieces for big parts (Thompson et al., 2016).

2. *Model by Ruffo, Tuck, and Hague*

Ruffo et al. improved the flexibility and authenticity of the model of *Hopkins and Dickens'* (Thompson et al., 2016). This model estimates the production cost of small and medium batch size systems. The cost model follows the cost-related tasks to direct and indirect costs, providing an accurate survey of the manufacturing costs. Accordingly, cost-related tasks include software, hardware, material, labor, machine, and process management (Schröder et al., 2015). *Ruffo et al.* assert that the major factors in the cost models are material and time (Ruffo, Tuck, & Hague, 2006). Their estimation is based on 100 h/week machine work for 50 weeks/year, which, in other words, is 57% utilization. *Ruffo et al.* introduced three different models for time calculations as mentioned below:

a) Laser scan and sintering time
b) Layers generation time
c) Preheating and cooling time for the powder bed

The *Hopkinson and Dickens's model* estimated the price of each produced part independently, while *Ruffo* model predicted the price of each part including the design cost, which is applicable for more products.

3. *Model by Gibson, Rosen, and Stucker*

According to this model, the AM total cost is the summation of the costs related to material, labor, machine, and manufacturing, while the production costs principally depend on the manufacturing time. The calculation of manufacturing time is based on the mathematical methods for measuring the vectors to build the product. Accordingly, layer-wise production can be utilized in modeling any type of process through the application of a welding torch, a nozzle, or laser (Fotovvati et al., 2018; Mehrpouya et al., 2019a; Mueller, 2012; Schröder et al., 2015).

4. *Model by Ingole*

This cost model was developed to enhance rapid tooling by making it quicker and less expensive utilizing the FDM process. It was applied to investigate the level of process/investment variations. In addition to the first three above-mentioned costs included in the model by Gibson, Rosen, and Stucker, this model includes the pre and post processing costs, as well. The calculation of the costs is according to dimensional and homogeneous equations including the advantages of utilizing AM. This model is based exclusively on a cost-benefit model and it is less qualified to apply only by considering the costs (Ingole et al., 2009; Schröder et al., 2015).

5. *Model by Lindemann*

This model takes a time-driven activity-based costing method. Accordingly, all activities are divided into four principal process steps which include preparation, manufacturing, post-processing, and material treatments to enhance the product characteristics (Lindemann et al., 2012; Schröder et al., 2015). Overall, cost models for different AM processes require considering (Schröder et al., 2015):

- Costs related to waste and recycling
- Support structures
- Manufacturing time
- Design optimization for effective use of the build volume
- The value and complexity of the product
- The time and cost of post-processing activities
- Quality control regulations

2.4.2 Environmental impacts evaluation

Advanced manufacturing processes and modern technologies provide numerous opportunities for designing and manufacturing new products. AM processes are the most advantageous processes which give the flexibility of fabricating different products with different materials. Thanks to these properties, in the past decade, these processes grew significantly. However, it is needed to investigate their suitability regarding their sustainability (Malshe et al., 2015). Shortly, after developing the first AM machines, the investigations on their environmental sustainability were conducted. Although it is considered that the AM processes are great replacements of conventional processes, there are some adverse properties to them (Bourell, Leu, & Rosen, 2009). As an instance, in AM of metals, the most used type of raw material is in powder form which needs precautions in handling and disposal (Dehghanghadikolaei, 2018). In addition, some types of raw materials, such as Mg powder (Dehghanghadikolaei et al., 2019a), are potentially ready to explode after an oxidization process which can impact the environmental sustainability of these processes. Therefore, the environmental impacts of AM processes are essential to be done based on the predefined regulations, considering the manufacturing by-products, emissions of transportation, material handling, end-of-life of the products, etc. while enhancing adoption of global standards of environmental safety, such as ISO 14000 (Maciel et al., 2019).

Numerous researches believe that AM processes can significantly decrease the environmental effects and energy consumption of manufacturing a product for each unit of gross domestic product (Baumers et al., 2011b; Campbell et al., 2011; Kreiger & Pearce, 2013; Peng et al., 2018). In addition, AM processes are able to reduce the carbon footprint of a production process via utilization of optimized designs which consequently results in less material use and less material waste. The latter ones, lead to a reduction in resource exploitation which reduces the amount of carbon dioxide (CO_2) (Frazier, 2014). Gebler et al. (2014), asserted that the CO_2 emissions should be in the range of 130.5−525.5 million tons. One effective way of decreasing these emissions is to produce lightweight parts which became possible using AM machines. Lightweight components mean a lighter structure and less material used for their fabrication. In a report by Hague (2010), design optimization can result in reduced material use up to 40%. This reduction in material usage comes into account when evaluations released that only a 100 kg decrease in the mass of a long-range aircraft results in $2.5 million and 1.3 million tons of CO_2 emission reduction over their lifespan (Frazier, 2014).

It is important to use proper Eco-designs and environmental evaluations prior to implementing an AM production process. However, these models are barely discussing all the aspects of AM processes due to their complexity and lack of data for a generalized solution. The majority of environmental assessments are based on a single product or a specific environment. In order to resolve this shortcoming, quantitative models must be implemented in order to perform a comprehensive sustainability evaluation of a process or a product, that is, environmental, economic, and social impacts (Liu et al., 2016). Knowing the environmental aspects facilitates determining the influence of a product on a global system (Burkhart & Aurich, 2015). According to environmental activities and the need for the evaluation of the influence, different measurement approaches have been presented (Hertwich & Peters, 2009). Peng et al. (2018), suggested the environmental impacts of AM processes be investigated in three different stages:

1. **Resource depletion:** the main resource consumption in AM processes are reflected in the energy and raw material consumption. Based on different analyses, conventional manufacturing processes in comparison to AM processes consume more raw material and as a result, leave more waste material behind. However, AM processes consume more energy for the fabrication of a similar part which needs to be optimized (Baumers et al., 2011a; Drizo & Pegna, 2006; Kellens et al., 2011; Meteyer et al., 2014). The reason for higher energy consumption in AM processes is that the products are more consumer-based and the number of productions is less comparatively. This problem increases in the AM processes which use high-temperature equipment such as laser, ion, and electron beams (Peng et al., 2018). In different studies by Nopparat and Kianian (Nopparat & Kianian, 2012) along with many other researchers (Zhou et al., 2019), the energy consumption and feedstock utilization were investigated on a product-service system (PSS) scheme in an industrial kit and the results revealed that AM processes utilize less raw material while the energy consumption is less efficient compared to the conventional manufacturing processes.
2. **Waste production and management:** utilization of AM processes could result in a 90% reduction of waste products takes place in subtractive processes for manufacturing the same component, implying that the AM processes produce less waste (Khajavi, Partanen, & Holmström, 2014). Many different applications need reliable mechanical properties of components such as strength, hardness, stiffness, and creep resistance at very high temperatures. In such applications, using intact raw materials, which might be costly, is important. For instance, in aerospace applications, a component must meet a limit of the mentioned properties and as a result, used powders could not be utilized in EBM or SLM (Gisario, Kazarian, Martina, & Mehrpouya, 2019). Since the amount of raw material consumption is less in AM processes, the overall cost reduces, compared to conventional fabrication methods

(Villamil et al., 2018). However, the quality of the used feedstock is not challenging in many applications, and the reuse and recycling reduce the cost of raw materials used for different parts (Nickels, 2019). In order to make sure of the quality of the used feedstock, the particle size, flowability, and other properties must be considered (Denti et al., 2019). In a study by Hiller & Lipson (2009), a new type of 3D fabrication is presented which is able to create components and the recycling of the feedstock is minimal since the input materials (i.e., powder particles) stay intact and keep their initial voxel. This new process utilizes digital matter fabrication which provides the opportunity of printing multiple materials together in the desired design. This type of manufacturing process presents the opportunity of developing desktop printing tools with multiple materials and the lowest needed equipment for feedstock treatment.

3. **Controlling the emissions/pollutions:** use of harmful chemicals, that is, coolants, lubricants, etc., are minimized in AM processes. These chemicals are abundantly used in conventional manufacturing processes, such as high-speed milling and turning processes of special alloys (Ghoreishi, Roohi, & Ghadikolaei, 2018; Mehrpouya, 2013, 2017; Mehrpouya). The applications of chemicals, input materials, and process consumables leave a high degree of uncertainty in the evaluation of the toxicity, end-of-life modes, carcinogenic effects, and human health hazards to individuals, who operate the conventional manufacturing machines. While AM methods are recognized as more material-efficient processes, wide-range data on the quantity of the material wastes (primary or secondary) is not available. The potential toxicity, environmental threads, and chemical solvents in AM can be considered as a subject for novel studies (Gao et al., 2015). Therefore, to realize and investigate the environmental impacts and sustainable potential of the AM processes, future studies are necessary to take into account the following (Hao et al., 2010):

- Quantification of the AM material waste and toxicity.
- Sustainable design optimization and AM process operation solution.
- AM sustainability measurement; and
- AM sustainable engineering materials recognition.

Environmental impacts of AM processes compose of feedstock extraction/preparation, energy consumption, material safety (toxicity, oxidation, etc.), recyclability, and the products end-of-life (i.e., landfilling, reusing, repairing) (Dehghanghadikolaei & Fotovvati, 2019; Fotovvati,

Namdari, & Dehghanghadikolaei, 2019; Luo et al., 1999). Comprehensive AM sustainability analysis results in enhancing the environmental impacts of these methods while in many cases, the adverse effects are mitigated, resulting in a more widely acceptable manufacturing technique (Faludi et al., 2019). Although many efforts have been taken in this matter, it is not straightforward to analyze the environmental impacts of interdisciplinary technologies by only considering the LCA of different processes and products. It is important to look at the processes from an industrial standpoint since many new challenges would be introduced. These challenges could be material characterization, development of new materials, controlling the different processes, identification of the different processes, different approaches of modeling a specific process, characteristics of different machines and tools used in fabrication, and software packages used from the beginning to the end of manufacturing processes. Combination of all of the mentioned sections arises the need to have an integrated means of measurement for the environmental impacts of an AM process (Gupta, Weber, & Newsome, 2012).

Based on Frazier (2014), the most used approaches of evaluating the environmental impacts are LCA, design for environment (DFE), and environmental impact scoring systems (EISS). Society of Environmental Toxicology and Chemistry (SETAC) first introduced LCA as an internationally approved means of systematic measurement of environmental impacts of processes, materials, and products, in the 1990s (Gungor & Gupta, 1999; Standardization, 2006). The reason for this success was that LCA was able to accurately quantify the measurements rather than qualitatively measuring the different impacts as other approaches such as Carbon Assessment and DFE were utilizing (Hernandez et al., 2012). Along with SETAC, United Nations Environment Program (UNEP) normalized the method under ISO 14044 standard (Le Bourhis et al., 2013). Considering ISO 14044 standard, LCA divides into four different sections: (1) specification of the study, the goal, the scope, and project specifications are identified, (2) life cycle inventory (LCI) in the form of identification and quantification for all input flows (raw material, energy consumption, manufacturing process, etc.) and all the outputs (waste material, emissions, etc.), (3) life cycle impact assessment (LCIA) is implemented to get the effect of quantification of data on human health and ecosystem (LCIA includes different techniques such as CML, Impacts 2002 +, Eco-Indicator 99, etc.), and (4) interpretation of the results which generally takes place by an automated series of calculations and comparisons (Burkhart & Aurich, 2015; Jolliet et al., 2003). A conceptual framework of LCA is presented in Fig. 2.4.

Up to now, LCA remains the most widely accepted and used analysis method from cradle to grave including materials, type of energy, type of manufacturing process, product, transportation, waste material, emissions, recycling, and end of life of a product (Burkhart & Aurich, 2015; Curran, 1996) (see Fig. 2.5). For evaluation of environmental performance, a process model can be applied based on the life cycle concept. Accordingly, the produced AM part can have the following principal steps (Malshe et al., 2015):

1. Preparation of building materials
2. Input the materials in the structures
3. Making the AM parts layer by layer
4. Post-treatment processes
5. Distribution of the built products
6. Disposal of the parts

In order to have a better view of the process, energy, and material consumption, and also process wastes and disposal can be taken into account in the production to estimate the environmental performance of the process (Malshe et al., 2015). Thanks to its accurate analysis, LCA is the broadly applicable method of analyzing the environmental effects of AM processes in digits. Despite the discussions in the scientific articles published on the advantages and disadvantages of AM processes, using LCA shows that CNC machining consumes a higher amount of energy compared to FDM for the same feedstock material and component. In addition, with a number of modifications inside the analysis window, it became possible to further reduce the energy consumption as low as half of the current values (Liu et al., 2016). In the metallic AM processes

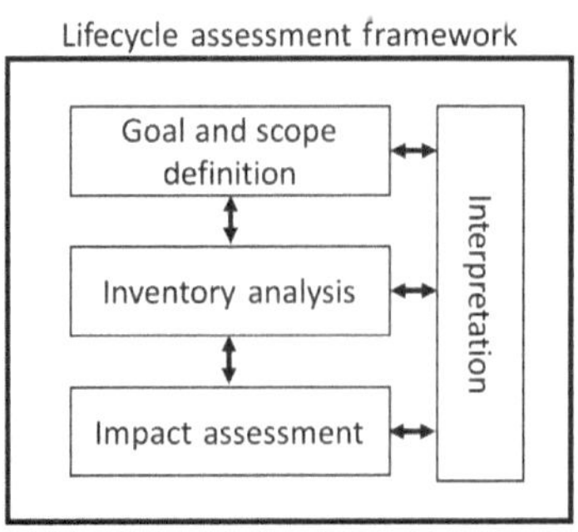

Figure 2.4 Conceptual framework of LCA.

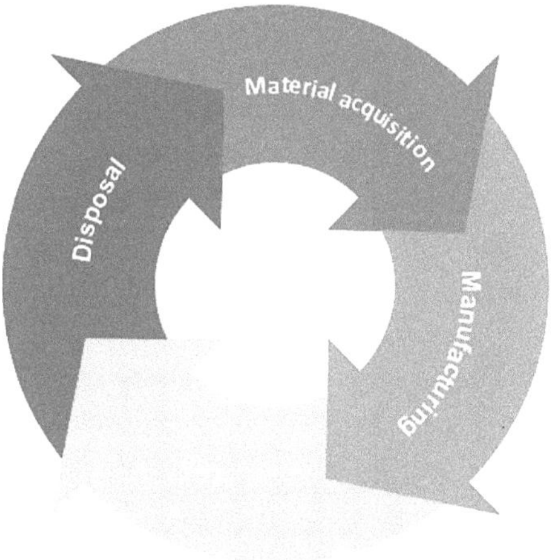

Figure 2.5 Life cycle phases (Burkhart & Aurich, 2015).

(the ones that use metallic feedstock) such as direct metal deposition (DMD), laser engineered net shaping (LENS), and construction laser additive direct (CLAD), it has been showed that through an LCA analysis, these methods are more cost-effective, less energy-consuming, and less waste generating processes compared to conventional processes on the same materials for the same components(Gao et al., 2015). Huang et al. (2016), predicted that the amount of greenhouse gas emissions (GHG) and energy consumption while using metallic airplane components manufactured by AM processes will result in a significant amount of 92.1–215 million metric ton, and 70–173 million GJ/year, respectively, by the year 2050. These results were only for the AM processes inside the United States of America.

Liu et al. (2016), proposed a framework of sustainability assessment for the parts fabricated by AM processes. The first step of an AM process is the preparation of a CAD file and at that stage, all the properties of the components such as materials, size, geometry, internal features, type of fabrication, etc. are introduced. All these decisions affect the sustainability of the component through and after the fabrication process and with an educated series of adjustments, it is possible to enhance the sustainability of the product in the primary phases of its design. A new scheme of sustainability analysis was proposed in a study by Faulkner et al. (2012), which utilized a stream mapping method to investigate the effect of water consumption, energy consumption, feedstock usage, hazards to the environment of production, and the labor share throughout the process and to successfully quantify the environmental, economic, and societal impacts. Morrow et al. (2007), put the stress on the potential of an environmentally friendly process which could be used for tool repair and proposed DMD as the most suitable option for this purpose. In addition, they emphasized that sustainability could be directly affected by maintenance, repair, and overhaul (MRO).

The majority of studies focus only on the electrical type of energy during the manufacturing process while in reality other energy types are consumed (Tapia & Elwany, 2014). In a study by Yosofi, Kerbrat, and Mognol (2019), AM processes were investigated for their sustainability and the inventory data during the manufacturing step of the analysis was gathered with a generic method of acquisition and characterization. The proposed method was focused on both i.e. electrical energy consumption and material usage throughout the manufacturing step in AM. In their method, all the aspects of environmental, technical, and economic assessments relating to the consumption of energy, material, and applicable fluids were included. In a similar study by Le Bourhis et al. (2013), a new approach was proposed which considered the overall effect of consumable flows, that is, energy, material, and fluids, and implemented a global view for a precise assessment of the flow consumption. Therefore, the main sections in the input data were the defined consumables in the manufacturing process and the CAD files that contained different design elements, that is, materials, manufacturing processes, geometries, internal features, etc.

Similarly, Kerbrat et al. (2016), presented a model that did not take into account the CAD file and the part properties but combined the effect of type of electricity consumed, used materials, and fluids throughout the process in order to accurately assess their influence on the environment. In addition, different process parameters that contributed to the different strategies of manufacturing were investigated from both analytical models and experimental results as a justification of the inputs. Cost-effectiveness is believed to have an equal effect on environmental sustainability during the design process. For this purpose, economic life cycle assessment could be implemented to evaluate the efficiency of decision making and different constraints related to the design and fabrication methods. Sustainability Life Cycle Assessment (SLCA) is also a qualitative technique to identify challenges and strengths in the complete life cycle of a part, evaluating both the social and environmental dimensions. The SLCA has a life cycle thinking perspective, evaluating all cycle stages, which is used to identify hot spots of sustainability impacts, and it can provide important information, which is associated during the product development processes (Hallstedt et al., 2013; Villamil et al., 2018). Other environmental measures should be taken for the less known byproducts such as emissions of CO_2 and NO_x, eco-toxicity, etc. They are considered to be the less investigated sectors of industrial sustainability while they could have a considerable impact on the environment. As an instance, many of the resin manufacturing companies simply landfill or incinerate their waste materials which could result in harmful consequences. In order to eliminate or mitigate this effect, it would be useful if biodegradable materials are used for manufacturing and chemical extraction to be applied to the disposed of liquid wastes. Although these might sound like good options, this field still needs a comprehensive investigation of alternatives (Malshe et al., 2015).

2.4.3 Social impacts of additive manufacturing

AM processes can be socially beneficial if the process is well-designed and well-organized. In this regard, the promising social impacts of AM processes are tied to the optimization of products, enhancement of product functions and defining new applications for the existing products, and a considerable reduction in energy and material consumption with minimal waste production (Garetti & Taisch, 2012). In addition to all of these, the social benefits of AM processes include the wellbeing of the labor force, a healthier and safer working environment, and a local impact on introducing new opportunities to the societies around them (Gao et al., 2015). Based on a research by Ford et al., AM processes can be potentially improving working conditions by less exposure time of the workers to a potentially hazardous working environment, higher quality of manufacturing since these processes are mostly done in enclosed environments and generally, they are equipped with fume exhausts, a better labor distribution, a higher potential to be creative in designing a manufacturing process and developing new products, and an opportunity for self-expression of the operators (Ford & Despeisse, 2016). Huang et al. (2013) reported that AM processes can provide improved health quality and life enhancement from a technical point of view. Although many other researchers are working in the assessment of the social impacts of AM processes, there is still a noticeable gap in this area of research which could be a potential topic for different research groups all around the world.

2.4.4 Sustainable manufacturing standards

Standards are the most crucial part of each design, manufacturing, marketing, and any other action that is conducted in a wide range of activities. Using standards, one can make sure of repeatability of a process with the same quality, and in the case of manufacturing, it lets the manufacturing processes be repeated with the same accuracy, order, and systematic process control for all the steps of the different processes (Henzold, 2006). Different internationally accepted and approved standards can be applied to any type of organization from a small machine shop to a multinational manufacturing company. In addition to that, another benefit of utilization of standards can be an easier integration of different processes in a smart manufacturing program which makes the labor share, fabrication, transportation, and marketing easier tasks. An instance is that several companies produce different components of a structure or a machine, and they are shipped to another location for assembly. Using standards, it is not a difficult task to make the parts according to the defined specifications and assemble them in a place where the manufacturing facilities are not in contact with the producers of the spare parts (Lu, Morris, & Frechette, 2016).

Similar to all the practices which use standards, sustainability analysis has standards for measurement and analysis which were established in 2008, by ASTM in order to have an integrated mean of interpreting sustainable manufacturing. The focus of this program is mainly on the sustainability assessment of different manufacturing processes and especially AM. Since sustainability is a naturally complicated series of measures, it is crucial to have a unified set of measures to assess the individual or global effect of different parameters on a product or a manufacturing process. It is known that the collection of experimental data from numerous different manufacturing corporations might seem impossible, but thanks to the quantifications applied and the standardization of the measures, there are many commercially available standards for sustainability assessments (Garetti & Taisch, 2012). For instance, ISO 22400 published an addendum standard for energy consumption/management on KPIs. Despite numerous efforts to define integrated standards for the sustainability of AM processes and manufacturing processes in general, there are still gaps in the fields of social and economic sustainability of AM and other manufacturing processes (Lu et al., 2016).

2.5 Conclusion

Industrial revolutions in recent centuries have changed the domain of manufacturing processes in various fields. AM technology has attained much attention in recent years due to providing a novel opportunity to manufacture customized and complex shape parts. Accordingly, the major roles of factories, designers, and customers are fundamentally redefined in this novel generation of manufacturing.

AM has provided a lot of innovations in numerous industrial fields such as design, medical/biomedical, aerospace, and automotive. In particular, it gives a great opportunity for rapid prototyping, free dimension, time and cost-saving production. In this way, AM is called as a sustainable production process which preserves the resources and significantly protects the environment. In product design, AM provides an amazing opportunity for manufacturing eco-design products so that environmental matters are considered in any design and fabrication process. In this way, AM helps to produce high quality and sustainable products. In general, sustainability issues are categorized in three sections including economic, environmental, and social, and there are various assessment tools, for example, life-cycle assessment tools (LCA), that can assess the sustainable influence of the AM products in various aspects.

On the economic side, cost models have a direct influence on the financial benefits of a production process and effect directly on the market. Therefore, sustainability measurement can efficiently reduce the cost and manufacturing time while it improves the design process. Besides, the advantages of AM processes for environments, including waste management or pollution reduction, are essential to be considered for manufacturing products and total product's life cycle. Accordingly, global standards and regulations are established, such as ISO 14000, for environmental safety. The social benefits of the AM process have a direct effect on both the quality of products and having a safer and healthier working condition as well.

In summary, AM industries are quickly growing, and all machines and materials are improving day-by-day. Therefore, the regulations and standards which are related to the sustainability of AM processes are still needed to be completed and improved in all aspects. There are still gaps between the economic and social sustainability of AM and the other manufacturing processes, that is expected to be filled by researcher by conducting further research, development and innovations in the field.

References

Achillas, C., et al. (2015). A methodological framework for the inclusion of modern additive manufacturing into the production portfolio of a focused factory. *Journal of Manufacturing Systems*, *37*, 328–339.

Al-Meslemi, Y., Anwer, N., & Mathieu, L. (2018). Environmental performance and key characteristics in additive manufacturing: a literature review. *Procedia CIRP*, *69*, 148–153.

Attaran, M. (2017). The rise of 3-D printing: the advantages of additive manufacturing over traditional manufacturing. *Business Horizons*, *60*(5), 677–688.

Baumers, M., et al. (2011a). Energy inputs to additive manufacturing: does capacity utilization matter. *EOS (Rome, Italy)*, *1000*(270), 30–40.

Baumers, M., et al., Sustainability of additive manufacturing: measuring the energy consumption of the laser sintering process. Proceedings of the Institution of Mechanical Engineers, Part B: Journal of Engineering Manufacture, 2011b. 225(12): p. 2228–2239.

Böckin, D., & Tillman, A.-M. (2019). Environmental assessment of additive manufacturing in the automotive industry. *Journal of Cleaner Production*, *226*, 977–987.

Bourell, D. L., Leu, M. C., & Rosen, D. W. (2009). *Roadmap for additive manufacturing: identifying the future of freeform processing* (pp. 11–15). Austin, TX: The University of Texas at Austin.

Brenken, B., et al. (2019). Development and validation of extrusion deposition additive manufacturing process simulations. *Additive Manufacturing*, *25*, 218–226.

Burkhart, M., & Aurich, J. C. (2015). Framework to predict the environmental impact of additive manufacturing in the life cycle of a commercial vehicle. *Procedia CIRP*, *29*, 408–413.

Caffrey, T. (2013). Additive manufacturing and 3D printing state of the industry annual worldwide progress report. *Engineering Management Research*, *2*(1), 209–222.

Campbell, T., et al. (2011). *Could 3D printing change the world? Technologies, Potential, and Implications of Additive Manufacturing* (p. 3) Washington, DC: Atlantic Council.

Charter, M., & Tischner, U. (2017). *Sustainable solutions: developing products and services for the future*. Routledge.

Chekurov, S., et al. (2018). The perceived value of additively manufactured digital spare parts in industry: an empirical investigation. *International Journal of Production Economics*, *205*, 87–97.

Chen, D., et al. (2015). Direct digital manufacturing: definition, evolution, and sustainability implications. *Journal of Cleaner Production*, *107*, 615–625.

Cunningham, C., et al. (2017). Cost modelling and sensitivity analysis of wire and arc additive manufacturing. *Procedia Manufacturing*, *11*, 650–657.

Curran, M. A. (1996). Environmental life-cycle assessment. *The International Journal of Life Cycle Assessment*, *1*(3), 179-179.

Dehghanghadikolaei, A. (2018). *Enhance its corrosion behavior of additively manufactured NiTi by micro arc oxidation coating*. ProQuest Dissertations Publishing.

Dehghanghadikolaei, A., & Fotovvati, B. (2019). Coating techniques for functional enhancement of metal implants for bone replacement: a review. *Materials*, *12*(11), 1795.

Dehghanghadikolaei, A., et al. (2018a). Abrasive flow finishing of stainless steel 304 biomedical devices. *Res. Dev. Mater. Sci*, *8*, 1–8.

Dehghanghadikolaei, A., et al. (2018b). Additive manufacturing methods: a brief overview. *J. Sci. Eng. Res*, *5*, 123–131.

Dehghanghadikolaei, A., et al. (2019a). *Biodegradable magnesium alloys. Metals for biomedical devices* (pp. 265–289). Elsevier.

Dehghanghadikolaei, A., et al. (2019b). Improving corrosion resistance of additively manufactured nickel–titanium biomedical devices by micro-arc oxidation process. *Journal of Materials Science, 54*(9), 7333–7355.

Denti, L., et al. (2019). Effect of powder recycling in laser-based powder bed fusion of Ti-6Al-4V. *Manufacturing Technology*.

Despeisse, M. and S. Ford. The role of additive manufacturing in improving resource efficiency and sustainability. in IFIP International Conference on Advances in Production Management Systems. 2015. Springer.

Diegel, O., et al., Tools for sustainable product design: additive manufacturing. 2010.

Diegel, O., et al. (2016). *Additive manufacturing and its effect on sustainable design. Handbook of sustainability in additive manufacturing* (pp. 73–99). Springer.

Drizo, A., & Pegna, J. (2006). Environmental impacts of rapid prototyping: an overview of research to date. *Rapid Prototyping Journal, 12*(2), 64–71.

Esmaeilian, B., Behdad, S., & Wang, B. (2016). The evolution and future of manufacturing: a review. *Journal of Manufacturing Systems, 39*, 79–100.

Eyers, D. R., & Potter, A. T. (2017). Industrial additive manufacturing: a manufacturing systems perspective. *Computers in Industry, 92*, 208–218.

Faludi, J., et al. (2019). Novel materials can radically improve whole-system environmental impacts of additive manufacturing. *Journal of Cleaner Production, 212*, 1580–1590.

Faulkner, W., et al. Visualizing sustainability performance of manufacturing systems using sustainable value stream mapping (Sus-VSM). in Proceedings of the 2012 international conference on industrial engineering and operations management. Istanbul, Turkey. 2012.

Ford, S., & Despeisse, M. (2016). Additive manufacturing and sustainability: an exploratory study of the advantages and challenges. *Journal of Cleaner Production, 137*, 1573–1587.

Fotovvati, B., & Asadi, E. (2019). Size effects on geometrical accuracy for additive manufacturing of Ti-6Al-4V ELI parts. *The International Journal of Advanced Manufacturing Technology, 104*(5-8), 2951–2959.

Fotovvati, B., Etesami, S. A., & Asadi, E. (2019). Process-property-geometry correlations for additively-manufactured Ti–6Al–4V sheets. *Materials Science and Engineering: A, 760*, 431–447.

Fotovvati, B., Namdari, N., & Dehghanghadikolaei, A. (2018). Fatigue performance of selective laser melted Ti6Al4V components: state of the art. *Materials research express, 6* (1), 012002.

Fotovvati, B., Namdari, N., & Dehghanghadikolaei, A. (2019). On coating techniques for surface protection: a review. *Journal of Manufacturing and Materials Processing, 3*(1), 28.

Fotovvati, B., et al. (2018). A review on melt-pool characteristics in laser welding of metals. *Advances in Materials Science and Engineering, 2018*.

Frazier, W. E. (2014). Metal additive manufacturing: a review. *Journal of Materials Engineering and Performance, 23*(6), 1917–1928.

Fuad-Luke, A. (2008). *The eco-travel handbook: a complete sourcebook for business and pleasure*. Thames & Hudson.

Gao, W., et al. (2015). The status, challenges, and future of additive manufacturing in engineering. *Computer-Aided Design, 69*, 65–89.

Garetti, M., & Taisch, M. (2012). Sustainable manufacturing: trends and research challenges. *Production planning & control, 23*(2-3), 83–104.

Gebler, M., Uiterkamp, A. J. S., & Visser, C. (2014). A global sustainability perspective on 3D printing technologies. *Energy Policy, 74*, 158–167.

Ghoreishi, R., Roohi, A. H., & Ghadikolaei, A. D. (2018). Analysis of the influence of cutting parameters on surface roughness and cutting forces in high speed face milling of Al/SiC MMC. *Materials Research Express, 5*(8), 086521.

Ghoreishi, R., Roohi, A. H., & Ghadikolaei, A. D. (2019). Evaluation of tool wear in high-speed face milling of Al/SiC metal matrix composites. *Journal of the Brazilian Society of Mechanical Sciences and Engineering*, *41*(3), 146.

Gisario, A., Kazarian, M., Martina, F., & Mehrpouya, M. (2019). Metal additive manufacturing in the commercial aviation industry: a review. *Journal of Manufacturing Systems*, *53*, 124–149.

Glavič, P., & Lukman, R. (2007). Review of sustainability terms and their definitions. *Journal of cleaner production*, *15*(18), 1875–1885.

Götze, U., Peças, P., & Richter, F. (2019). Design for eco-efficiency—a system of indicators and their application to the case of moulds for injection moulding. *Procedia Manufacturing*, *33*, 304–311.

Gungor, A., & Gupta, S. M. (1999). Issues in environmentally conscious manufacturing and product recovery: a survey. *Computers & Industrial Engineering*, *36*(4), 811–853.

Gupta, N., Weber, C., & Newsome, S. (2012). *Additive manufacturing: status and opportunities*. Washington: Science and Technology Policy Institute.

Hague, G. (2010). *ATKINS: manufacturing a low carbon footprint*. UK: Southborough University.

Hallstedt, S., et al. A decision support approach for modeling sustainability consequences in an aerospace value chain. in ASME 2013 International Design Engineering Technical Conferences and Computers and Information in Engineering Conference. 2013. American Society of Mechanical Engineers.

Hao, L., et al. (2010). *Enhancing the sustainability of additive manufacturing*.

Henzold, G. (2006). *Geometrical dimensioning and tolerancing for design, manufacturing and inspection: a handbook for geometrical product specification using ISO and ASME standards*. Elsevier.

Hernandez, N. V., et al. (2012). Development of an expert system to aid engineers in the selection of design for environment methods and tools. *Expert Systems with Applications*, *39*(10), 9543–9553.

Hertwich, E. G., & Peters, G. P. (2009). Carbon footprint of nations: a global, trade-linked analysis. *Environmental Science & Technology*, *43*(16), 6414–6420.

Hiller, J.D. and H. Lipson. Fully recyclable multi-material printing. in Solid Freeform Fabrication Proceedings. 2009. Citeseer.

Hopkinson, N., & Dicknes, P. (2003). Analysis of rapid manufacturing—using layer manufacturing processes for production. *Proceedings of the Institution of Mechanical Engineers, Part C: Journal of Mechanical Engineering Science*, *217*(1), 31–39.

Huang, R., et al. (2016). Energy and emissions saving potential of additive manufacturing: the case of lightweight aircraft components. *Journal of Cleaner Production*, *135*, 1559–1570.

Huang, S. H., et al. (2013). Additive manufacturing and its societal impact: a literature review. *The International Journal of Advanced Manufacturing Technology*, *67*(5-8), 1191–1203.

Hyvarinen, J. (2006). The 2005 world summit: UN reform, security, environment and development. *Review of European Community & International Environmental Law*, *15*(1), 1–10.

Ibrahim, H., et al. (2018). In vitro corrosion assessment of additively manufactured porous NiTi structures for bone fixation applications. *Metals*, *8*(3), 164.

Ibrahim, H., et al. (2019). Ceramic coating for delayed degradation of Mg-1.2 Zn-0.5 Ca-0.5 Mn bone fixation and instrumentation. *Thin Solid Films*, *687*, 137456.

Ingole, D. S., et al. (2009). Rapid prototyping—a technology transfer approach for development of rapid tooling. *Rapid Prototyping Journal*, *15*(4), 280–290.

Jolliet, O., et al. (2003). IMPACT 2002 + : a new life cycle impact assessment methodology. *The international journal of life cycle assessment*, *8*(6), 324.

Karunakaran, K., et al. (2012). Rapid manufacturing of metallic objects. *Rapid Prototyping Journal*, *18*(4), 264–280.

Kellens, K., et al. Energy and resource efficiency of SLS/SLM processes (keynote paper). in SFF Symposium 2011. 2011.

Kellens, K., et al. (2017). Environmental dimensions of additive manufacturing: Mapping application domains and their environmental implications. *Journal of Industrial Ecology*, *21*(S1), S49–S68.

Kerbrat, O., et al. (2016). *Environmental impact assessment studies in additive manufacturing. Handbook of Sustainability in Additive Manufacturing* (pp. 31–63). Springer.

Khajavi, S. H., Partanen, J., & Holmström, J. (2014). Additive manufacturing in the spare parts supply chain. *Computers in industry*, *65*(1), 50–63.

Khorram Niaki, M., & Nonino, F. (2017). Additive manufacturing management: a review and future research agenda. *International Journal of Production Research*, *55*(5), 1419–1439.

Kiely, R. (2005). *Industrialization & development*. Routledge.

Kim, B., & Park, C. (2013). Firms' integrating efforts to mitigate the tradeoff between controllability and flexibility. *International Journal of Production Research*, *51*(4), 1258–1278.

Kreiger, M., & Pearce, J. M. (2013). Environmental life cycle analysis of distributed three-dimensional printing and conventional manufacturing of polymer products. *ACS Sustainable Chemistry & Engineering*, *1*(12), 1511–1519.

Le Bourhis, F., et al. (2013). Sustainable manufacturing: evaluation and modeling of environmental impacts in additive manufacturing. *The International Journal of Advanced Manufacturing Technology*, *69*(9-12), 1927–1939.

Lindemann, C., et al. Analyzing product lifecycle costs for a better understanding of cost drivers in additive manufacturing. in 23th Annual International Solid Freeform Fabrication Symposium—An Additive Manufacturing Conference. Austin Texas USA 6th–8th August. 2012.

Liu, Z., et al. Sustainability of 3D Printing: A Critical Review and Recommendations. in ASME 2016 11th International Manufacturing Science and Engineering Conference. 2016. American Society of Mechanical Engineers.

Lu, Y., Morris, K. C., & Frechette, S. (2016). *Current standards landscape for smart manufacturing systems* (8107, p. 39). National Institute of Standards and Technology, NISTIR.

Luo, Y., et al. Environmental performance analysis of solid freedom fabrication processes. in Proceedings of the 1999 IEEE International Symposium on Electronics and the Environment (Cat. No. 99CH36357). 1999. IEEE.

Maciel, V. G., et al. (2019). Environmental performance of 3D-printing polymerisable ionic liquids. *Journal of cleaner production*, *214*, 29–40.

Malshe, H., et al. Profile of sustainability in additive manufacturing and environmental assessment of a novel stereolithography process. in ASME 2015 International Manufacturing Science and Engineering Conference. 2015. American Society of Mechanical Engineers.

Mani, M., Lyons, K. W., & Gupta, S. (2014). Sustainability characterization for additive manufacturing. *Journal of Research of the National Institute of Standards and Technology*, *119*, 419.

Mehrpouya, M. (2013). Modeling of machining process of nickel-titanium based shape memory alloy. *Department of Mechanical and Manufacturing Engineering, Faculty of Engineering*. Universiti Putra Malaysia.

Mehrpouya, M., Laser welding of NiTi shape memory sheets: experimental analysis and numerical modeling. Department of Mechanical and Aerospace Engineering, Sapienza University of Rome, 2017: p. 134.

Mehrpouya, M., & Emamian, S. (2017). Recent advantages in laser fabrication of micro-channel heat exchangers: fortschritte in der Herstellung von Mikrokanal Wärmetauschern mittels Laserherstellung. *Materialwissenschaft und Werkstofftechnik, 48*(3-4), 205–209.

Mehrpouya, M., et al. (2017). An investigation on the optimum machinability of NiTi based shape memory alloy. *Materials and Manufacturing Processes, 32*(13), 1497–1504.

Mehrpouya, M., Lavvafi, H., & Darafsheh, A. (2018). *Microstructural characterization and mechanical reliability of laser-machined structures. Advances in Laser Materials Processing* (pp. 731–761). Elsevier.

Mehrpouya, M., et al. (2018). Laser welding of NiTi shape memory sheets using a diode laser. *Optics & Laser Technology, 108*, 142–149.

Mehrpouya, M., et al. (2019a). A prediction model for finding the optimal laser parameters in additive manufacturing of NiTi shape memory alloy. *The International Journal of Advanced Manufacturing Technology, 105*(11), 4691–4699.

Mehrpouya, M., et al. (2019b). Numerical study for prediction of optimum operational parameters in laser welding of NiTi alloy. *Optics & Laser Technology, 118*, 159–169.

Mehrpouya, M., et al. (2019c). The potential of additive manufacturing in the smart factory industrial 4.0: A review. *Applied Sciences, 9*(18), 3865.

Mehrpouya, M., et al. (2020a). Investigation on the Functionality of Thermo-Responsive Origami Structures. *Advanced Engineering Materials.*

Mehrpouya, M., et al. (2020b). Investigation on shape recovery of 3D printed honeycomb sandwich structure. *Polymers for Advanced Technologies.*

Mellor, S., Hao, L., & Zhang, D. (2014). Additive manufacturing: a framework for implementation. *International Journal of Production Economics, 149*, 194–201.

Meteyer, S., et al. (2014). Energy and material flow analysis of binder-jetting additive manufacturing processes. *Procedia CIRP, 15*, 19–25.

Mihelcic, J. R., et al. (2003). Sustainability science and engineering: the emergence of a new metadiscipline. *Environmental Science & Technology, 37*(23), 5314–5324.

Morrow, W., et al. (2007). Environmental aspects of laser-based and conventional tool and die manufacturing. *Journal of Cleaner Production, 15*(10), 932–943.

Mueller, B. (2012). Additive manufacturing technologies–Rapid prototyping to direct digital manufacturing. *Assembly Automation, 32*(2).

Nickels, L., AM cleans up its act. Metal Powder Report, 2019.

Nopparat, N. and B. Kianian, Resource Consumption of Additive Manufacturing Technology. 2012.

Peng, T., et al. (2018). Sustainability of additive manufacturing: an overview on its energy demand and environmental impact. *Additive Manufacturing.*

Pinkerton, A. J. (2016). Lasers in additive manufacturing. *Optics & Laser Technology, 78*, 25–32.

Reeves, P., Additive Manufacturing—a supply chain wide response to economic uncertainty and environmental sustainability. Econolyst Ltd., Derbyshire, UK. 2008.

Ruffo, M., Tuck, C., & Hague, R. (2006). Cost estimation for rapid manufacturing-laser sintering production for low to medium volumes. *Proceedings of the Institution of Mechanical Engineers, Part B: Journal of Engineering Manufacture, 220*(9), 1417–1427.

Sadooghi, A., et al. (2019). Bending strength and notched-sample fatigue life of hBN/TiC-reinforced steel 316 L: a numerical and experimental analysis. *Journal of Composite Materials*, 0021998319875211.

Schröder, M., Falk, B., & Schmitt, R. (2015). Evaluation of cost structures of additive manufacturing processes using a new business model. *Procedia CIRP, 30*, 311–316.

Standardization, I.O.f, Environmental Management: Life Cycle Assessment; Principles and Framework. 2006: ISO.

Tang, Y., & Zhao, Y. F. (2016). A survey of the design methods for additive manufacturing to improve functional performance. *Rapid Prototyping Journal*, *22*(3), 569–590.
Tang, Y., Mak, K., & Zhao, Y. F. (2016). A framework to reduce product environmental impact through design optimization for additive manufacturing. *Journal of Cleaner Production*, *137*, 1560–1572.
Tang, Y., Yang, S., & Zhao, Y. F. (2016). *Sustainable design for additive manufacturing through functionality integration and part consolidation. Handbook of sustainability in additive manufacturing* (pp. 101–144). Springer.
Tapia, G., & Elwany, A. (2014). A review on process monitoring and control in metal-based additive manufacturing. *Journal of Manufacturing Science and Engineering*, *136*(6), 060801.
Telenko, C., C.C. Seepersad, and M.E. Webber, A compilation of design for environment principles and guidelines. ASME Paper No. DETC2008-49651, 2008.
Thomas, D. (2016). Costs, benefits, and adoption of additive manufacturing: a supply chain perspective. *The International Journal of Advanced Manufacturing Technology*, *85*(5-8), 1857–1876.
Thompson, M. K., et al. (2016). Design for additive manufacturing: trends, opportunities, considerations, and constraints. *CIRP annals*, *65*(2), 737–760.
Thornton, A. C. (1999). A mathematical framework for the key characteristic process. *Research in engineering design*, *11*(3), 145–157.
Tuck, C. J., et al. (2008). Rapid manufacturing facilitated customization. *International Journal of Computer Integrated Manufacturing*, *21*(3), 245–258.
Villamil, C., et al. ADDITIVE MANUFACTURING FROM A STRATEGIC SUSTAINABILITY PERSPECTIVE. in DS92: Proceedings of the DESIGN 2018 15th International Design Conference. 2018.
Wohlers, T., Wohlers report 2016. 2016: Wohlers Associates, Inc.
Wohlers, T., & Caffrey, T. (2014). *Wohlers report 2014: 3D printing and additive manufacturing state of the industry*. Fort Collins, CO, USA: Wohlers Associates. Inc.
Yosofi, M., Kerbrat, O., & Mognol, P. (2019). Additive manufacturing processes from an environmental point of view: a new methodology for combining technical, economic, and environmental predictive models. *The International Journal of Advanced Manufacturing Technology*, 1–13.
Zhou, Y., et al. (2019). Selective laser melting of typical metallic materials: an effective process prediction model developed by energy absorption and consumption analysis. *Additive Manufacturing*, *25*, 204–217.

CHAPTER 3

Experimental study on feasibility of a novel vegetable oil-based pressurized spray cooling system for sustainable surface grinding process

Janak Valaki
Mechanical Engineering Department, Government Engineering College, Bhavnagar, India

3.1 Introduction

Metal cutting fluid, also known as metalworking fluid (MWF), are indispensable integral element of metal cutting processes. MWFs increase productivity and quality of machining by providing efficient cooling and stable lubrication. MWFs play an effective role in controlling friction, amount of heat generation, and cutting force development in machining. Selection of appropriate cutting fluid type, optimum amount, and flow conditions largely affects machinability and helps to extend tool life, enhance work surface quality, and attain sustainability especially when cutting difficult-to-machine (DTM) materials. The main functions of metal cutting fluids are manifolds (Gupta & Laubscher, 2017; Gupta, Laubscher, Davim, & Jain, 2016):

- To provide cooling between tool-work interfaces by taking away the heat generated.
- To lubricate the tool-chip interface and prevents the heat to generate by decreasing the amount of friction.
- To reduce tool wear by protecting tool surface from abrasion, adhesion, and chipping.
- To protect work surface by ensuring smooth removal of chips.
- To flush away chips, debris, and residues from the machining zone.

Apart from the indispensable role of MWFs, their cost of the application in metal cutting contributes significantly in the production

Sustainable Manufacturing
DOI: https://doi.org/10.1016/B978-0-12-818115-7.00005-5

economics. It is reported that the cost associated with cutting fluid preparation, supply, usage, maintenance, end-of-life waste treatment, and safe disposal ranges from 15% to 20% of the total machining cost (Gupta & Laubscher, 2017; Klocke & Eisenblätter, 1997; Marksberry & Jawahir, 2008; Pusavec, Krajnik, & Kopac, 2010; Pusavec, Kramar, Krajnik, & Kopac, 2010). In some cases, it may rise as high as two times the cost of machining. Moreover, application of cutting fluids in machining affects economics, personnel health, and environment as it involves high disposal cost, serious health issues due to their toxicity, and nonbiodegradable nature. The amount of cutting fluid required depends upon the amount of heat generated during machining process. High energy-intensive processes, that is, machining and finishing of DTM materials, require a huge quantity of cutting fluid to dissipate the heat generated at the tool-chip interface in the machining zone.

Grinding is a unique metal cutting (finishing) process used to manufacture semifinished and mostly finished parts due to its ability to produce high surface quality and surface integrity. However, it is the most energy, resource, and time-intensive machining process. Therefore, Grinding process has a very high environment, economic and technological impacts (Klocke & Kuchle, 2009). Abrasive impregnated or embedded wheel is used in grinding to precisely remove a very small amount of material from the work surface. Grinding can produce high dimensional accuracies, geometrical tolerances, and precise surface finish in metallic, ceramics, composites, and difficult to cut materials with relatively higher rates (Hadad, 2015). Unlike other conventional machining processes, grinding employs intermittent and non-homogeneous shearing action along with rubbing and plowing that generates extremely high cutting forces and friction (Malkin & Guo, 2008; Shaw, 1996). Higher cutting speeds employed and thereby cutting forces developed during grinding consumes significantly higher energy per unit volume of material removed, that results into higher temperatures generation compared to its counterparts (Malkin, 1989; Wang et al., 2016). Grinding process has an energy distribution pattern different than other machining processes. In grinding, approximately 65%–85% of the total heat is transferred into workpiece surface (Komanduri & Hou, 2001; Shaw & Vyas, 1994). The heat accumulated in workpiece may cause thermal, mechanical, and metallurgical damages to the surface and subsurface integrity by cracking, microstructural alterations, dimensional changes, oxidation, and corrosion (de Souza Ruzzi et al., 2017; Guo & Malkin, 2000; Sieniawski & Nadolny, 2016).

The heat accumulated in the grinding wheel also affects the grindability of abrasives grits and reduces the wheel life (Jermolajev, Epp, Heinzel, & Brinksmeier, 2016).

In order to ensure sufficient heat dissipation from grinding wheel and workpiece surface to affirm better surface quality of the workpiece with longer service life of the grinding wheel, huge amount of fluid is required to be supplied at work-wheel contact zone. The fluids must have very effective lubrication and cooling capability to keep the temperature at the zone below acceptable values to prevent thermal and other damages, and residual stresses in the work and wheel materials.

Generally, mineral oil-based water-soluble oils in the form of emulsions are used as cutting fluids for grinding in most of the applications. These fluids are mixtures of water and mineral oils mixed with emulsifiers in different proportions depending upon the materials and cutting parameters. In water-soluble emulsions, water cools the surface and oil lubricates the interface (Cambiella et al., 2007). Even though flood cooling is preferable for grinding, it has several drawbacks like high cost of fluid preparation, maintenance and disposal, high pumping power requirement, operator's health risks due to misting and vaporization (Cearsolo, Cabanes, Sanchez, Pombo, & Portillo, 2016). The enormous quantity of used cutting fluid after its end of life has been a major threat due to its toxicity, adverse impact on operator health and environment (Lawal, Choudhury, & Nukman, 2012; Shashidhara & Jayaram, 2010). It is reported that over 80% of all occupational diseases to industrial workers were due to skin contact of cutting fluids (Bennett, 1983; Shokoohi & Shekarian, 2015).

These issues pertaining to the use of huge quantity of MWFs in machining processes that include grinding as well, make it a highly unsustainable machining process from economics, environment, and health point of view. This sustainbility issue of grinding process has compelled research community to explore and develop new cutting fluids and with efficient supply techniques for effective cooling and lubrication for metal cutting processes to attain economy, safety, and sustainability.

In order to improve sustainability of grinding process, the major thrust area is to minimize the usage of harmful cutting fluids and opt sustainable cutting fluids. In order to minimize the usage of cutting fluids, various methods namely dry grinding (Tawakoli & Rabiey, 2007), near dry grinding or minimum quantity lubrication (MQL) grinding (Sadeghi, Haddad, Tawakoli, & Emami, 2009; Tawakoli, Hadad, Sadeghi, Daneshi, & Sadeghi, 2011), and cryogenic grinding (Fredj & Sidhom, 2006) have been attempted by researchers. All the alternate methods are effective to perform both the functions of cooling and lubrication in wheel–workpiece zone efficiently at laboratory and specialised research conditions, however, there is a need of more significant attempts for commercial acceptance of these methods.

3.1.1 Vegetable oils as cutting fluids

Due to the strict environmental regulations and the increased awareness to adopt the sustainable manufacturing practices, the machining sector has started replacing petroleum or mineral-based cutting fluids with more sustainable vegetable oil-based metal cutting fluids, which are less toxic, highly biodegradable, and safe for operator's health (Lawal et al., 2012; Shashidhara & Jayaram, 2010). Vegetable oils are basically consisting of triglycerides, which are glycerol molecules with three long-chain fatty acids attached at the hydroxyl groups via ester linkages (Davim, 2010; Debnath, Reddy, & Yi, 2014). Major advantages of vegetable-based oils compared to conventional MWFs as stated by Khan and Dhar (2006) are: long, heavy, and dipolar molecules of vegetable oils create a dense and strong lubricating film for greater capacity to bear cutting pressure; effective lubrication film helps to reduce friction and prevent heat generation; lower smoke formation due to high flash points, higher boiling point, and greater molecular weight prevents fluid loss, nontoxic, safe, environment-friendly, and biocompatible.

Vegetable oils are comparable to the mineral oil-based metal cutting fluids in terms of fluid properties. Vegetable oils are generally used as water-soluble emulsion by mixing vegetable oil into the water. Its oil part contributes to lubricity index of the emulsion due to high viscosity, and water part contributes to a better cooling effect due to high thermal conductivity.

Higher rotational speed of the wheel during grinding tends to form tiny mists of the emulsion fluid. Moreover, due to lower viscosity and flash point, the conventional MWFs tend to exhibit higher evaporation and recondensation forming tiny mist particles in the surrounding atmosphere. It is reported that vegetable oil-based MWFs produced 30%–90% less mist during machining than mineral oil-based cutting fluids due to evaporation and recondensation (Raynor, Kim, & Bhattacharya, 2005).

3.1.2 Minimum quantity lubrication for grinding

Since the first dedicated attempt to reduce the amount of cutting fluids for metal cutting processes in 1997, Near Dry Machining (NDM) or Minimum Quantity Lubrication (MQL) has gained limited attention among the researchers for grinding process due to the technological limitations, like high wheel wear (high G ratio), high tangential forces, and high surface roughness (SR). Due to this, the application of MQL for grinding was limited for the parts with loose tolerances, and the prime objective was to minimize the usage of cutting fluid (up to 99% less than flood cooling) and hence cost. Some of the major limitations of MQL applications in grinding are as below (Lawal, Choudhury, & Nukman, 2013; Pervaiz, Kannan, & Kishawy, 2018; Sharma, Tiwari, & Dixit, 2016):

- Limited choice of fluids for MQL, as cutting fluids with low viscosity tend to form mist and high viscosity fluids tend to clog the system.
- Cutting fluids with high oxidative stability are to be used.
- Lack of proper cooling ability for hard materials.
- Higher grinding wheel temperature results in high wheel wear and lower grinding ratio.
- Crucial requirement of optimum nozzle position, that is, angle and stand-off distance, flow pressure, and mist control.
- Lacks the ability to evacuate chips from cutting zone.

MQL based on vegetable oil cutting fluids has emerged as an important sustainable grinding practice. Seven different vegetable oils in terms of their lubricating properties for MQL grinding can be ranked as follows: maize oil < rapeseed oil < soybean oil < sunflower oil < peanut oil < palm oil < castor oil (Wang et al., 2016). However, vegetable oils possess some limitations for both flood cooling and MQL, such as poor thermal stability, poor oxidation stability, nozzle clogging, and poor chip evacuation ability (Pervaiz et al., 2018; Sharma et al., 2016).

3.2 Research motivation

3.2.1 Rationale for selection of feedstocks for potential metal cutting fluid

Table 3.1 shows the fatty acid contents of some of the vegetable oils (Giakoumis, 2013, 2018). Selection of the vegetable oils has been done based on the oil chemistry for fatty acid contents. Higher amount of

Table 3.1 Proportion of fatty acids contents in different vegetable oils.

Fatty acid contents	Jatropha oil	Sunflower oil	Soybean oil	Palm oil	Castor oil	Coconut oil
Saturated	21.6	11.3	15.1	49.9	4.5	89.1
Monounsaturated	45.4	21.1	23.4	39.2	91.4	6.3
Polyunsaturated	33	67.6	61.5	10.9	4.1	4.6
Total	100	100	100	100	100	100

saturated fatty acid content contributes to the better oxidative stability of the oil and higher degree of unsaturation is important to form water-soluble emulsion (Al-Sabagh et al., 2012). So, to ensure that the emulsion formed should have higher oxidative stability and form water-based emulsion easily, base oils must have a high percentage of monosaturated fatty acid contents. Oils with high polyunsaturated contents are not desirable, as it is likely to exhibit poor oxidative stability of the emulsion. Castor and palm oils contain more than 90% saturated and monosaturated contents, availability in abundance, and comparatively cheaper. Castor oil possesses better low-temperature viscosity and high-temperature lubrication viscosity and other lubricant properties of the Castor oils are much better than the vegetable oils. Unique properties of castor oils such as high viscosity and specific gravity are due to the 80% ricinoleic acid content, which is the only source of C18:1-OH. It gives the castor oil a unique combination of physical properties such as relatively high viscosity and specific gravity as well as solubility in alcohol in any proportions.

3.2.2 Rationale for proposed pressurized spray cooling method

Grinding is one of the most energy-intensive and unsustainable conventional machining processes. In conventional flood cooling, effectively less fluid is available at wheel–work interface zone due to high rejected flowrate, which tends to deteriorate the cooling and lubrication actions (Marinescu, Rowe, Dimitrov, & Ohmori, 2013). Even though MQL has emerged as a potential technique for reducing the usage of cutting fluids and thereby increasing the sustainability of machining processes, it has limitations to ensure proper cooling and lubrication at the wheel–work interface zone. Cutting fluid jets of high velocity increases the fluid entry in the interface and also increases the coefficient of heat transfer

(Lavisse et al., 2018). In this research, authors have proposed a hybrid method of cutting fluid supply of pressurized spray cooling that is intended to simultaneously utilize the merits of conventional flood cooling and MQL lubrication/cooling, and minimize the limitations of both the techniques. Vegetable oil-based cutting fluids are used in place of mineral oil-based fluids with the increased fluid entry in the wheel—workpiece zone for effective cooling and lubrication. Pressurized droplets of cutting fluid are sprayed exactly at the entry point of wheel—workpiece contact zone, which is expected to increase heat transfer coefficient, evacuate the chips effectively, ensure fresh wheel surface to perform the grinding operation by minimizing the sticking of the chips on the workpiece and wheel surfaces.

3.3 Experimental methodology

3.3.1 Grinding wheel and workpiece materials

Silicon-based grinding wheel with the grade AA46/54 of the dimension 200 × 20 mm, capable of grinding at maximum 3350 RPM has been used in this experimental research. High carbon alloy steel, EN31 grade steel has been used as workpiece material having composition as shown in Table 3.2.

3.3.2 Metal cutting fluids preparation and properties

Three types of cutting fluids such as water-miscible semisynthetic oil (Ecocool 650), castor oil (sulfated castor oil-50% or Turkey Red Oil), and palm oil are used as base oils for preparing cutting fluids. Some of the important properties of the cutting fluids are as listed in Table 3.3. Emulsions of the selected oils were prepared by mixing 80% base oil, 3% anticorrosive agent, 5% antioxidant, and 12% emulsifier, and sodium tripolyphosphate. Cutting fluids were prepared in mixing ratio of 1:12, that is, 1 L of emulsion of oil mixed with 12 L of demineralized water.

Table 3.2 Composition of EN31 steel.

EN31	C	Mn	Si	S	P	Cr
Composition by (%)	0.95	0.45	0.27	0.040	0.040	1.26

Table 3.3 Important properties of cutting fluid.

Sr. no.	Properties	Semisynthetic oil	Castor oil	Palm oil
1	Kinematic viscosity (cS) at 40°C	30	220	41
2	Kinematic viscosity (cS) at 100°C	5.5	19.5	8.21
3	Density (g/cm^3)	0.985	0.955	0.925
4	Thermal conductivity (W/m K)	0.610	0.180	0.170
5	Flash point (°C)	150	250	260
6	Specific heat (J/kg K)	4.118	1.80	1.902
7	Pour point (°C)	18	−27	20
8	pH	9.4	7.0	6.2
9	Refractometer factor, % (v/v)	1.6	1.47	1.45

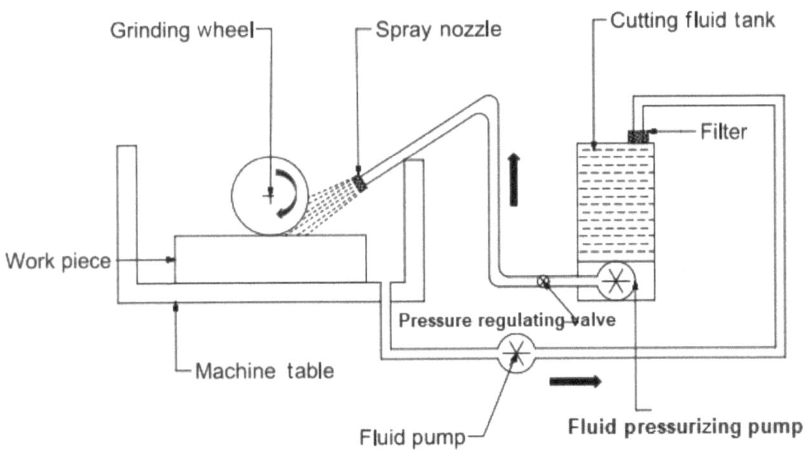

Figure 3.1 Experimental set up for pressurized spray cooling.

3.3.3 Cutting fluid supply

The proposed pressurized spray cooling system is shown in Fig. 3.1. Pressurization of the fluid (being supplied via classical flood cooling method) is obtained through an electrical pressure regulating pump with pressure ratings of 1.53 kg/cm^2 (1.5 bar) and 2.55 kg/cm^2 (2.5 bar), where 1.5 bar is used as low-pressure mode (LP) and 2.5 bar as high-pressure mode (HP) for the experimentation. Cutting fluid flow rates of 10, 1.8,

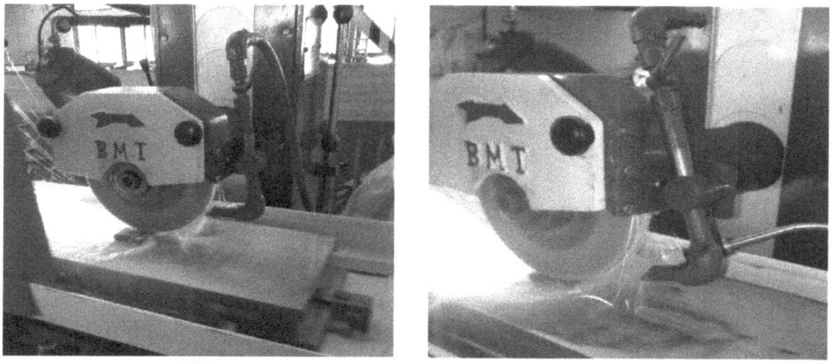

(A) (B)

Figure 3.2 Actual photographs of the (A) flood cooling, and (B) pressurized spray cooling.

Table 3.4 Details of machining parameters.

Machining parameter	Level 1	Level 2	Level 3
Table feed (mm/rev)	0.1	0.2	0.3
Depth of cut (mm)	0.1	0.2	0.3
Table speed (m/min)	22	27	32
Type of cutting fluid	SSO	CO	PO
Fluid supply method	CF	LP	HP

CF, conventional flooding; *CO*, castor oil-based MWF; *HP*, high-pressure spraying at 2.55 kg/cm^2 (2.5 bar); *LP*, low pressure spraying at 1.53 kg/cm^2 (1.5 bar); *PO*, palm oil-based MWF; *SSO*, semisynthetic oil-based MWF.

and 2.5 L/min have been used for conventional flood cooling, LP and HP systems, respectively. The spray mists are formed using swirl generating nozzles. Fig. 3.2 shows the actual photographs of conventional flooding and pressurized spray cooling systems used for the experiments.

3.3.4 Experimental plan

In this work, five input parameters viz. feed rate (f), depth of cut (d), table speed (t), type of cutting fluid (fl), and cutting fluid supply method (m) are used with three levels of each as shown in Table 3.4. SR, SH, and material removal rate (MRR) are measured as output parameters to evaluate the performance and effectiveness of oils and cooling system in grinding process. Experiments are performed as per Taguchi robust design of experiment technique with L_{27} orthogonal array as illustrated in Table 3.5.

Table 3.5 Experimental plan and corresponding results.

Sr. no.	Feed (mm/min)	Depth of cut (mm)	Table speed (m/min)	Fluid	Method	SR (μm)	SH (HV)	MRR (mm^3/min)
1	0.1	0.1	22	SSO	CF	0.366	183.33	28
2	0.1	0.1	22	SSO	LP	0.434	166.67	66.65
3	0.1	0.1	22	SSO	HP	0.474	204.67	50.61
4	0.1	0.2	27	CO	CF	0.325	167.67	106.3
5	0.1	0.2	27	CO	LP	0.3	201.33	53.78
6	0.1	0.2	27	CO	HP	0.41	172.33	103.79
7	0.1	0.3	32	PO	CF	0.431	180.67	171.79
8	0.1	0.3	32	PO	LP	0.498	208.33	159.53
9	0.1	0.3	32	PO	HP	0.6	204.33	168.22
10	0.2	0.1	27	PO	CF	0.476	183.33	74.25
11	0.2	0.1	27	PO	LP	0.626	198.67	45.99
12	0.2	0.1	27	PO	HP	0.362	201.67	67.29
13	0.2	0.2	32	SSO	CF	0.296	155.33	99.12
14	0.2	0.2	32	SSO	LP	0.33	196	132.54
15	0.2	0.2	32	SSO	HP	0.427	196.33	121.82
16	0.2	0.3	22	CO	CF	0.429	185.33	166.74
17	0.2	0.3	22	CO	LP	0.381	187	201.22
18	0.2	0.3	22	CO	HP	0.456	194.33	208.52
19	0.3	0.1	32	CO	CF	0.305	192	72.96
20	0.3	0.1	32	CO	LP	0.421	186.33	106.03
21	0.3	0.1	32	CO	HP	0.606	186.33	179.29
22	0.3	0.2	22	PO	CF	0.601	204.33	148.77
23	0.3	0.2	22	PO	LP	0.48	193.67	114.9
24	0.3	0.2	22	PO	HP	0.557	201	101.65
25	0.3	0.3	27	SSO	CF	0.466	190.67	189.21
26	0.3	0.3	27	SSO	LP	0.429	213.33	152.16
27	0.3	0.3	27	SSO	HP	0.229	191.33	182.27

CF, conventional flooding; *HP*, high-pressure mode; *LP*, low-pressure mode; *MRR*, material removal rate; *SH*, surface hardness; *SR*, surface roughness; *SSO*, semisynthetic oil; *CO*, castor oil; *PO*, palm oil.

3.3.5 Measurement of response parameters

This research study investigates the machining characteristics of SR, SH, and MRR. SR is one of the most important machinability indicators that governs the surface quality of the machined or finished surface and largely affects the performance characteristics of the machined part. Average roughness R_a is the arithmetic mean of the roughness profiles measured along the evaluation length (Bhushan, 2013; Valaki, Rathod, & Sankhavara, 2016). Lower values of SR, that is, precision finish is desirable for high functional performance of the machined parts. SH is also an important property that affects tribological performance. In order to avoid

Figure 3.3 Actual workpiece samples finished by surface grinding process.

premature failure, wear and tear, and other defects, the surface of the machined samples should be hard enough (Neale, 2001; Valaki et al., 2016). MRR is one of the important grinding process parameters from process economics point of view and deals with the productivity of the process. A higher value of MRR is desirable (Valaki et al., 2016).

MGW SRT-1, Taiwan-made SR tester is used for the measurement of average SR. Yamayo Bliss TH-330, make Portable Leeb Metal hardness tester is used for the measurement of Vickers hardness of the machined samples. Average of the five measurements is taken as the final value for each sample for roughness and hardness measurements. Fig. 3.3 shows the actual pictures of the machined samples.

3.4 Results and discussions

Experimental results obtained are shown in Table 3.5. Effects of machining parameters, cutting fluids, fluid supply method, and table speed on SR, SH, and MRR have been investigated and discussed below.

3.4.1 Analysis of variance

Analysis of variance (ANOVA) is a statistical technique, used to identify the significantly contributing parameters from the set of process parameters. Various statistical parameters like the sum of squares, degree of freedom, variance, and percentage contribution can be calculated for each experimental factor. General linear model of ANOVA is used with a 95% confidence level. ANOVA results computed for SR, SH, and MRR are shown in Tables 3.6, 3.7, and 3.8, respectively.

Table 3.6 Analysis of variance for SR, using adjusted SS for tests.

Source	DF	Seq SS	Adj SS	Adj MS	*F*	*P*	% Contribution
Feed	2	0.0082	0.0082	0.0041	1.37	0.282	2.84
Depth of cut	2	0.03961	0.03961	0.01981	6.62	0.088	13.75
Table speed	2	0.00474	0.00474	0.00237	0.79	0.47	1.66
Fluid	2	0.11241	0.11241	0.05621	18.79	**0**	**39.02**
Method	2	0.0752	0.0752	0.0376	12.57	**0.001**	**26.11**
Error	16	0.04787	0.04787	0.00299			
Total	26	0.28803					

Bold values indicate that the parameters having *P* values less than .05 are significant parameters.
SR, surface roughness; *SS*, sum of squares.

Table 3.7 Analysis of variance for SH, using adjusted SS for tests.

Source	DF	Seq SS	Adj SS	Adj MS	*F*	*P*	% Contribution
Feed	2	137.31	276.07	138.04	1.58	0.241	2.85
Depth of cut	2	322.94	501.66	250.83	2.87	0.09	6.7
Table speed	2	472.4	557.78	278.89	3.19	0.072	9.8
Fluid	2	1901.13	1584.77	792.39	9.06	**0.003**	**39.53**
Method	2	749.99	749.99	374.99	4.29	**0.035**	**15.6**
Error	14	1224.35	1224.35	87.45			
Total	24	4808.12					

Bold values indicate that the parameters having *P* values less than .05 are significant parameters.
SR, surface roughness; *SS*, sum of squares.

Table 3.8 Analysis of variance for MRR, using adjusted SS for tests.

Source	DF	Seq SS	Adj SS	Adj MS	*F*	*P*	Contribution (%)
Feed	1	2837.2	1564.4	1564.4	4.09	0.057	**4.44**
Depth of cut	1	53,742.6	52,958.3	52,958.3	138.43	0	**84.05**
Table speed	1	18.6	21.8	21.8	0.06	0.814	0.03
Fluid	1	50	50	50	0.13	0.722	0.08
Method	1	26.4	26.4	26.4	0.07	0.796	0.04
Error	19	7268.6	7268.6	382.6			
Total	24	63943.3					

MRR, material removal rate; *SS*, sum of squares.

It is noted from the ANOVA results that type of cutting fluid and cutting fluid supply method are the most significant contributors to SR, SH, and MRR characteristics. Type of cutting fluid has 39% contribution for both SR and SH, while cutting fluid supply method has 26% and 15% contribution to SR and SH, respectively. Moreover, depth of cut and feed rate are the significant parameters with 84% and 4.44% contribution on MRR. It can be inferred from the ANOVA results that pressurized spray cooling system with vegetable oil-based MWFs significantly governs the surface characteristics of the surface grinding process.

3.4.2 Taguchi optimization

The variations caused in the SR, SH, and MRR during experiments are reported by calculating the S/N ratio values. Based on the process characteristic requirements, minimization objective is selected as "smaller the better" for SR and maximization objectives are selected as "higher the better" for SH and MRR. Following equation gives the S/N ratio and it is denoted by "η" and expressed in dB.

$$\text{S/N ratio for SR is expressed as, } \eta_{\text{SR}} = -10 \log \left(1/\text{n } \Sigma\text{yi2}\right)$$

$$\text{S/N ratio for SH and MRR is expressed as, } \eta_{\text{SH}} = -10 \log \left(1/\text{n } \Sigma 1/\text{yi } 2\right)$$

The mean S/N ratio values of SR, SH, and MRR responses for each level are tabulated in Tables 3.9–3.11.

From the above S/N ratio tables, optimal set of combinations can be identified. For optimal combination for the SR is *A3B1C1D3E3* (Table 3.9), SH is *A3B3C1D3E3* (Table 3.10), and MRR is A3B3C3D2E3 (Table 3.11). With these optimal parametric combinations, the confirmation tests have been conducted and results obtained are tabulated in Table 3.12.

Table 3.9 S/N ratios for surface roughness using Taguchi analysis (smaller is better).

Level	Feed	Depth of cut	Table speed	Fluid	Method
1	7.581	7.11[a]	6.767[a]	8.539	7.949
2	7.721	7.938	8.233	8.077	7.454
3	7.205[a]	7.46	7.509	5.892[a]	7.104[a]
Delta	0.516	0.828	1.466	2.647	0.845
Rank	5	4	2	1	3

[a]Desirable level of S/N ratios for the given parameter.

Table 3.10 S/N ratios for surface hardness using Taguchi analysis (larger is better).

Level	Feed	Depth of cut	Table speed	Fluid	Method
1	45.44	45.52	45.61[a]	45.48	45.2
2	45.49	45.43	45.6	45.37	45.76
3	45.81[a]	45.79[a]	45.53	45.89[a]	45.78[a]
Delta	0.37	0.36	0.09	0.52	0.57
Rank	3	4	5	2	1

[a]Desirable level of S/N ratios for the given parameter.

Table 3.11 S/N Ratios for MRR using Taguchi analysis (larger is better).

Level	Feed	Depth of cut	Table speed	Fluid	Method
1	38.69	36.62	40.13	39.78	40.24
2	40.91	40.48	39.68	41.69[a]	40.25
3	42.45[a]	44.95[a]	42.24[a]	40.59	41.56[a]
Delta	3.76	8.33	2.57	1.91	1.31
Rank	2	1	3	4	5

[a]Desirable level of S/N ratios for the given parameter.

The results of the confirmation tests indicate that for the identified set of optimum combinations, the predicted values of the responses SR, SH, and MRR have shown very good conformance to the experimental values with 93%, 95%, and 97% accuracies respectively. It can also be noticed that high-pressure spray cooling system with palm oil-based MWF has produced optimal SR of 0.68 μm and SH of 195.40 HV. Castor oil-based MWF resulted optimal MRR of 232.55 mm^3/min. It can be concluded that vegetable oil-based MWFs with pressurized spray cooling system have generated optimal results for the responses under consideration than semisynthetic MWF with conventional flood cooling system.

3.4.3 Regression analysis

Regression analysis is a well-known statistical technique to establish the relationship between input process parameters and output parameter or responses. Multiple regression analysis is the best tool for regression modeling. In the present work, least squares method has been used to estimate regression parameters with 99% confidence level. The coefficient of determination (R^2) measures the amount of variation in the dependent variables. The value closer to 1 is desirable. Value of R^2 can be increased

Table 3.12 Confirmation results for optimization.

Response	Objective	Optimum set of process parameters					Mathematical value	Experimental value	%Error
		f	*d*	*t*	*fl*	*m*			
SR (μm)	Minimize	0.3	0.1	22	PO	HP	0.73	0.68	7.35
SH (HV)	Maximize	0.3	0.3	22	PO	HP	186.22	195.40	4.70
MRR (mm^3/min)	Maximize	0.3	0.3	32	CO	HP	241.24	232.55	2.72

HP, high-pressure mode; *MRR*, material removal rate; *SR*, surface roughness; *SS*, sum of squares.

by adding more predictor variables to the regression model, known as R^2 (adj) (Valaki et al., 2016). In this research, regression analysis has been performed using Minitab 17.0 statistical analysis software. General regression is performed using a linear regression model with 95% confidence level. The regression models developed for surface grinding process to fit the SR, SH, and MRR parameters are as shown in Eqs. (3.1), (3.2), and (3.3), respectively.

$$\begin{aligned} SR = & -0.053 + 1.004 * f - 2.852 * d + 0.0694 * t - 0.2849 * fl \\ & - 0.1692 * m - 2.48 * f^2 + 8.85 * d^2 - 0.00149 * 1t^2 + 0.0710 * fl^2 \\ & + 0.0266 * m^2 + 0.0758 * f * m - 0.4382 * d * m + 0.00520 * t * m \\ & + 0.0331 * fl * m \end{aligned} \quad (3.1)$$

$$\begin{aligned} SH = & 36 - 399 * f - 432 * d + 5.2 * t + 54.8 * fl + 5 * m + 2212 * f^2 \\ & + 1078 * d^2 - 0.158 * t^2 - 3.8 * fl^2 - 45.2 * m^2 + 1057 * f * m \\ & - 335 * d * m + 9.85 * t * m - 56.9 * fl * m - 795 fl^2 * m - 645 * f * d * m \\ & - 41.5 * * f * t * m + 115.2 * f * m^2 + 892 * d^{2*} m + 0.8 * d * m^2 \\ & - 0.42 * t * m^2 + 10.33 * fl * m^2 \end{aligned} \quad (3.2)$$

$$\begin{aligned} MRR = & 735 + 703f - 67d - 58.8t + 129.2fl - 105.7m - 1436f^2 \\ & + 2306d^2 + 0.973t^2 - 27.89fl^2 + 14.71m^2 + 29.8f.m - 150.4d.m \\ & + 3.30t.m - 8.00fl.m \end{aligned} \quad (3.3)$$

(Where f = feed, d-depth of cut, t = table speed, fl = cutting oil, m = method of cutting oil supply)

Table 3.13 shows the summary of R-sq and R-sq (adj) for the developed regression models for SR, SH, and MRR. These values indicate that the developed regression models are best fit to the experimental responses (more than 90% accuracy) and the predicted values of all three responses have shown good conformance with their experimental values as shown in Table 3.14.

Table 3.13 Summary table of R-Sq, R-Sq (adj) for the regression models.

Response parameters	SR	SH	MRR
R-Sq	93.03	93.29	90.62
R-Sq (adjusted)	80.66	91.26	87.86

SR, surface roughness; *SH*, surface hardness; *MRR*, material removal rate.

Table 3.14 Experimental and predicted values of parameters with percentage of errors.

Sr. no.	Feed	Depth of cut	Table speed	Fluid	Method	SR			SH			MRR		
	mm/min	mm	m/min			Exp. values	Predicted values	% Error	Exp. values	Predicted values	% Error	Exp. values	Predicted values	% Error
1	0.1	0.1	22	SSO	CF	0.366	0.39	−6.56	183.33	182.44	0.49	28	47.49	−69.61
2	0.1	0.1	22	SSO	LP	0.434	0.41	5.53	216.67	225.13	−3.9	66.65	38.46	42.3
3	0.1	0.1	22	SSO	HP	0.474	0.48	−1.27	204.67	202.8	0.91	50.61	58.85	−16.28
4	0.1	0.2	27	CO	CF	0.325	0.29	10.77	167.67	168.99	−0.79	106.3	93.35	12.18
5	0.1	0.2	27	CO	LP	0.3	0.33	−10	201.33	195.02	3.13	53.78	77.78	−44.63
6	0.1	0.2	27	CO	HP	0.41	0.42	−2.44	172.33	172.65	−0.19	103.79	91.63	11.72
7	0.1	0.3	32	PO	CF	0.431	0.44	−2.09	180.67	179.44	0.68	171.79	178.19	−3.73
8	0.1	0.3	32	PO	LP	0.498	0.49	1.61	208.33	206.65	0.81	159.53	156.08	2.16
9	0.1	0.3	32	PO	HP	0.6	0.6	0	204.33	202.08	1.1	168.22	163.39	2.87
10	0.2	0.1	27	PO	CF	0.476	0.49	−2.94	183.33	183.52	−0.1	74.25	57.86	22.07
11	0.2	0.1	27	PO	LP	0.626	0.61	2.56	198.67	194.5	2.1	45.99	52.31	−13.74
12	0.2	0.1	27	PO	HP	0.732	0.79	−7.92	201.67	200.62	0.52	67.29	76.18	−13.21
13	0.2	0.2	32	SSO	CF	0.296	0.3	−1.35	155.33	152.08	2.09	99.12	95.55	3.6
14	0.2	0.2	32	SSO	LP	0.33	0.33	0	196	196.93	−0.47	132.54	107.46	18.92
15	0.2	0.2	32	SSO	HP	0.427	0.43	−0.7	196.33	191.56	2.43	121.82	148.79	−22.14
16	0.2	0.3	22	CO	CF	0.429	0.43	−0.23	185.33	179.91	2.92	166.74	256.22	−53.66
17	0.2	0.3	22	CO	LP	0.381	0.41	−7.61	187	194.39	−3.95	201.22	212.09	−5.4
18	0.2	0.3	22	CO	HP	0.456	0.43	5.7	194.33	187.87	3.32	208.52	197.38	5.34
19	0.3	0.1	32	CO	CF	0.305	0.31	−1.64	192	188.25	1.95	72.96	87.12	−19.41
20	0.3	0.1	32	CO	LP	0.421	0.43	−2.14	186.33	187.85	−0.82	106.03	109.05	−2.85
21	0.3	0.1	32	CO	HP	0.606	0.6	0.99	186.33	180.77	2.98	179.29	160.4	10.54
22	0.3	0.2	22	PO	CF	0.4801	0.4	16.68	204.33	204.13	0.1	148.77	145.89	1.94
23	0.3	0.2	22	PO	LP	0.48	0.46	4.17	193.67	190.57	1.6	114.9	111.78	2.72
24	0.3	0.2	22	PO	HP	0.557	0.58	−4.13	201	199.55	0.72	101.65	107.09	−5.35
25	0.3	0.3	27	SSO	CF	0.466	0.46	1.29	190.67	192.79	−1.11	189.21	181.06	4.31
26	0.3	0.3	27	SSO	LP	0.429	0.44	−2.56	213.33	203.74	4.5	152.16	164.41	−8.05
27	0.3	0.3	27	SSO	HP	0.442	0.47	−6.33	191.33	191.87	−0.28	182.27	177.18	2.79
								−0.39			**0.77**			**−4.99**

MRR, material removal rate; *SR*, surface roughness; *SH*, surface hardness.

The mean errors have been calculated for SR, SH, and MRR by taking absolute difference between experimental and predicted values of the responses. It can be noticed from the results that the mean error for SR, SH, and MRR found to be 0.39%, 0.77%, and 4.99%, respectively.

Results of the experimental and predicted values are superimposed in the form of trend plots for SR, SH, and MRR as shown in Figs. 3.4, 3.5, and 3.6,

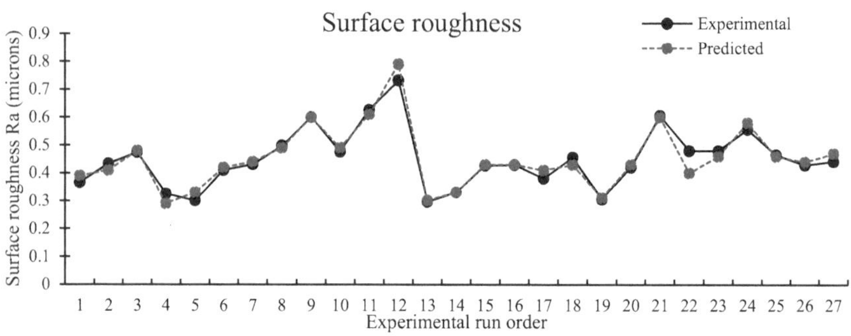

Figure 3.4 Trend of experimental and predicted values of surface roughness.

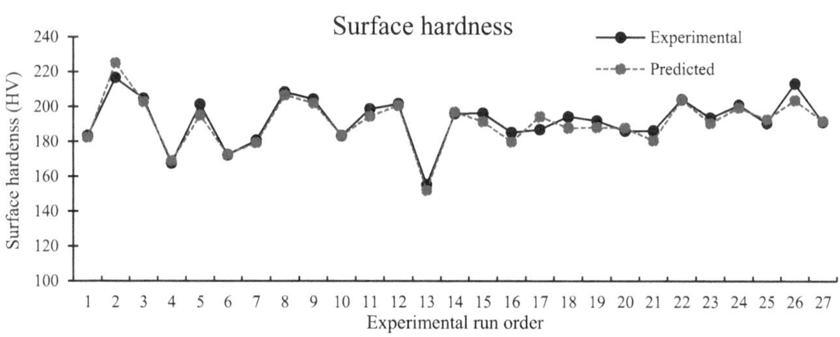

Figure 3.5 Trend of experimental and predicted values of surface hardness.

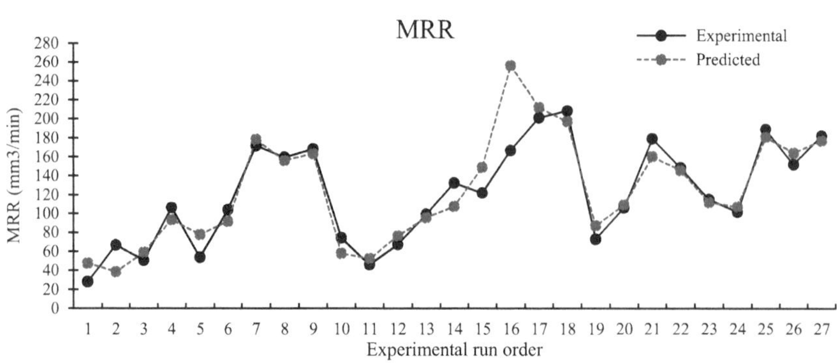

Figure 3.6 Trend of experimental and predicted values of MRR.

respectively. The trend plots indicate that the regression models have good predication power to the actual trends of the dataset. The close agreement between experimental and mathematical values indicates that the developed models can be used for the selection of process parameters for the desired response values.

3.4.3.1 Validation of regression models

In order to assess the validity of the regression models developed, the experiments are performed with different levels of feed rate and depth of cut values to change the combinations for the experiments. The results obtained for the validation/confirmation tests for SR and SH are presented in Tables 3.15–3.17. It can be seen that the regression models developed have good predictability of the responses with mean error of less than 5% over wide range of parameter values.

Table 3.15 Confirmation test results for SR (surface roughness).

Sr. no.	Feed	Depth of cut	Table speed	Fluid	Method	SR-PR	SR-Ex	%Error
1	0.15	0.25	27	1	3	0.41	0.346	−18.5
2	0.25	0.15	27	1	1	0.37	0.31	−19.35
3	0.15	0.25	27	3	3	0.61	0.543	−12.34
4	0.25	0.15	27	3	1	0.44	0.498	11.65
5	0.25	0.15	27	2	3	0.53	0.505	−4.95
6	0.15	0.25	27	2	1	0.35	0.416	15.87
Mean error								**4.66**

Table 3.16 Confirmation test results for SH (surface hardness).

Sr. no.	Feed	Depth of cut	Table speed	Fluid	Method	SH-PR	SH-Ex	% Error
1	0.15	0.25	27	1	3	206.29	190.67	−8.19
2	0.25	0.15	27	1	1	187	192.67	2.94
3	0.15	0.25	27	3	3	130.03	121	−7.46
4	0.25	0.15	27	3	1	173.06	186.33	7.12
5	0.25	0.15	27	2	3	202.59	214.33	5.48
6	0.15	0.25	27	2	1	164.06	211.67	22.49
Mean error								**3.73**

Table 3.17 Confirmation test results for MRR (material removal rate).

Sr. no.	Feed	Depth of cut	Table speed	Fluid	Method	MRR-PR	MRR-Ex	% Error
1	0.15	0.25	27	1	3	117.74	101.94	−15.5
2	0.25	0.15	27	1	1	60.86	70.42	13.58
3	0.15	0.25	27	3	3	105.02	105.72	0.66
4	0.25	0.15	27	3	1	80.14	95.12	15.75
5	0.25	0.15	27	2	3	120.65	122.16	1.24
6	0.15	0.25	27	2	1	153.05	159.71	4.17
Mean error								**3.32**

3.4.4 Effects on response parameters

3.4.4.1 Effects on surface roughness

In surface grinding, lower SR is a desirable surface characteristic. SR generated during grinding is mostly attributed to uneven grinding forces which subsequently contribute to the detachment of burrs through shearing, rubbing or plowing of the work material due to improper lubrication.

Fig. 3.7A shows the 3D bar plots for effects of type of cutting fluid and cutting fluid supply method on SR for depth of cut of 0.1 mm. It can be observed from Fig. 3.7B that palm oil-based MWF with HP system has produced low SR compared to other combinations of fluid and method. The reason might be attributed to the fact that better viscosity of vegetable oils might have generated better lubrication at wheel−work interface and produced uniform and continuous shearing action. Moreover, better lubrication would have also produced uniform wheel wear to get uniform grinding action in order to obtain better surface finishes. In addition, high-pressure spray of tiny droplets of low-density vegetable oils would have produced effective flushing due to better cleaning of resolidified burrs from work and wheel surfaces, which in turn might have produced uniform and better surface finish. Fig. 3.7B represents cumulative response behavior of the effects of type of fluid and fluid supply method on SR. It can be noticed that the combination of castor oil-based MWF with low-pressure spray cooling results into lower SR.

3.4.4.2 Effects on surface hardness

High SH is a desirable surface characteristic and it is directly affected by the cooling capacity of the cutting fluid used for grinding. Change in

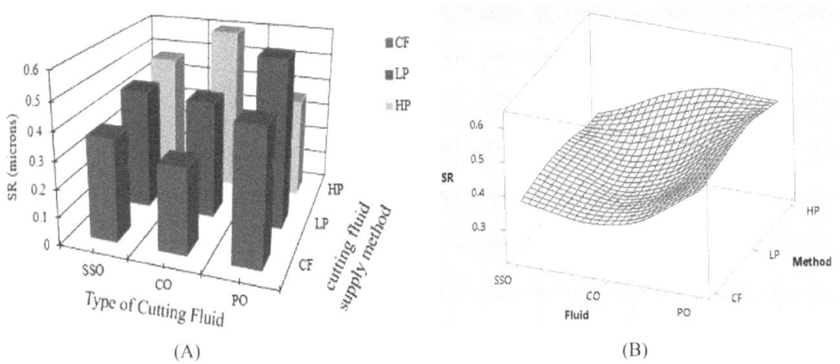

Figure 3.7 Effects on SR versus fluid and method. (A) SR versus fluid and supply method (for $d = 0.1$ mm); (B) SR versus fluid and supply method (for $d = 0.1$ mm). *SR*, surface roughness.

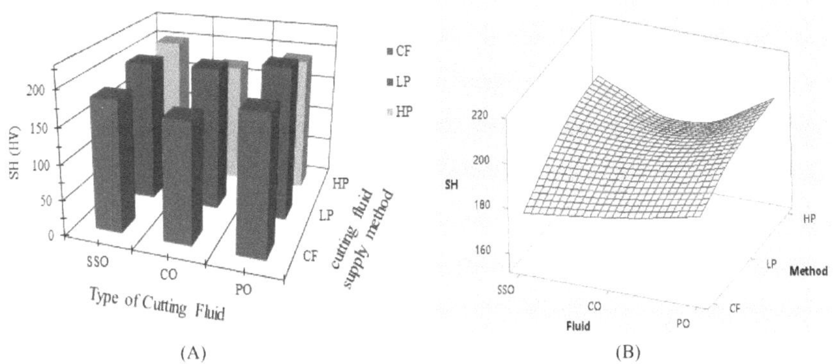

Figure 3.8 Effects on SH versus fluid and method. (A) SH versus fluid and supply method (for $d = 0.3$ mm); (B) SH versus fluid and supply method (for $d =$ 0.3 mm). *SH*, surface hardness.

material hardness during grinding is an indication that material must have experienced quenching, annealing, or severe strain hardening due to plastic deformation of the work material before it leaves the material surface in the form of burrs.

Fig. 3.8A shows the bar plots for effects of type of cutting fluid and cutting fluid supply method on SH for depth of cut 0.3 mm. It can be observed from the plot that vegetable oil-based MWFs with pressurized spray cooling system have produced comparable or higher SH compared to conventional flooding system. It might be attributed to the fact that impingement of high-pressure spray in discrete droplet form must have broken the continuous fluid layer formed on the wheel and work surfaces, which in turn would have increased heat transfer coefficient. Improved heat dissipation capability with spray cooling might have resulted into efficient cooling of the work surfaces, causing tempering and hardening effects to produce hard surfaces. It can also be noticed from Fig. 3.8D that combination of vegetable oil-based MWFs and pressurized spray cooling system results into the better surface finish. In addition, experimental results are similar to the data generated by statistical and optimization analyses, where it has been reported that palm oil with HP system generated optimal SH.

3.4.4.3 Effects on material removal rate

High MRR is desirable in the surface grinding process. It indicates unit amount of material removed for a certain set of parameters. Chips in the form of burrs generated in surface grinding are too small for direct removal from the material surface. It tends to get entrapped in the deep surface valleys and get rewelded due to high temperature. Change in MRR is due to the effectiveness of the burr evacuation tendency of the cooling method used apart from parametric influence.

Fig. 3.9A represents the 3D bar plots for effects of type of cutting fluid and cutting fluid supply method on MRR for depth of cut of 0.3 mm. It can be observed from the plots that vegetable oil-based MWFs with pressurized spray cooling system produced high MRR compared to conventional flooding system. It is attributed to the fact that high-velocity

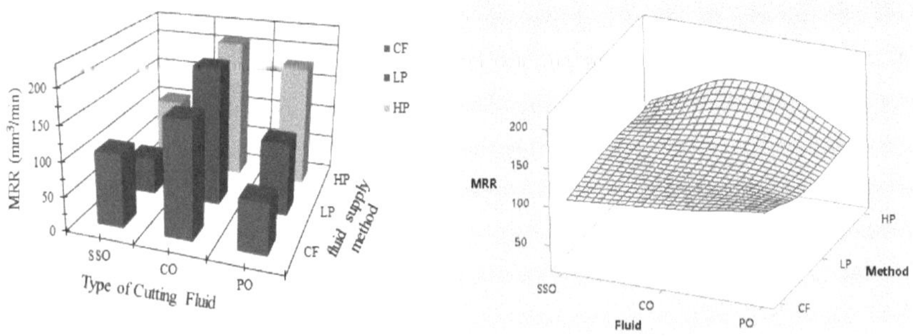

(A) (B)

Figure 3.9 Effects on MRR versus fluid and method. (A) MRR versus fluid and supply method ($d = 0.3$ mm); (B) MRR versus fluid and supply method (for $d = 0.3$ mm). *MRR*, material removal rate.

droplets upon impingement on the material surface tend to remove the entrapped and rewelded burrs through effective splashing effect due to lower densities of vegetable oils. Availability of fresh surface for grinding results into improved MRR. Fig. 3.9B represents the cumulative response behavior of MRR under the influence of fluid and method. It can be noticed that castor oil-based MWF with high-pressure system, produced highest MRR. Moreover, the role of vegetable oil-based MWFs and pressurized spray cooling system becomes very imperative to obtain better MRR. In addition, these results are well inline to the data generated by statistical and optimization analyses, where it has been reported that castor oil with HP system generated optimal MRR.

From the above discussions, it is clear that pressurized spray cooling method with castor and palm oil-based MWF have the potential to produce lower SR, and higher SH and MRR for sustainable surface grinding process.

3.5 Conclusions

The present research summarizes the detailed investigations conducted on green vegetable oil-based cutting fluids and sustainable pressurized spray cooling system-based surface grinding process. The following specific conclusions can be drawn from this work:

- Type of cutting fluid has been identified as the significant parameter for SR and SH with 39% contribution for each response. While type of cutting fluid supply method has been found as the significant parameter for SR and SH with 26% and 15.6% contribution, respectively. However, depth of cut and feed rate have been found as significant parameters for MRR with 84% and 4.5% contribution, respectively.
- Pressurized spray system with palm oil resulted optimum value for SR of 0.68 μm and SH of 195.4 HV. Moreover, pressurized spray system with castor oil generated optimum MRR of 232.55 mm^3/min.
- The developed regression models have shown good predictability of the responses SR, SH, and MRR with mean error of 0.39%, 0.77%, and 4.99%, respectively.
- High-pressure spray cooling system resulted better performance for SR, SH, and MRR than conventional flood cooling method. The pressurized spray system with palm oil-based MWF resulted high machinability in context of SR and SH for surface grinding process.

However, pressurized spray cooling with castor oil-based MWF produced high machinability for MRR.

- Pressurized spray cooling system can reduce the amount of cutting fluid requirement in surface grinding process by 70%–80% and the produce better machinability compared to conventional flood cooling system.
- Pressurized spray cooling system can be employed as a sustainable alternate in place of two extreme methods for cutting fluids supply for surface grinding process, that is, conventional flood cooling and MQL.

The experimental studies carried out in this research study suggest that pressurized spray cooling system with vegetable oil-based green cutting fluids is a feasible and sustainable solution for improved machinability and sustainability of surface grinding process.

Acknowledgment

This research did not receive any specific grant from any funding agency. However, the authors wish to express their gratitude to the Principal, Government Engineering College, Bhavnagar, India and Commissioner of technical education, Gujarat state, India for providing infrastructural and moral support to carry out this research.

References

Al-Sabagh, A. M., Khalil, S. A., Abdelrahman, A., Nasser, N. M., Noor Eldin, M. R., Mishrif, M. R., & El-Shafie, M. (2012). Investigation of oil and emulsion stability of locally prepared metalworking fluids. *Industrial Lubrication and Tribology*, *64*(6), 346–358.

Bennett, E. O. (1983). Water based cutting fluids and human health. *Tribology International*, *16*(3), 133–136.

Bhushan, B. (2013). *Introduction to tribology*. John Wiley & Sons.

Cambiella, A., Benito, J. M., Pazos, C., Coca, J., Hernández, A., & Fernández, J. E. (2007). Formulation of emulsifiable cutting fluids and extreme pressure behaviour. *Journal of Materials Processing Technology*, *184*(1–3), 139–145.

Cearsolo, X., Cabanes, I., Sanchez, J. A., Pombo, I., & Portillo, E. (2016). Dry-dressing for ecological grinding. *Journal of Cleaner Production*, *135*, 633–643.

Davim, J. P. (2010). *Surface integrity in machining*. London: Springer.

Debnath, S., Reddy, M. M., & Yi, Q. S. (2014). Environmental friendly cutting fluids and cooling techniques in machining: a review. *Journal of Cleaner Production*, *83*, 33–47.

de Souza Ruzzi, R., de Mello Belentani, R., de Mello, H. J., Canarim, R. C., D'Addona, D. M., Diniz, A. E., . . . Bianchi, E. C. (2017). MQL with water in cylindrical plunge grinding of hardened steels using CBN wheels, with and without wheel cleaning by compressed air. *International Journal of Advanced Manufacturing Technology*, *90*(1–4), 329–338.

Fredj, N. B., & Sidhom, H. (2006). Effects of the cryogenic cooling on the fatigue strength of the AISI 304 stainless steel ground components. *Cryogenics*, *46*(6), 439–448.

Giakoumis, E. G. (2013). A statistical investigation of biodiesel physical and chemical properties, and their correlation with the degree of unsaturation. *Renewable Energy*, *50*, 858–878.

Giakoumis, E. G. (2018). Analysis of 22 vegetable oils' physico-chemical properties and fatty acid composition on a statistical basis, and correlation with the degree of unsaturation. *Renewable Energy*, *126*, 403–419.

Guo, C., & Malkin, S. (2000). Energy partition and cooling during grinding. *Journal of Manufacturing Processes*, *2*(3), 151–157.

Gupta, K., & Laubscher, R. F. (2017). Sustainable machining of titanium alloys: A critical review. *Proceedings of the Institution of Mechanical Engineers, Part B: Journal of Engineering Manufacture*, *231*(14), 2543–2560.

Gupta, K., Laubscher, R. F., Davim, J. P., & Jain, N. K. (2016). Recent developments in sustainable manufacturing of gears: A review. *Journal of Cleaner Production*, *112*, 3320–3330.

Hadad, M. (2015). An experimental investigation of the effects of machining parameters on environmentally friendly grinding process. *Journal of Cleaner Production*, *108*, 217–231.

Jermolajev, S., Epp, J., Heinzel, C., & Brinksmeier, E. (2016). Material modifications caused by thermal and mechanical load during grinding. *Procedia CIRP*, *45*, 43–46.

Khan, M. M., & Dhar, N. R. (2006). Performance evaluation of minimum quantity lubrication by vegetable oil in terms of cutting force, cutting zone temperature, tool wear, job dimension and surface finish in turning AISI-1060 steel. *Journal of Zhejiang University-SCIENCE A*, 7(11), 1790–1799.

Klocke, F., & Kuchle, A. (2009). *Manufacturing processes*. Heidelberg: Springer.

Klocke, F. A., & Eisenblätter, G. (1997). Dry cutting. *CIRP Annals*, *46*(2), 519–526.

Komanduri, R., & Hou, Z. B. (2001). A review of the experimental techniques for the measurement of heat and temperatures generated in some manufacturing processes and tribology. *Tribology International*, *34*(10), 653–682.

Lavisse, B., Lefebvre, A., Torrance, A. A., Sinot, O., Henrion, E., Lemarié, S., & Tidu, A. (2018). The effects of the flow rate and speed of lubricoolant jets on heat transfer in the contact zone when grinding a nitrided steel. *Journal of Manufacturing Processes*, *35*, 233–243.

Lawal, S. A., Choudhury, I. A., & Nukman, Y. (2012). Application of vegetable oil-based metalworking fluids in machining ferrous metals—A review. *International Journal of Machine Tools and Manufacture*, *52*(1), 1–2.

Lawal, S. A., Choudhury, I. A., & Nukman, Y. (2013). A critical assessment of lubrication techniques in machining processes: A case for minimum quantity lubrication using vegetable oil-based lubricant. *Journal of Cleaner Production*, *41*, 210–221.

Malkin, S. (1989). Grinding Technology, Theory and Applications of Machining with Abrasives, Soc. of Manuf. Eng., Dearborn, MI.

Malkin, S., & Guo, C. (2008). *Grinding technology: Theory and application of machining with abrasives*. Industrial Press Inc.

Marinescu, I., Rowe, W.B., Dimitrov, B., Ohmori, H. (2013). Tribology of abrasive machining processes, Chapter 9 (Abrasives and Abrasive Tools, pp. 243–311).

Marksberry, P. W., & Jawahir, I. S. (2008). A comprehensive tool-wear/tool-life performance model in the evaluation of NDM (near dry machining) for sustainable manufacturing. *International Journal of Machine Tools and Manufacture*, *48*(7–8), 878–886.

Neale, M. J. (2001). *Lubrication and reliability handbook*. Newnes.

Pervaiz, S., Kannan, S., & Kishawy, H. (2018). An extensive review of the water consumption and cutting fluid based sustainability concerns in the metal cutting sector. *Journal of Cleaner Production*. Available from 10.1016/j.jclepro.2018.06.190.

Pusavec, F., Krajnik, P., & Kopac, J. (2010). Transitioning to sustainable production−Part I: Application on machining technologies. *Journal of Cleaner Production*, *18*(2), 171 104.

Pusavec, F., Kramar, D., Krajnik, P., & Kopac, J. (2010). Transitioning to sustainable production−part II: Evaluation of sustainable machining technologies. *Journal of Cleaner Production*, *18*(12), 1211−1221.

Raynor, P. C., Kim, S. W., & Bhattacharya, M. (2005). Mist generation from metalworking fluids formulated using vegetable oils. *Annals of Occupational Hygiene*, *49*(4), 283−293.

Sadeghi, M. H., Haddad, M. J., Tawakoli, T., & Emami, M. (2009). Minimal quantity lubrication-MQL in grinding of Ti−6Al−4V titanium alloy. *International Journal of Advanced Manufacturing Technology*, *44*(5−6), 487−500.

Sharma, A. K., Tiwari, A. K., & Dixit, A. R. (2016). Effects of Minimum Quantity Lubrication (MQL) in machining processes using conventional and nanofluid based cutting fluids: A comprehensive review. *Journal of Cleaner Production*, *127*, 1−8.

Shashidhara, Y. M., & Jayaram, S. R. (2010). Vegetable oils as a potential cutting fluid—An evolution. *Tribology International*, *43*(5−6), 1073−1081.

Shaw, M.C. (1996). Principles of Abrasive Processing (Oxford Series on Advanced Manufacturing).

Shaw, M. C., & Vyas, A. (1994). Heat-affected zones in grinding steel. *CIRP Annals*, *43* (1), 279−282.

Shokoohi, Y., & Shekarian, E. (2015). Application of nanofluids in machining processes—A review. *Journal of Nanoscience and Technology*, 59−63.

Sieniawski, J., & Nadolny, K. (2016). The effect upon grinding fluid demand and workpiece quality when an innovative zonal centrifugal provision method is implemented in the surface grinding of steel CrV12. *Journal of Cleaner Production*, *113*, 960−972.

Tawakoli, T., Hadad, M., Sadeghi, M. H., Daneshi, A., & Sadeghi, B. (2011). Minimum quantity lubrication in grinding: Effects of abrasive and coolant−lubricant types. *Journal of Cleaner Production*, *19*(17−18), 2088−2099.

Tawakoli T., Rabiey M. (2007). An innovative concept for dry grinding with resin and vitrified bond CBN wheel. In International Grinding Conference ISSAT.

Valaki, J. B., Rathod, P. P., & Sankhavara, C. D. (2016). Investigations on technical feasibility of *Jatropha curcas* oil-based bio dielectric fluid for sustainable electric discharge machining (EDM). *Journal of Manufacturing Processes*, *22*, 151−160.

Wang, Y., Li, C., Zhang, Y., Li, B., Yang, M., Zhang, X., ... Liu, G. (2016). Experimental evaluation of the lubrication properties of the wheel/workpiece interface in MQL grinding with different nanofluids. *Tribology International*, *99*, 198−210.

Wang, Y., Li, C., Zhang, Y., Yang, M., Li, B., Jia, D., ... Mao, C. (2016). Experimental evaluation of the lubrication properties of the wheel/workpiece interface in minimum quantity lubrication (MQL) grinding using different types of vegetable oils. *Journal of Cleaner Production*, *127*, 487−499.

CHAPTER 4

Remanufacturing and refurbishment in the age of Industry 4.0: an integrated research agenda

Okechukwu Okorie[1], Konstantinos Salonitis[2] and Fiona Charnley[1]
[1]Exeter Center for Circular Economy, Science, Innovation, Technology and Entrepreneurship Department, The Business School, University of Exeter, Exeter, United Kingdom
[2]Sustainable Manufacturing Systems Centre, School of Aerospace, Transport and Manufacturing, Cranfield University, Bedfordshire, United Kingdom

4.1 Introduction

Research into resource efficiency and circular strategies due to rising environmental concerns, increased competition in the market, constantly changing customer needs, business model innovation, and the emergence of new regulations concerning end-of-product-life has increased in the last decade (Okorie et al., 2018) Thus concepts like sustainability (Etzion, 2018) and more recently, the circular economy (CE) (Circle Economy, 2016; Kalmykova, Sadagopan, & Rosado, 2017; Korhonen, Nuur, Feldmann, & Birkie, 2018) have had growing use among the academia, industry, and policy-making circles. Most academic literature on the CE (Kirchherr, Reike, & Hekkert, 2017) draw their definition from the Ellen MacArthur Foundation's definition. They define the CE as an "industrial system which is restorative and regenerative by intention and design, replacing the end-of-life concept with restoration, a shift toward the use of renewable energy, and aims for the elimination of waste through the superior design of components, ranging from materials to business models" (Ellen MacArthur Foundation, 2013a). In the academia, CE research has rapidly grown; more than 100 articles were published on the topic in 2016, as opposed to about 30 articles in 2014. In 2018, these had increased to over 400. At the national and regional level of discourse, countries such as China has embraced resource efficiency regulations through the implementation of the CE law in 2008 (CIRAIG, 2015)

Sustainable Manufacturing
DOI: https://doi.org/10.1016/B978-0-12-818115-7.00001-8

while the European Union has followed up with a CE package which is an extension of its waste directive (European Commission, 2015).

According to Potting, Hekkert, Worrell, and Hanemaaijer (2017), the 9R's effectively capture the circular strategies important for resource efficiency. These include Refuse, Rethink, Reduce, Repair, Remanufacture, Refurbish, Repurpose, Recycle, and Recover. Of interest to this study is remanufacturing and refurbishment. These circular strategies have been classified as focused on "extending lifespan of products and its parts," along with reuse, repair, and refurbish (Potting et al., 2017). Product remanufacturing is an industrial process that transforms used products or cores into useful life, by taking it through a sequence of operations (Sundin, 2004). As described in a recent UNEP study, refurbishment refers to the modification of an object that is a waste of a product that takes place within maintenance or intermediate maintenance operations (Unep, 2018a). The traditional goal of refurbishment is to increase or restore performance or/and functionality of the refurbished product (Unep, 2018a). A remanufactured product is made to be as new in condition or performance or better, hence the often-used phrase, "like-new" (Ijomah, McMahon, Hammond, & Newman, 2007).

Thus several literature have argued that the similarity in definition and practice of remanufacturing and refurbishment remains a challenge for manufacturers and customers (Abbey, Meloy, Blackburn, & Guide, 2015; Okorie, Salonitis, Charnley, & Turner, 2018a); remanufacturing, for example, has been described in several names such as "reconditioned, refurbished, recertified, certified preowned, and more" (Abbey et al., 2015). Understanding the difference has only continued in the literature, where it is argued that while a refurbished item is usually priced approximately 30% lower than the new product (Mugge, Jockin, & Bocken, 2017), a remanufactured product which is restored to a like-new or higher condition than the new product. Thus differences in quality and price (Rathore, Kota, & Chakrabarti, 2011). However, this lack of clarity and inconsistencies between the definitions and their practical descriptions have only continued in research and industrial practice (Gharfalkar, Ali, & Hillier, 2016).

As with the concept of the CE, Industry 4.0 (INDUSTRY 4.0)—also known as smart manufacturing—has been the focus of research, industry, and policymakers since it originated from the high-tech strategy of the German government promoting the computerization of manufacturing (Stankovic & Gupta, 2017). It has been defined in terms of full-scale automation (Lasi, Fettke, Kemper, Feld, & Hoffmann, 2014) as well as in relation to emerging technologies (Stankovic & Gupta, 2017). Its influence and integration in research have given rise to various new terms such as Supply Chain 4.0 (Makris, Hansen, & Khan, 2019), remanufacturing 4.0 (Yang, Aravind, Kaminski, & Pepin, 2018), and construction 4.0 (Craveiro, Duarte, Bartolo, & Bartolo, 2019). Thus the argument can be made for the integration of Industry 4.0 technologies and CE principles (Beatriz et al., 2018), for example, recovering components through digital tracking of products after their use stage. However, these topics have been largely analyzed independently of each other (Okorie et al., 2018; UNEP, 2018b). Gharfalkar et al. (2016) suggest that a more systematic review of the literature is needed in order to bring clarity and consistency to "refurbishment" and "remanufacturing." We, however, propose an integrated research study in order to bring clarity to remanufacturing and refurbishment based on the two considerations:

1. Remanufacturing and refurbishment, while traditionally resource efficiency processes, will be influenced by Industry 4.0 paradigm in the current Fourth Industrial Revolution.
2. Value creation within smart manufacturing can only be attained when there is a clear knowledge of how it affects resource efficiency concepts such as remanufacturing and refurbishment.

Therefore this study aims to propose a research agenda for future integrated research of remanufacturing and refurbishment in the age of Industry 4.0 through a review of the literature.

The organization of this chapter is as follows: Section 4.2 provides a theoretical background of the relevant subject areas; the review of the literature from SCOPUS will follow in Section 4.3; we conclude with the proposed research agenda as well as important discussions and conclusion in Section 4.4.

4.2 Theoretical background

4.2.1 Remanufacturing

Remanufacturing has become an important process in automotive, aerospace, ICT equipment, ink, and toner cartridges as well as rail industry sectors due to its strategic importance for economic and sustainable benefits and reduction in environmental impacts (Xia, Govindan, & Zhu, 2015). This importance has also been highlighted by government and nongovernment organizations such as the National Development and

Reform Commission (NDRC) in China (NDRC, 2010), the European Remanufacturing Network (ERN) (ERN, 2015) as well as the United Nations Environmental Program (UNEP) (UNEP, 2018c) among others. Consequently, there has been an uptake in remanufacturing research as can be seen in Okorie et al. (2018). Finally, as captured in Ellen MacArthur Foundation's widely cited "butterfly diagram," remanufacturing is observed to be an important component of the CE model, which conserves value within a restorative and regenerative circular loop (Ellen MacArthur Foundation, 2013b).

Product remanufacturing is an industrial process whereby used products or broken down components normally described as "cores" are restored to useful life through a sequence of processes (Östlin, Sundin, & Björkman, 2009). These used products go through inspection, disassembly, cleaning, inspection, reconditioning, reassembly, inspection, and final testing (Östlin et al., 2009; Okorie, Salonitis, Charnley, & Turner, 2018b) as shown in Fig. 4.1; thus a thorough process which ensures that the final product or remanufactured product attains a state of being "like-new or better" in quality and warranty to a newly manufactured product (Matsumoto & Ijomah, 2013).

Fig. 4.2 shows a used product and its final remanufactured state. Remanufacturing has been described as the "ultimate form of recycling" in several literature (Giuntini & Gaudette, 2003; Steinhilper, 1998) as it is focused on value conservation. It conserves the raw material content of the product, much of the value added during the remanufacturing process (Giuntini & Gaudette, 2003) as well as it is energy-saving in comparison to manufacturing of new product. Cross-country industry research suggests that remanufacturing achieves energy savings of 50% when compared to new products (Steinhilper & Weiland, 2015).

Remanufacturing research has been identified to be varied across science and business-related disciplines (Okorie et al., 2018). Using citation network analysis approach, Lee and Kwak (2018) identified the following areas as the top 12 research topics in remanufacturing: cost and profit in market, logistics and inventory, design for disassembly, quality

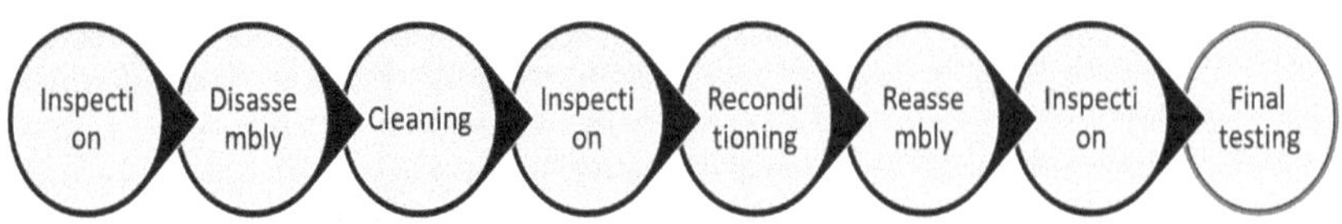

Figure 4.1 Stages in the remanufacturing process.

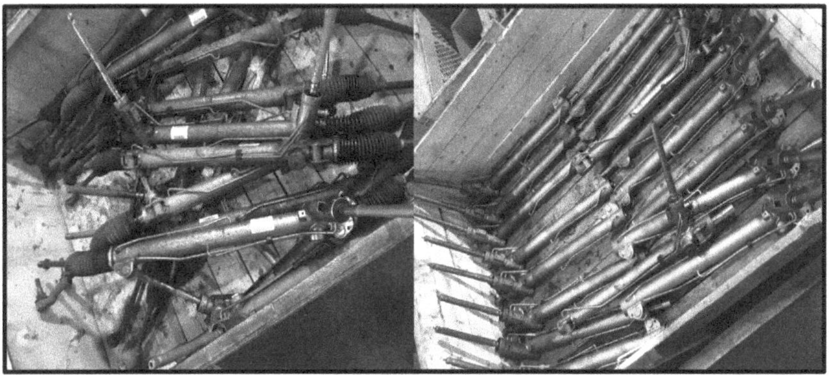

Figure 4.2 (L-R): Used products (automobile steering racks). Same products after being remanufactured. *Courtesy PSS, North Walsham, United Kingdom.*

management, capacity optimization, green design, material efficiency, reverse logistics, closed-loop supply chains, electronic waste recycling, product service system business model, and network optimization models. Growth in remanufacturing and remanufacturing research is expected to persist given: (1) the recognized economic and social advantages of remanufacturing, (2) the increasing regulatory pressures felt by manufacturers to pursue more environmentally friendly manufacturing processes. Carter and Ellram (1998) estimate that over $124 billion is spent in the United States in order to comply with regulations and statures related to the environment and sustainability. Third, the presence of nongovernmental institutions focused on research areas such as the CE and sustainability such as the Ellen MacArthur Foundation (founded in 2009). Lastly, the market for "environmentally friendly" products has grown to over $200 billion, according to a 1998 study on reverse logistics (Carter & Ellram, 1998). This brings in an extra incentive for manufacturers and remanufacturers.

This work explores remanufacturing and refurbishment in the age of Industry 4.0; it is important to state that remanufacturing and refurbishment as end-of-life practice [although remanufacturing has been argued to be a middle-of-life practice (Sundin, 2019)] predates Industry 4.0. While Industry 4.0—which was coined in 2011 by a German initiative of the German government, universities, and private institutions (Frank, Dalenogare, & Ayala, 2019)—sets itself in the heart of the Fourth Industrial Revolution, the early remanufacturing industry is said to have formed in the 1950s or earlier (Xu, 2013). Thus established concepts for remanufacturing which include the remanufacturing processes, research, and benefits can be said to have less influence from INDUSTRY 4.0 compared to the previous three eras of industrialization of mechanization, mass production, and automation (Stankovic & Gupta, 2017). Several papers, while noting that the integrated area of the CE and industry 4.0 has been unexplored (Beatriz et al., 2018; Hunt, 2018; Okorie et al., 2018; Rajput & Singh, 2019), argue that sustainability can be achieved by an integration of these two research areas. This study advances this argument by stating that understanding the barriers and enablers of Industry 4.0 can establish a link between remanufacturing and refurbishment and hence support the development of a future research framework. The next section presents a review on refurbishment.

4.2.2 Refurbishment

Similar to remanufacturing, refurbishment has been identified and described as a value-retention process (VRPs) (UNEP, 2018d). These are complementary processes (remanufacturing, repair, and refurbishment) that can enable faster achievement of the CE if pursued strategically with recycling by manufacturers. UNEP studies have shown that VRPs can create net-positive outcomes for CE as they enable product-level efficiency gains in material and energy use (UNEP, 2018d) and in emissions and waste generation. Comprehensive refurbishment with remanufacturing can contribute to reduction in Green-House Gases emissions by between 79% and 99% in appropriate sectors (Ekins, Hughes, & UNEP, 2017). These studies also show that comprehensive refurbishment can contribute to material savings of between 82% and 99% on average (Ekins et al., 2017).

Refurbishment has been described as the modification of an object that is waste[1] or a product to increase or restore its performance and/or functionality or to meet applicable technical standards or regulatory requirements, with the result of making a fully functional product to be used for a purpose that is at least the one that was originally intended[2]. Thus the refurbished item must be returned back to a good working condition. Similarly, the Ellen MacArthur Foundation and the European Remanufacturing Network, ERN in their respective definitions states that

[1] In this study "Waste" does not infer or refer to materials or waste products that do not have value, thus aligned with the definitions of the Basel Convention (http://www.basel.int/TheConvention/Overview/tabid/1271/Default.aspx).

[2] This definition comes from Document UNEP/CHW.13/4/Add.20.

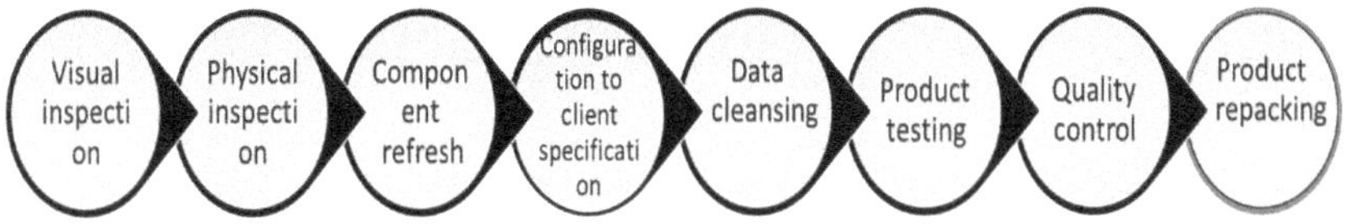

Figure 4.3 The product refurbishment process of an electronic product.

a refurbished product must go through cosmetic changes that ensure that the final product has the appearance of a new product (ERN, 2015; MacArthur, 2013). Within the context of warranty, refurbishment shares a similarity with remanufacturing as a refurbished product must come with an expected warranty. For example in the European Union, the refurbished smartphones come with a 1-year warranty, as set by the EU law[3]. Generally, the refurbishment end-of-life approach follows a sequence of operations from visual inspection to final product repackaging of the refurbished product as in Fig. 4.3.

This study focuses on refurbishment of manufacturing products. It should be noted that refurbishment extends beyond products such as the smartphones or the electronic switchboards as in Fig. 4.4 below. Building refurbishment (also called renovation), where the decision is made to demolish or refurbish older housing, is a key study in refurbishment, and are the highest cited articles on refurbishment on the SCOPUS database. However, this is outside the context of this chapter.

There is consensus in academic literature and among industry that refurbishment is an environmentally beneficial end-of-life strategy that allows companies to retain the value of products and materials (MacArthur, 2013). Furthermore, through the process of refurbishment, significant energy and raw materials while CO_2 emissions are avoided (Andrae, 2016) thereby contributing to the CE. Studies have also shown that like remanufacturing, refurbishment is a commercially economically viable business for companies (Atasu & Wassenhove, 2010; Daniel, Guide, & Van, 2001), as the total number of components needed for the refurbishment process can be reduced (and this enables easier refurbishing). Prior research on refurbishing has included the examination of issues on customer responsiveness to refurbished products (Mugge et al., 2017); aspects such as pricing (Subramanian, 2012); sales cannibalization for refurbishment (Atasu & Wassenhove, 2010); building refurbishment (Kaklauskas, Zavadskas, & Raslanas, 2005); and design for refurbishment (Hatcher, Ijomah, & Windmill, 2011).

[3] EU Law Directive, 1999/44/EC Article 7.1.

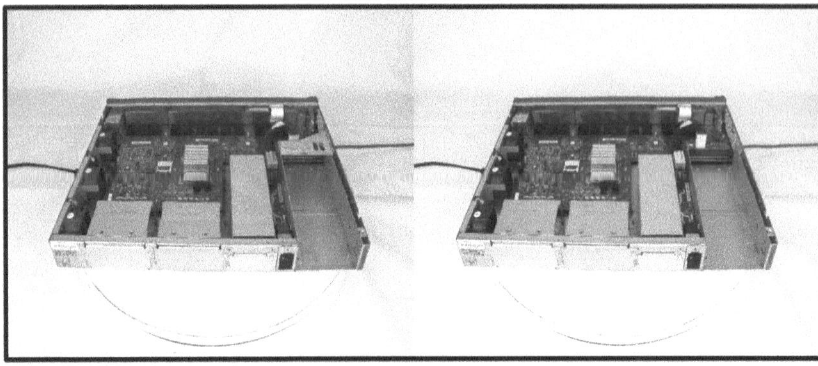

Figure 4.4 (L-R): Used switchboard and a refurbished switched board. The used switchboard appears darker, thus showing obvious signs of wear. *Courtesy Techbuyer, United Kingdom.*

While these studies discuss refurbishment, it should be noted that remanufacturing is also discussed in the same terminology as refurbishment. We argue that in the context of Industry 4.0, there should be some clear distinction between remanufacturing and refurbishment if Industry 4.0 is to achieve the aim of "smart" and "intelligent" manufacturing (Stankovic & Gupta, 2017). The next section reviews remanufacturing and refurbishment as discussed in the articles and publications available on SCOPUS and Web of Science.

4.3 Review of related concepts

4.3.1 Industry 4.0: perspectives

The stated objectives of Industry 4.0 are (1) to achieve a higher level of operational efficiency and productivity, and (2) to achieve a higher level of automatization (Thames & Schaefer, 2016). Industry 4.0 is rooted in the smart manufacturing concept, which is an adaptable system where flexible lines are able to automatically adjust its production processes for multiple types of products and a variability in conditions (Wang, Tãrngren, & Onori, 2015). Several definitions have been postulated as to what constitute Industry 4.0, but it is generally accepted that Industry 4.0 is a combination of (1) cyber-physical systems (CPSs) and (2) the internet of things (IoT), and (3) cognitive computing (Beatriz et al., 2018; Rajput & Singh, 2019; Stankovic & Gupta, 2017; Thames & Schaefer, 2016). Cognitive computing includes cloud manufacturing and additive manufacturing, which are exemplified by the internet and 3D printing, respectively. At its core, is the connectivity between machines, orders, employees, suppliers, and customers (Shrouf, Ordieres, & Miragliotta, 2014) driven by the IoT and other electronic devices. Lu (2017) summarizes the characteristics of Industry 4.0 as being able to ensure that manufacturing processes are integrated, adapted, optimized, and interoperable.

As remanufacturing and refurbishing are respective circular approaches, it is also important to understand the intersection of the CE and Industry 4.0 wherever applicable. These two areas have been largely independently studied, as noted in several studies (Beatriz et al., 2018; Okorie et al., 2018; Tseng, Tan, Chiu, Chien, & Kuo, 2018), however, it is largely agreed that Industry 4.0 through digital technologies can enable the CE. For example, quantitative tools and metrics derived from big data can support CE decision-making process; combining CPS, big data, data analytics, and new circular business models could provide important opportunities toward more sustainable industrial value creation, value capture, and value deployment (Ellen MacArthur Foundation, 2016).

With specific reference to remanufacturing and refurbishment, knowledge of the product location in real time enables increased product accessibility and enhances the likelihood for end-of-life strategies such as remanufacturing, refurbishment, recycling as well as the important aspect of waste collection (Bressanelli, Adrodegari, Perona, & Saccani, 2018). Thus in Table 4.1, we present an overview of the functional technologies of Industry 4.0 as well as possible implications for refurbishment and remanufacturing processes to support a faster CE transition. The next section shall review the research progression of Industry 4.0 with respect to remanufacturing and refurbishment.

4.3.2 Links between Industry 4.0, remanufacturing, and refurbishment: an SCOPUS overview

SCOPUS and Web of Science databases are generally regarded as important databases in the science community and have been consistently found to have a greater overall coverage of academic journals (Mongeon, 2016; Waltman et al., 2011). However, the comprehensiveness of SCOPUS has been highlighted as a database of choice for multidisciplinary research. SCOPUS also allows for a single search to be done on multiple qualifiers, unlike the Web of Science database. For example, we can search through

Table 4.1 Overview of the functional technologies of Industry 4.0 (Kang et al., 2016).

Industry 4.0 technology	Description	Resources in use	Reman applications	Refurb applications
Cyber-physical systems	Enables automation, monitoring, and control of processes and objects in real time (Wang et al., 2015)	Controllers and sensor systems (Wang et al., 2015; Yu, Xu, & Lu, 2015)	Cleaning, disassembly, inspection	Product testing, quality control
Cloud manufacturing	Virtual portals which create a shared network of manufacturing resources and capabilities offered as services (Yu et al., 2015)	The Internet	Disassembly, inspection	Product repackaging, data cleansing
Internet of things (IoT)	Computational system which collects and exchanges data (Kang et al., 2016)	Radio-frequency Identification (RFID), sensors embedded in products, technology tags, barcodes, and smartphones (Xu, He, & Li, 2014; Atzori, Iera, & Morabito, 2014)	Inspection	Inspection
Additive manufacturing	Represents agile and connected prototyping of parts of products on a large scale, enabling customization (Holmström, Holweg, Khajavi, & Partanen, 2016)	3D printers	Reconditioning, reassembly	Reconditioning

Source: With kind permission from Springer Nature.

the title, abstract, and keywords of an article at the same time using a single string on SCOPUS. Web of Science does not offer this advantage.

We adopt a similar review protocol as detailed in Okorie et al. (2018). As the term Industry 4.0 based literature started publishing in 2011 with the launch in Germany (Frank et al., 2019), we restrict the year of review to 2010 and 2019 inclusive. In order to keep this review simple, we restrict the keywords to "Industry 4.0," "Remanu*," and "Refurbish*." We review these keywords on the title, abstract, and keywords section in articles. The following are run on SCOPUS:

1. (TITLE-ABS-KEY ("remanufactur*") AND TITLE-ABS-KEY ("industry 4.0"))
2. (TITLE-ABS-KEY ("refurbish*") AND TITLE-ABS-KEY ("industry 4.0"))

The figures below show the results by Year of Publication, Institutional Affiliation of Publication, and documents by Subject Area.

4.3.2.1 Remanufacturing and Industry 4.0 on SCOPUS

Fig. 4.5 shows the number of related documents which we find on SCOPUS when remanufacturing is integrated with Industry 4.0 as a search string. While there is an increase in studies in this area of research, the total papers are less than academic studies in the individual areas of remanufacturing and Industry 4.0.

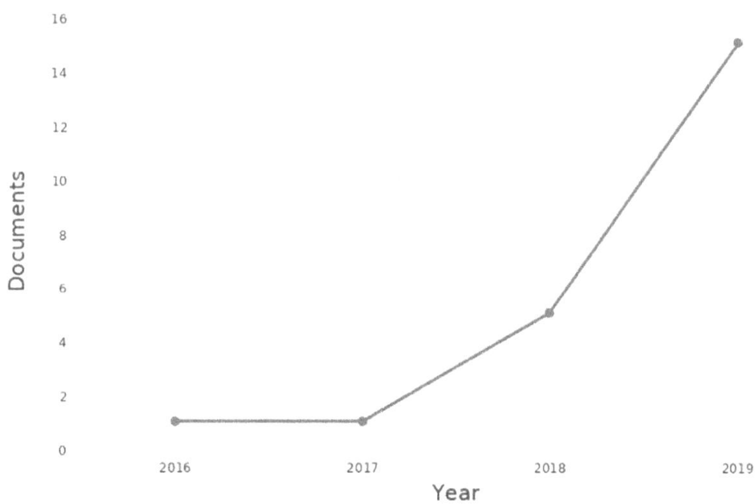

Figure 4.5 Number of documents versus year of publication.

In Fig. 4.6, we capture the affiliated universities and research centers invested in the integrated area of remanufacturing and Industry 4.0 as obtained from SCOPUS. Researchers publishing in this area are affiliated with these institutions.

The pie-chart in Fig. 4.7 gives the distribution of the subject area in the articles/documents published during 2016–2019 as presented in Fig. 4.5.

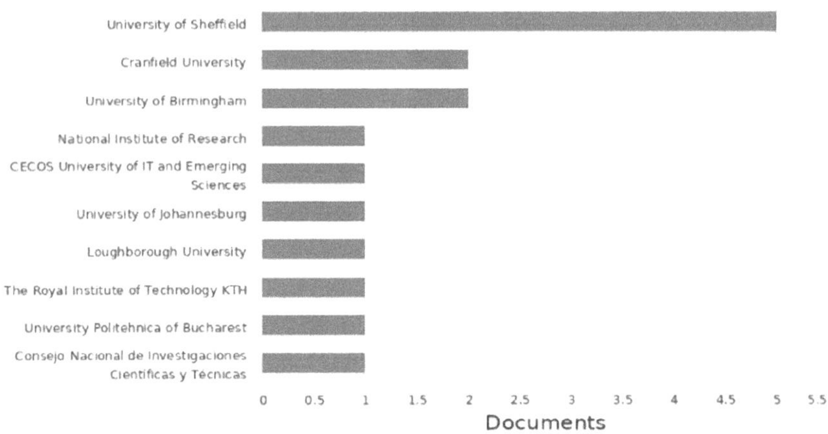

Figure 4.6 Institutional affiliation versus number of documents published.

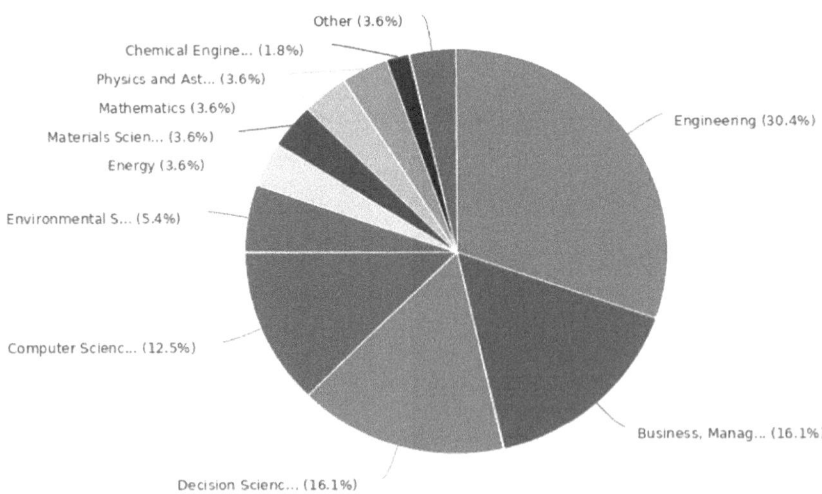

Figure 4.7 Documents by subject area.

4.3.2.2 Refurbishment and Industry 4.0 on SCOPUS

The search string (TITLE-ABS-KEY ("refurbish*") AND TITLE-ABS-KEY ("industry 4.0")) was used on SCOPUS and this produced a single article published in 2019 in the Journal Computers in Industry, "*An industrial evaluation of an Industry 4.0 reference architecture demonstrating the need for the inclusion of security and human components*" (Sharpe et al., 2019).

4.3.3 Results from SCOPUS

The descriptive results from SCOPUS clearly indicate that the integration of remanufacturing with Industry 4.0 and the integration of refurbishment with Industry 4.0 are still a novel, developing area with published research in this area only starting from 2016. With refurbishment, there is only one published paper in this area. This may imply that research in refurbishment is being replicated in remanufacturing and Industry 4.0 research (due to the inconsistencies in their terms). It does imply, however, that research in the integrated area of refurbishment and Industry 4.0 is very novel, as we expect Industry 4.0 to influence product refurbishment.

By Fig. 4.6, we conclude that this area of research is primarily being driven by institutions in the United Kingdom, as they make up the top three institutions publishing in this area of research. From the results on the subject area, it is quite clear that this research area is focused within the space of interdisciplinary research. This is important in defining what an integrated research agenda may consist. The next section presents an integrated research agenda

4.4 Discussion and agenda for integrated research

4.4.1 Lessons from similar studies

Several studies have been undertaken in research where an agenda for future research was proposed after an investigation of integrated studies. These include the study of Jabbour et al. (Beatriz et al., 2018) where a roadmap for sustainable operations and a research agenda was proposed by the authors after an integration of CE and industry 4.0 paradigms. They develop a matrix of the relationships between CE, Industry 4.0, and sustainable operations management, by focusing on the ReSOLVE[4] CE

[4] The ReSOLVE framework stands for REgenerate, Share, Optimize, Loop, Virtualize, and Exchange. These are six business actions developed by the Ellen MacArthur Foundation to guide organizations through implementing the principles of the CE (Ellen MacArthur Foundation, 2015).

framework, design of products, production of products, and reverse logistics. Their work argues that CE and Industry 4.0 integration should be focused on addressing the current threats to adoption of Industry 4.0 technologies, addressing the implementation of Industry 4.0 technologies as well as conducting in-depth case studies in order to understand the "soft side" of this integrated research area (Beatriz et al., 2018). In a similar integrative study, Mariale Moreno and Fiona Charnley (2014) focused on the integration of redistributed manufacturing and digital intelligence in order to enable a regenerative economy. This study identified multiple similarities between the drivers of redistributed and circular models of production and consumption that could be supported by the use of digital intelligence. They conclude that understanding the position and type of value capture and delivery is important to unlock significant opportunities for this integrative research study in the consumer goods sector. Finally, in a highly cited research Jabbour and De Sousa Jabbour (2016) explores the integrated and emerging research area between green human resource management and green supply chain management. They propose a synergistic framework with implications for academia, managers, research practitioners, and teachers. This framework is focused on the reduction of human barriers when implementing green supply chain management.

Thus we find out that integrative studies are useful in driving prevalent agendas such as sustainable development agenda, useful for academic, industrial, and policy-making practice as well as useful in understanding existing barriers and new insights into these studies, either as integrative research areas or as individual research areas. For example, Jabbour and De Sousa Jabbour (2016) suggest that a future research on green human resource management and green supply chain management area can focus on overcoming the main integration barriers as well as investigating the main benefits for employees and organization performance provided by the adoption of green purchasing. This has implications for academia, managers, and policymakers. Integrative studies also identify technological opportunities, as in the case of Yang et al. (2018). They review Industry 4.0 and remanufacturing and identify that with increased digitalization across the supply chain and improved cyber-physical intelligence within a remanufacturing factory, major concerns that remanufacturers encounter are addressed. Thus in this study, we recognize that implications for Industry 4.0 and refurbishment as well as Industry 4.0 and remanufacturing exists for these stakeholders as identified in Beatriz et al. (2018), Jabbour and De Sousa Jabbour (2016), and Mariale Moreno and Fiona

Charnley (2014) and may well identify new technology or profitability opportunities. The next section shall propose an agenda for integrated research by analyzing the literature review in Section 4.3.

4.4.2 Agenda for integrated research

From the review of the literature conducted in the previous sections, we identify that there is ample scope for research and publication contribution on the integration between remanufacturing and Industry 4.0 and refurbishment and Industry 4.0. We hypothesize that this integration will help to improve the clarity and distinction between remanufacturing and refurbishment as VRPs.

In order to assess what should constitute a research agenda, it is important to understand the outcomes from the SCOPUS review. When we merge remanufacturing and Industry 4.0 on SCOPUS, we observe that research in this area has been focused on identification, mapping, and application of Industry 4.0 technologies to remanufacturing processes (Stock & Seliger, 2016; Yang et al., 2018; Tao, Qi, Liu, & Kusiak, 2018; French, Benakis, & Marin-Reyes, 2018); remanufacturing as analyzed through the integration of Industry 4.0 and the CE (Manavalan & Jayakrishna, 2019; Cezarino, Liboni, Oliveira Stefanelli, Oliveira, & Stocco, 2019). Business models to support Industry 4.0 implementation ranked third in the reviewed papers on Industry 4.0–remanufacturing integration. Research integration between Industry 4.0 and refurbishment is still very novel as identified in Section 3.2.2. We also observe that the CE will remain an important core of this research agenda. We suggest that refurbishment is likely to develop in research with close similarity to remanufacturing.

We recognize that this novelty is also motivated by the fact that Industry 4.0 is a predominately new area of research and its integration with CE research (through remanufacturing and refurbishment) will be predominately technical and practitioner-oriented. Thus we propose the integrated research agenda as explained in Fig. 4.8.

As outlined in Fig. 4.8, the first step for researchers is the identification of suitable Industry 4.0 technologies for remanufacturing and refurbishment, their benefits, barriers to implementation, and so on. A suggestion for this is to review these technologies under available Industry 4.0 perspectives [such as Smart Life Cycle Data, Smart Factory, Smart Service (Yang et al., 2018)]. Following this, we propose the identification and

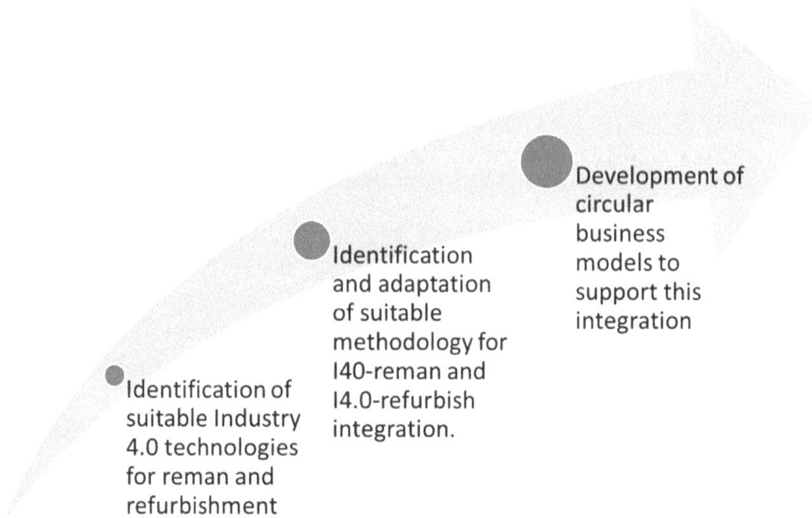

Figure 4.8 Future agenda for integrated research.

adaptation of suitable methodologies for Industry 4.0-remanufacturing and Industry 4.0-refurbishment integration as a future research agenda. Such methodology can be adapted in order to measure circularity (for companies implementing Industry 4.0 in remanufacturing or refurbishment) and can help identify key performance indicators for Industry 4.0 adoption.

Finally, we identify the development of circular business models to support the adoption of Industry 4.0 in remanufacturing and refurbishment. From the literature reviewed in this study, this is the research area with the least of research conducted. However, we recognize that remanufacturing and refurbishing companies may have to adapt or completely change to new business models in order to assimilate Industry 4.0 in their practice. This study can be a potential boost to CE practices.

4.5 Conclusion

This study aimed to add to the growing research area of Industry 4.0 integration with the CE having a specific focus on remanufacturing and refurbishment through a review of the literature. A number of new results are highlighted in this study. These include the overview of functional Industry 4.0 technologies which maps into remanufacturing and refurbishment processes and applications. From the review of existing literature, we propose areas of future research. These include the clear identification of suitable Industry 4.0 technologies for remanufacturing and refurbishment; the identification and adaptation of suitable methodology for integration of Industry 4.0 technologies to remanufacturing and refurbishment. Finally, the identification of business models for the CE to support Industry 4.0 adoption in remanufacturing and refurbishment practice is identified as a potential area for future research.

Industry 4.0 adoption in remanufacturing and refurbishment is still a very new area, with published research in this area started a few years back. Remanufacturing and refurbishment currently suffer from a disadvantage of ambiguity in definition and practice. We expect this ambiguity to diminish causing a greater understanding of both VRPs once the clear integrated research agenda is followed. As this is a very recent area of research, we believe that academic scholars should play a leading role in understanding the implications, methodology, and applications of this integrated area, where the support of practitioners and policymakers is desirable.

Acknowledgments

The authors are grateful to PSS Steering and Hydraulics, Norfolk, England and to Techbuyer, Harrogate, England for the permission to use photographs from their respective companies.

References

Abbey, J. D., Meloy, M. G., Blackburn, J., & Guide, V. D. R. (2015). Consumer markets for remanufactured and refurbished products. *California Management Review*, *57*(4), 26–42. Available from 10.1525/cmr.2015.57.4.26.

Andrae, A. S. G. (2016). Life-cycle assessment of consumer electronics. *IEEE Consumer Electronics Magazine*, *5*, 51–60. Available from 10.1109/MCE.2015.2484639.

Atasu, A., & Wassenhove, L. N. Van (2010). So what if remanufacturing cannibalizes my new product sales. *California Management Review*, *52*(2).

Atzori, L., Iera, A., & Morabito, G. (2014). From 'smart objects' to 'social objects': The next evolutionary step of the internet of things. *IEEE Communications Magazine*, *52*(1), 97–105. Available from 10.1109/MCOM.2014.6710070.

Beatriz, A., Sousa, L. De, Charbel, J., Chiappetta, J., Godinho, M., & David, F. (2018). *Industry 4. 0 and the circular economy: A proposed research agenda and original roadmap for sustainable operations. Annals of Operations Research* (270, pp. 273–286). Available from 10.1007/s10479-018-2772-8.

Bressanelli, G., Adrodegari, F., Perona, M., & Saccani, N. (2018). Exploring how usage-focused business models enable circular economy through digital technologies. *Sustainability (Switzerland)*, *10*(3), 639. Available from 10.3390/su10030639.

Carter, C. R., & Ellram, L. M. (1998). Reverse logistics: A review of the literature and framework for future investigation. *Journal of Business Logistics, 19*(1).
Cezarino, L. O., Liboni, L. B., Oliveira Stefanelli, N., Oliveira, B. G., & Stocco, L. C. (2019). Diving into emerging economies bottleneck: Industry 4.0 and implications for circular economy. *Management Decision*. Available from 10.1108/MD-10-2018-1084.
CIRAIG. (2015). International reference centre for the life cycle of products P and S. circular economy. A critical literature review of concepts. CIRAIG. https://doi.org/10.1017/CBO9781107415324.004.
Circle Economy. (2016). Circular economy. *Nature Sustainability, 531*, 6–9. Available from 10.1038/531435a.
Craveiro, F., Duarte, J. P., Bartolo, H., & Bartolo, P. J. (2019). Additive manufacturing as an enabling technology for digital construction: A perspective on Construction 4.0. *Automation in Construction., 103*(2018), 251–267. Available from 10.1016/j.autcon.2019.03.011.
Daniel, V., Guide, R., J., & Van, W. L. N. (2001). Managing product returns for remanufacturing. *Production and Operations Management, 10*(2), 142–155.
Ekins P., Hughes N., UNEP (2017) Resource efficiency: Potential and economic implications. A report of the International Resource Panel. 2017. Retrieved from https://doi.org/10.1017/S1368980007000365.
Ellen MacArthur Foundation. (2013a). Towards the circular economy. 596/09-08-2013.
Ellen MacArthur Foundation. (2013b). Towards the circular economy: Economic and business rationale for an accelerated transition. https://doi.org/10.1007/b116400.
Ellen MacArthur Foundation. (2015). *Growth within: A circular economy vision for a competitive Europe* (p. 100) Ellen MacArthur Foundation.
Ellen MacArthur Foundation. (2016). Intelligent assets: Unlocking the circular economy potential. Ellen MacArthur Foundation (December) (pp. 1–25). Retrieved from http://www.ellenmacarthurfoundation.org/assets/downloads/publications/EllenMacArthurFoundation_Intelligent_Assets_080216.pdf.
ERN. (2015) Remanufacturing market study. Retrieved from www.remanufacturing.eu.
Etzion, D. (2018). Management for sustainability. *Nature Sustainability, 1*(12), 744–749. Available from 10.1038/s41893-018-0184-z.
European Commission. (2015). Communication from the Commission to the European Parliament, the Council, the European Economic and Social Committee and the Committee of the Regions: Closing the Loop—An EU Action Plan for the Circular Economy. Brussels.
Frank, A. G., Dalenogare, L. S., & Ayala, N. F. (2019). Industry 4.0 technologies: Implementation patterns in manufacturing companies. *International Journal of Production Economics, 210*, 15–26. Available from 10.1016/j.ijpe.2019.01.004.
French, R., Benakis, M., & Marin-Reyes, H. (2018). Intelligent sensing for robotic remanufacturing in aerospace—An industry 4.0 design based prototype. In *Proceedings—2017 IEEE 5th international symposium on robotics and intelligent sensors, IRIS 2017* (pp. 272–277). https://doi.org/10.1109/IRIS.2017.8250134.
Gharfalkar, M., Ali, Z., & Hillier, G. (2016). Clarifying the disagreements on various reuse options: Repair, recondition, refurbish and remanufacture. *Waste Management and Research, 34*(10), 995–1005. Available from 10.1177/0734242X16628981.
Giuntini, R., & Gaudette, K. (2003). Remanufacturing: The next great opportunity for boosting US productivity. *Business Horizons, 46*(6), 41–48. Available from 10.1016/S0007-6813(03)00087-9.
Hatcher, G. D., Ijomah, W. L., & Windmill, J. F. C. (2011). Design for remanufacture: A literature review and future research needs. *Journal of Cleaner Production., 19*(17–18), 2004–2014. Available from 10.1016/j.jclepro.2011.06.019.

Holmström, J., Holweg, M., Khajavi, S. H., & Partanen, J. (2016). The direct digital manufacturing (r)evolution: definition of a research agenda. *Operations Management Research*, *9*(1–2), 1–10. Available from 10.1007/s12063-016-0106-z.

Hunt, R. (2018). Design for a circular economy in Industry 4.0. In M. Charter (Ed.), *Designing for the circular economy* (pp. 221–231). Routledge.

Ijomah, W. L., McMahon, C. A., Hammond, G. P., & Newman, S. T. (2007). Development of design for remanufacturing guidelines to support sustainable manufacturing. *Robotics and Computer-Integrated Manufacturing*, *23*(6), 712–719. Available from 10.1016/j.rcim.2007.02.017.

Jabbour, C. J. C., & De Sousa Jabbour, A. B. L. (2016). Green human resource management and green supply chain management: Linking two emerging agendas. *Journal of Cleaner Production*, *112*, 1824–1833. Available from 10.1016/j.jclepro.2015.01.052.

Kaklauskas, A., Zavadskas, E. K., & Raslanas, S. (2005). *Multivariant design and multiple criteria analysis of building refurbishments*, *37*, 361–372. Available from 10.1016/j.enbuild.2004.07.005.

Kalmykova, Y., Sadagopan, M., & Rosado, L. (2017). Circular economy—From review of theories and practices to development of implementation tools. *Resources, Conservation and Recycling.*, *135*, 190–201. Available from 10.1016/j.resconrec.2017.10.034.

Kang, H. S., Lee, J. Y., Choi, S., Kim, H., Park, J. H., Son, J. Y., et al. (2016). Smart manufacturing: Past research, present findings, and future directions. *International Journal of Precision Engineering and Manufacturing—Green Technology*, *3*(1), 111–128. Available from 10.1007/s40684-016-0015-5.

Kirchherr, J., Reike, D., & Hekkert, M. (2017). Conceptualizing the circular economy: An analysis of 114 definitions. *Resources, Conservation and Recycling*, *127*, 221–232. Available from 10.1016/j.resconrec.2017.09.005.

Korhonen, J., Nuur, C., Feldmann, A., & Birkie, S. E. (2018). *Circular economy as an essentially contested concept. Journal of Cleaner Production* (175, pp. 544–552). Available from 10.1016/j.jclepro.2017.12.111.

Lasi, H., Fettke, P., Kemper, H. G., Feld, T., & Hoffmann, M. (2014). Industry 4.0. *Business & Information Systems Engineering*, *6*(4), 239–242.

Lee, S., & Kwak, M. (2018). A review of the research on remanufacturing using the citation network analysis. *ICIC Express Letters*, *12*(4), 345–351. Available from https://www.researchgate.net/requests/r51557416.

Lu, Y. (2017). *Industry 4.0: A survey on technologies, applications and open research issues. Journal of Industrial Information Integration* (6, pp. 1–10). Available from 10.1016/j.jii.2017.04.005.

MacArthur, E. (2013). Towards the circular economy: A opportunities for the consumer goods sector. *Journal of Industrial Ecology*, *10*, 4–8. Available from 10.1162/108819806775545321.

Makris, D., Hansen, Z. N. L., & Khan, O. (2019). Adapting to supply chain 4.0: An explorative study of multinational companies. *Supply Chain Forum.*, *20*(2), 116–131. Available from 10.1080/16258312.2019.1577114.

Manavalan, E., & Jayakrishna, K. (2019). *An analysis on sustainable supply chain for circular economy. Procedia Manufacturing* (33, pp. 477–484). Available from 10.1016/j.promfg.2019.04.059.

Moreno, M., & Charnley, F. (2014). Can re-distributed manufacturing and digital intelligence enable a regenerative economy? An integrative literature review. In P. Theobald (Ed.), *Design and manufacturing* (pp. 18–22). https://doi.org/10.1007/978-3-319-32098-4.

Matsumoto, M., & Ijomah, W. (2013). *Remanufacturing. Handbook of Sustainable Engineering* (62, pp. 389–408). Dordrecht: Springer.

Mongeon, P. (2016). The journal coverage of web of science and scopus. *Scientometrics*, 213–228. Available from 10.1007/s11192-015-1765-5.
Mugge, R., Jockin, B., & Bocken, N. (2017). *How to sell refurbished smartphones? An investigation of different customer groups and appropriate incentives. Journal of Cleaner Production* (147, pp. 284–296). Available from 10.1016/j.jclepro.2017.01.111.
NDRC. (2010). *National Development and Reform Commission: An overview of the 13th Five-Year Plan*. Retrieved from http://en.ndrc.gov.cn/newsrelease/201612/t20161207_829923.html.
Okorie, O., Salonitis, K., Charnley, F., Moreno, M., Turner, C., & Tiwari, A. (2018). Digitisation and the circular economy: A review of current research and future trends. *Energies.*, *11*(11), 3009. Available from 10.3390/en11113009.
Okorie, O., Salonitis, K., Charnley, F., & Turner, C. (2018a). A systems dynamics enabled real-time efficiency for fuel cell data-driven remanufacturing. *Journal of Manufacturing and Materials Processing*, *2*(4), 77. Available from 10.3390/jmmp2040077.
Okorie, O., Salonitis, K., Charnley, F., & Turner, C. (2018b). A systems dynamics enabled real-time efficiency for fuel cell data-driven remanufacturing. *Journal of Manufacturing and Materials Processing*, *2*(77). Available from 10.3390/jmmp2040077.
Östlin, J., Sundin, E., & Björkman, M. (2009). Product life-cycle implications for remanufacturing strategies. *Journal of Cleaner Production*, *17*(11), 999–1009. Available from 10.1016/j.jclepro.2009.02.021.
Potting, J., Hekkert, M., Worrell, E., & Hanemaaijer, A. (2017) Circular economy: Measuring innovation in the policy report (January).
Rajput, S., & Singh, S. P. (2019). Connecting circular economy and industry 4.0. *International Journal of Information Management*, *49*, 98–113. Available from 10.1016/j.ijinfomgt.2019.03.002.
Rathore, P., Kota, S., & Chakrabarti, A. (2011). Sustainability through remanufacturing in India: A case study on mobile handsets. *Journal of Cleaner Production.*, *19*(15), 1709–1722. Available from 10.1016/j.jclepro.2011.06.016.
Sharpe, R., van Lopik, K., Neal, A., Goodall, P., Conway, P. P., & West, A. A. (2019). *An industrial evaluation of an Industry 4.0 reference architecture demonstrating the need for the inclusion of security and human components. Computers in Industry* (108, pp. 37–44). Elsevier B.V. Available from 10.1016/j.compind.2019.02.007.
Shrouf, F., Ordieres, J., & Miragliotta, G. (2014) Smart factories in Industry 4.0: A review of the concept and of energy management approached in production based on the Internet of Things paradigm. In *IEEE international conference on industrial engineering and engineering management* (pp. 697–701). IEEE. https://doi.org/10.1109/IEEM.2014.7058728.
Stankovic, M., & Gupta, R. F. J. I. Industry 4.0—Opportunities behind the challenge. UNIDO Background paper. 2017. Retrieved from https://www.unido.org/gc17/industry40. (Accessed 7 August 2018).
Steinhilper, R. (1998) Remanufacturing—The ultimate form of recycling.
Steinhilper, R., & Weiland, F. (2015). *Exploring new horizons for remanufacturing an up-to-date overview of industries, products and technologies. Procedia CIRP* (29, pp. 769–773). Available from 10.1016/j.procir.2015.02.041.
Stock, T., & Seliger, G. (2016). Opportunities of sustainable manufacturing in Industry 4.0. *Procedia CIRP*, *40*, 536–541. Available from 10.1016/j.procir.2016.01.129.
Subramanian, R. (2012). Key factors in the market for remanufactured products. *Manufacturing & Service Operations Management*, *14*(2), 315–326.
Sundin, E. (2004). Product and process design for successful remanufacturing. *Science and Technology*.
Sundin, E. (2019). Circular economy and design for remanufacturing. In M. Charter (Ed.), *Designing for the circular economy* (1st Edition, pp. 186–199). Routledge.

Tao, F., Qi, Q., Liu, A., & Kusiak, A. (2018). Data-driven smart manufacturing. *Journal of Manufacturing Systems*, *48*, 157–169. Available from 10.1016/j.jmsy.2018.01.006.

Thames, L., & Schaefer, D. (2016). Software-defined cloud manufacturing for Industry 4.0. *Procedia CIRP.*, *52*, 12–17. Available from 10.1016/j.procir.2016.07.041.

Tseng, M. L., Tan, R. R., Chiu, A. S. F., Chien, C. F., & Kuo, T. C. (2018). Circular economy meets industry 4.0: Can big data drive industrial symbiosis? Resources, conservation and recycling. *Resources, Conservation and Recycling*, *131*, 146–147. Available from 10.1016/j.resconrec.2017.12.028.

UNEP. (2018a). Redefining value: The manufacturing revolution. Remanufacturing, refurbishment, repair and direct reuse in the circular economy.

UNEP. (2018b). Redefining value: The manufacturing revolution. Remanufacturing, refurbishment, repair and direct reuse in the circular economy. Summary for policy makers.

UNEP. (2018c). Redefining value: The manufacturing revolution. Remanufacturing, refurbishment, repair and direct in the circular economy. Summary for business leaders (p. 52).

UNEP. (2018d). Redefining value: The manufacturing revolution. Remanufacturing, refurbishment, repair and direct reuse in the circular economy. Key insights for policy makers.

Waltman, L., Eck, N. Jv, Leeuwen, T. Nv, Visser, M. S., & Raan, A. F. Jv (2011). Towards a new crown indicator: Some theoretical considerations. *Journal of Informetrics.*, *5*(1), 37–47. Available from 10.1016/j.joi.2010.08.001.

Wang, L., Tãrngren, M., & Onori, M. (2015). Current status and advancement of cyber-physical systems in manufacturing. *Journal of Manufacturing Systems*, *37*, 517–527. Available from 10.1016/j.jmsy.2015.04.008.

Xia, X., Govindan, K., & Zhu, Q. (2015). Analyzing internal barriers for automotive parts remanufacturers in China using grey-DEMATEL approach. *Journal of Cleaner Production*, *87*(1), 811–825. Available from 10.1016/j.jclepro.2014.09.044.

Xu, B. (2013). Progress of remanufacturing engineering and future technology expectation. *Advances in Manufacturing*, *1*, 8–12. Available from 10.1007/s40436-013-0003-8.

Xu, L. D., He, W., & Li, S. (2014). Internet of things in industries: A survey. *IEEE Transactions on Industrial Informatics*, *10*(4), 2233–2243. Available from 10.1109/TII.2014.2300753.

Yang, S., Aravind, M. R., Kaminski, J., & Pepin, H. (2018). Opportunities for Industry 4.0 to support remanufacturing. *Applied Sciences*, *8*(7), 1177. Available from 10.3390/app8071177.

Yu, C., Xu, X., & Lu, Y. (2015). Computer-integrated manufacturing, cyber-physical systems and cloud manufacturing—Concepts and relationships. *Manufacturing Letters*, *6*, 5–9. Available from 10.1016/j.mfglet.2015.11.005.

CHAPTER 5

Nanofluid-minimum quantity lubrication system in machining: towards clean manufacturing

Sujan Debnath[1], Mahmood Anwar[1], Alokesh Pramanik[2] and Animesh Kumar Basak[3]
[1]School of Engineering and Science, CDT250, Curtin University, Miri, Malaysia
[2]School of Civil and Mechanical Engineering, Curtin University, Bentley, WA, Australia
[3]Adelaide Microscopy, University of Adelaide, Adelaide, SA, Australia

5.1 Introduction

Owing to environmental concerns and growing regulations over contamination and pollution, the demand for renewable and biodegradable cutting fluids is rising. Machinability is an important index in the manufacturing industry to measure the machining performance for clean manufacturing. Machining processes such as milling and turning are widely used in manufacturing industries to obtain desired surface, shape, and size of engineering parts/products. During machining, high temperature and friction generated at the cutting zone highly influence the tool life and surface quality of the work piece (Debnath, Reddy, & Yi, 2014; Gupta, Laubscher, Davim, & Jain, 2016). Development of high temperature generates compressive stress at the workpiece—tool interface resulting plastic deformation at the tool edge. Thus the high cutting temperature at the cutting zone adversely affects the efficiency of the tool surface that consequently lead to the reduced tool life, poor work surface quality as well as high-power consumption. Therefore improved surface quality and longer tool life largely depend on the reduction of temperature at the cutting zone during machining.

The temperature at the cutting zone can be reduced by applying cutting fluids as a source of coolant and lubricant. The conventional synthetic cutting fluids used in large quantities in machining are harmful to the environment as well as to the health of the workplace personnels. It is revealed that prolonged exposures to these conventional synthetic cutting fluids during machining may cause several respiratory illnesses namely

Sustainable Manufacturing
DOI: https://doi.org/10.1016/B978-0-12-818115-7.00014-6

asthma, bronchitis, respiratory irritations, and loss of lung function. Apart from environment and health impacts, cutting fluids are also responsible for additional manufacturing cost (Debnath et al., 2014; Gupta et al., 2016).

In order to get rid of the adverse impact of conventional cutting fluids, researchers are showing keen interest toward developing environmentally friendly cutting fluids and innovating lubrication techniques with the aim to reduce or even completely remove the need for any cutting fluid. Dry machining is one of the modern lubricating concepts where machining is performed without the application of cutting fluid or lubricant. The other well-known lubricating concept is called near-dry machining where cryogenic coolants or a small quantity of green lubricants are used to produce sort of mist or gaseous environment in the form of micro droplets. The near-dry machining is more popularly known as minimum quantity lubrication (MQL) technique. MQL reduces the amount of cutting fluids quite significantly yet provides excellent cooling and lubricating which is often comparable with the traditional wet machining performance in terms of surface finish, tool life, cutting forces, and temperature reduction (Debnath et al., 2014; Gupta et al., 2016).

In recent years, there has been series of innovations as regards to the improvisation of MQL. Combination of MQL with cryogenic cooling, MQL with different green oils and lubricants, and use of nanoparticles in MQL fluids are some of them. In nanofluid-based MQL, nanoparticles due to their significantly large surface area compared to density are well dispersed in the base fluid and hence capable to interact with the tiny contact surfaces of cutting zone more effectively in order to reduce friction and better cooling and lubrication effects.

This chapter introduces machinability, cutting fluids, ecological problems with cutting fluids and subsequently demand for environmentally friendly cutting fluid, the development toward clean manufacturing, MQL, and nanofluid-assisted MQL system. Some of the recent research works conducted by the authors on nanofluid-MQL assisted machining are also highlighted.

5.2 Cutting fluids

Improvement of machining performance by the application of cutting fluids has long been practiced in manufacturing industries. Water was used as the cutting fluid by Taylor in 1907 which resulted in a 40% increment of cutting speed for machining steel (Vieira, Machado, & Ezugwu, 2001). Thus water possess great cooling capabilities. However, water lacks lubricity, and it is highly corrosive to the metallic workpiece and cutting tools. A wide variety of cutting fluids with improved cooling and lubricating properties have been developed since then. Nevertheless, the type of machining process, the workpiece, and cutting tool materials have great influence on the selection of cutting fluid (Çakōr, Yardimeden, Ozben, & Kilickap, 2007).

The primary functions of cutting fluid are to provide cooling and lubricating properties during machining process (El Baradie, 1996). The cooling property of a cutting fluid is the capacity to absorb heat and carry away from the cutting zone. This helps to improve the tool life due to the reduction of tool wear at the workpiece—tool interface. The lubricating properties of a cutting fluid can be defined as its ability to reduce friction at the tool—chip interface. A good lubricant will ensure reduction of plastic deformation and smooth chip removal from the cutting zone and thus further reduction of heat generation (El Baradie, 1996). A good lubricant will also contribute in reduction of cutting forces, which helps to reduce the power requirement for machining. The other important functions of cutting fluid is to provide the media to the developed chips to be flushed away from the cutting zone preventing entanglement of chips at the tool tip as well as to prevent built-up edge (Bruni, Forcellese, Gabrielli, & Simoncini, 2006). Moreover, cutting fluid functions as an insulating material which protects tool and workpiece from corrosion (Bruni et al., 2006).

5.2.1 Ecological problems of conventional cutting fluid

The conventional cutting fluid, such as hydrocarbon-based mineral oil, is effective for cooling the cutting zone; however, it results in various ecological problems (Debnath et al., 2014; Gupta et al., 2016). It often generates airborne mists, smoke, and other particles in the workshop's atmosphere. Due to its toxicity, conventional cutting fluid may lead to environmental pollution as well as health problems.

Exposure to conventional cutting fluid had resulted in 80% of occupational infection among contractors (Shashidhara & Jayaram, 2010). According to International Agency for Research on Cancer (IARC), the petroleum (hydrocarbon)-based cutting fluid contains carcinogenic heterocyclic and polyaromatic rings which could lead to occupational skin cancer (Debnath et al., 2014). Moreover, frequently expose to conventional cutting fluid may cause respiratory disease, lung cancer, genetic disease, and dermatological (Ozcelik, Kuram, Cetin, & Demirbas, 2011). Therefore a proper disposal plan is needed for conventional cutting fluid, and therefore its disposal cost is higher than the biodegradable cutting fluid. In fact, additional space is required for storing and processing the used conventional cutting fluid (Davoodi & Tazehkandi, 2014).

5.2.2 Environmentally friendly cutting fluids

Due to rising awareness of the negative impacts associated with the application of cutting fluids, the demand for environmentally friendly cutting fluids has been increasing. In recent years, new laws and restrictions have been passed that requires the harmful contents of cutting oils to be minimized or reduced to an acceptable level in order to alleviate the adverse environmental impact. Sustainability and biodegradability are important aspects to be considered for environmentally friendliness.

Biodegradable substance is capable of being broken down by bacteria or micro-organism without causing harm to the environment (Aydin, 2017). Due to biological action, the substance which changes in the chemical structure experiences the loss of its specific properties during biotransformation. The substance which achieves ultimate biodegradation can completely be decomposed to form water, carbon dioxide, biomass, and mineral salts.

The researchers utilized vegetable oils to develop a variety of bio-based cutting fluids. Ni et al. (2018) indicated that the application of vegetable oil in machining operation was more biodegradable and less toxic than using the synthetic or semisynthetic oil. The researchers are seeking for an eco-friendlier cutting fluid to replace the conventional mineral oil-based cutting fluid which resulted in pollutions and various health problems. Wang et al. (2017) studied the properties of various vegetable-based nanofluid by conducting the frictional test. The vegetable oil was chosen as base oil due to its renewability, high viscosity, and low volatilization. Besides, Padmini, Krishna, and Rao (2016) dispersed different concentration of nano MoS_2 to coconut, sesame, and canola oil to produce nanofluids for MQL turning of AISI 1040 steel. The machining performance of vegetable-based oils was compared with pure oil and conventional oil. The experiment found that the canola oil exhibited the best machining performance among the selected oils.

Synthetic esters, fatty acids, alcohols, etc. are also some sustainable or environmentally friendly cutting fluids and are highly biodegradable and ecological.

5.2.3 Types of cutting fluids

Cutting fluids are mainly classified into three types namely oil-based, water-based, and gas. Oil-based cutting fluids exhibit excellent lubricating properties, whereas water-based cutting fluids are superior as a coolant (Vieira et al., 2001). The gas cutting fluids are considered as a clean manufacturing option since disposal is not a concern.

5.2.3.1 Oil-based cutting fluids

Oil-based cutting fluids or neat oils consist of mineral, vegetable, animal, and synthetic oils. Oil-based cutting fluids are known for their excellent lubricating, antiseizure, and corrosion prevention properties. Most neat oils are petroleum-based mineral oils. Oil-based cutting fluids lubricate the tool–chip interface and reducing friction, cutting forces, and heat generation, which also lead to reduce tool wear. Mineral oil properties are enhanced by the addition of fatty oils from vegetable and animal sources. The fatty oil forms a thin layer at the interface of tool, chip, and workpiece during machining process. This developed thin layer continues to function as a lubricant even after the normal oil film has broken down, thus improving the antifriction characteristics of the mineral oils (El Baradie, 1996). However, mineral oils are highly flammable and produce mist and smoke at higher loads and temperatures (Shokrani, Dhokia, & Newman, 2012). As such, the application of highly flammable mineral oils are mainly applied at low-velocity operating conditions where temperature does not increase significantly (Debnath et al., 2014).

5.2.3.2 Water-based cutting fluids

Water-based cutting fluids are classified into three categories namely emulsions, synthetic, and semisynthetic fluids (Shokrani et al., 2012). Emulsions are prepared by mixing oil in water where oil droplets are suspended in water by blending the oil with emulsifying agents. The oil drops are broken down into tiny droplets and kept suspended in water with the aid of the emulsifying agent. Thus emulsions, as a cutting fluid, have the both capacities of oil and water. The oil droplets work as the lubricant and corrosion resistant, and the water works as an excellent coolant (El Baradie, 1996).

Apart from superior lubricating and cooling properties, emulsions are less expensive since only one part of oil in 100 parts water to 1:5 oil to water ratio required for emulsion (El Baradie, 1996). Due to the water dominancy in the solution, emulsions possess minimum fire hazards. However, emulsions are susceptible to the growth of microorganisms such as bacteria, fungi, and algae which lead to separation in the emulsion deteriorating the cooling and lubricating properties of the cutting fluid (Kuram, Ozcelik, & Demirbas, 2013).

Synthetic fluids are mixture of water and favorable chemical agents without containing any mineral oils (Avila & Abrao, 2001). The role of chemical agents in synthetic fluids are to provide lubrication, corrosion resistance, water softening as well as to reduce surface tension (El Baradie, 1996). One good advantage of synthetic cutting fluid is its color transparency which greatly assists monitoring the machining operation. Apart from that, synthetic fluids are resistant to micro-organism growth so that the cooling and lubricating properties are not degraded (Kuram et al., 2013). Semisynthetic fluids fall in between emulsion and synthetic fluids with 5%–30% of mineral oil content in the matrix of base oil (Kuram et al., 2013). Thus semisynthetic fluids show the compromised properties of both emulsions and synthetic fluids with better lubrication than synthetic fluids and better corrosion resistance than emulsions (Shokrani et al., 2012).

5.2.3.3 Gas cutting fluids

Gas cutting fluids include air, argon, helium, nitrogen, and carbon dioxide. These gases are mostly environment friendly options compared to the other cooling techniques since disposal is not required for gases. Gaseous fluids are also more effective in conditions where traditional liquid coolants fail to penetrate the tool–chip interface such as in heavy duty machining. Inert gases also prevent oxidation at high cutting temperature. Nevertheless, gaseous cutting fluids have poorer cooling capabilities compared to its liquid fluid counterpart (Shokrani et al., 2012). The properties of neat oil, soluble oil, semisynthetics, synthetics, and gas-based lubricant are listed in Table 5.1.

5.2.4 Vegetable oil as an environmentally friendly cutting fluid

The awareness of the health and environmental impacts toward clean cutting fluid in machining is ever increasing. Cutting fluids are recycled or

Table 5.1 Types of cutting fluids.

	Advantages (Debnath et al., 2014)	Disadvantages (Kuram et al., 2013)
Oil-based lubricant (neat oil)	• Excellent in lubrication • Excellent corrosion resistance	• Flammable • Poor coolant • May produce smoke or mist
Soluble oil (emulsified oil)	• Good corrosion resistance • Good in lubricity effect • Good cooling effect • Low rate of oil misting	• Rusting problem • Bacteria infestation
Semisynthetics	• Good rust control • Cooling effect	• Foaming problem • Easily polluted by other cutting fluid • Water hardness will influence the fluid's stability
Synthetics	• Excellent cooling effect • Good corrosion control • Nonflammable	• Poor lubricity • Easily polluted by other cutting fluid
Gas-based lubricant	• Biodegradable • High-corrosion resistance • High-oxidation resistance	• Poor thermal conductivity • Low cooling capacity

reused and easily gets contaminated over time due to the mixing with metal particles generated from chips as well as trump oil from various sources which affects its performance (Debnath et al., 2014). This causes disposal of these cutting fluids inevitable. According to the statistics nearly 85% mineral oil-based cutting fluids are currently in use and disposal of these huge volume of contaminated fluids possess serious environmental and health concern (Pop, Puşcaş, Bandur, Vlase, & Nuţiu, 2008). To address this, scientists and engineers proposed vegetable-based cutting fluids as a clean alternative to mineral oils due to its high biodegradability, renewability, and low toxicity (Debnath et al., 2014). Vegetable oils have a biodegradability of over 95% (Pop et al., 2008) compared to mineral oils with only 20%–30% (Kuram et al., 2013).

The long, polar fatty acid chains ion in vegetable oil forms a high-strength lubricant film that reduces friction and wear. Furthermore, vegetable oils outperform mineral oils in antiwear and friction, scuffing load capacity, and fatigue resistance (Lawal, Choudhury, & Nukman, 2013). In addition, due to high viscosity index, the viscosity of vegetable oil drops slower than that of mineral oils as temperature rises thus maintaining a more stable lubricity across the operating temperature range (Gajrani and Ravi Shankar, 2017). However, the disadvantage of vegetable oils are that it has low-thermal and oxidative stability, high-freezing point, and poor corrosion protection (Lawal et al., 2013). To overcome these drawbacks, various methods can be employed such as reformulation of additives, chemical modification, or genetic modification of the oil seed crop (Shashidhara & Jayaram, 2010). Table 5.2 summarizes the relative advantages and disadvantages of vegetable oil as a cutting fluid.

5.2.5 Clean manufacturing

5.2.5.1 Dry machining

Dry machining requires no cutting fluid thus it is considered the cleanest manufacturing technique. In dry machining, the machining components like the workpiece material, cutting tool as well as machining process parameters must be considered very carefully for successful operation (Najiha, Rahman, & Yusoff, 2016). For instance, high temperature hardness and wear resistant materials such as ceramics, cermet, cemented carbides, cubic boron nitride, and polycrystalline diamond are most suitable for dry machining. The cutting tool used must be optimized with respect to geometry and compatible to the workpiece materials. Coated

Table 5.2 Advantages and disadvantages of vegetable oil.

Advantages	Disadvantages
High biodegradability	Low-oxidation stability
Low toxicity	Low-thermal stability
High flash point	High-freezing point
Compatibility with additives	Poor corrosion protection
Higher viscosity index	Flammable
Greater lubricity	

tools are often utilized to improve the tool performance and tool life. Coating on tools works as a solid lubricant which reduces friction and adhesion (Klocke & Eisenblätter, 1998). Interrupted machining process such as milling operation is the best suited for dry machining. In interrupted machining, there is no thermal shock which consequently provides longer tool life. Soft materials like aluminum are not recommended for dry machining because they induces built-up edge during machining which affects surface finish (Kuram et al., 2013). Though an environment friendly option, dry machining incurs several disadvantages in machining. The main disadvantage associated with dry machining is that the tendency to produce high temperatures which induces excessive tool wear. The other disadvantage is that the generated chips in dry machining are likely to tangle at the tool tip and the cutting zone which may deteriorate tool life and generate poor surface finish. As such, there is a need to optimize the cutting process in terms of cutting parameters, cutting tool, machining process, and workpiece material.

5.2.5.2 Minimum quantity lubrication

MQL is a clean manufacturing technique which is also known as near-dry machining. In MQL, a drastically reduced amount of cutting fluid is sprayed in the cutting zone. A flow rate in the range of 10–100 mL/h are commonly applied for most industrial applications (Najiha et al., 2016). Due to low consumption of cutting fluids, MQL is considered as an environmentally friendly cooling technique. Moreover, several researchers observed that MQL machining has the potential to offer comparable or even better machinability compared to traditional dry and wet machining (Sharma, Dogra, & Suri, 2009). Due to its low consumption of cutting fluid and machining performance, MQL machining has been discussed and researched worldwide as an clean alternative to traditional wet machining (Boswell, Islam, Davies, Ginting, & Ong, 2017).

In MQL machining, cutting fluids are delivered without or with a medium such as air. In airless system, cutting fluid is supplied using a pump and fluids in the form of micro droplets which is injected to the tool–chip interface through a nozzle. In the system with air, a tiny amount of cutting fluid is mixed with air and delivered in the form of an aerosol spray through a diffuser or nozzle (Weinert, Inasaki, Sutherland, & Wakabayashi, 2004). The cutting fluid and air may either mix inside the nozzle or outside the nozzle. For the inside nozzle mixing process, extremely fine droplets of atomized cutting fluids mixes with air in the nozzle and sprayed to the cutting zone as shown in Fig. 5.1. In this process, harmful mists andvapors are reduced. Moreover, the mixture setting is very easy to control (Attanasio, Gelfi, Giardini, & Remino, 2006). This technique has benefits such as reduction of manufacturing cost and impact on the environment as well as human life.

Kamata and Obikawa (2007) applied MQL using biodegradable synthetic ester with a flow rate of 16.8 mL/h to perform turning operations on an Inconel 718 bar. Three kinds of coated carbide tools were employed to measure the performance. The results showed that MQL turning resulted massively increased tool life for two tool inserts compared to flood cooling and MQL turning resulted in the best surface finish compared to all other techniques. MQL and dry turning of AISI D2 steel using vegetable oil was conducted by Sharma and Sidhu (2014). The outcome showed that MQL produced lower temperature and surface roughness compared to dry turning. MQL also produced better surface finish at higher feed rates. It was observed that the chips produced by MQL cooling did not contain much residual oil which makes them easier to recycle.

Rahman, Kumar, and Salam (2002) observed catastrophic failure of tool inserts in dry and flood cooling, whereas the MQL performed milling

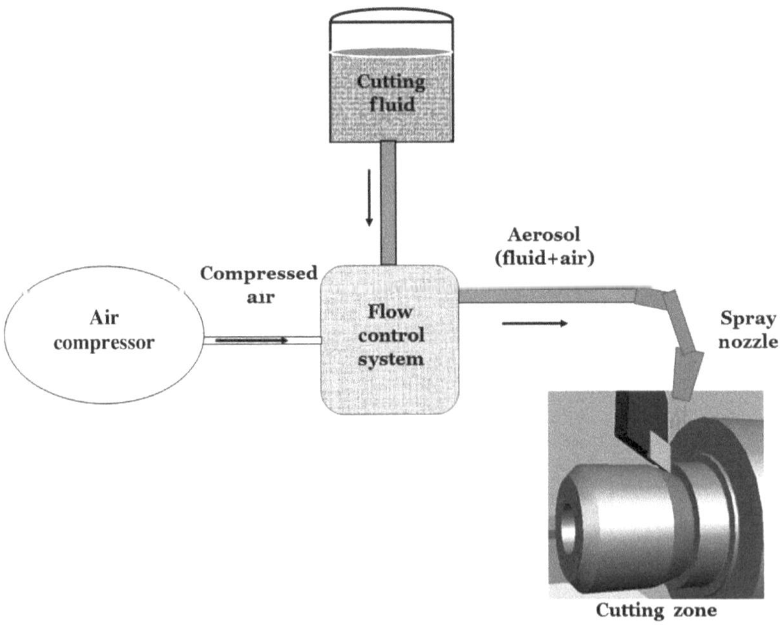

Figure 5.1 Schematic representation of MQL. *MQL*, minimum quantity lubrication.

of ASSAB 718 HH steel successfully without any damage of the tool insert. MQL also experienced lower fracturing or chipping compared to dry machining and flood cooling. Moreover, MQL produced the lowest cutting forces and generated comparable surface finish when compared to that of flood cooling and dry machining. However, all the above superior outcomes are only effectively realized for low machining parameters including low depth of cut, feed rate, and cutting speed.

Khan, Mithu, and Dhar (2009) used vegetable oil-based MQL to examine the machining performance on the turning of AISI 9310 alloy steel. The experimental findings showed that MQL performed significantly better compared to dry and wet cooling in terms of tool wear, surface finish, and tool/chip interface temperature. MQL reduced average tool/chip interface temperatures by up to 10% compared to flood cooling.

5.2.5.3 Nanoparticle-enhanced minimum quantity lubrication

Nanofluid is a fluid that combines nanometer-size of solid particle and fibers (less than 100 nm) in base fluids. This fluid can be considered as the next generation of heat transfer due to their excellent heat conductive properties (Brousseau, Dimov, & Pham, 2010). Some nanoparticles such as aluminum oxide, graphite, zirconium dioxide, boric, molybdenum disulfide, and diamond are frequently used as lubricants and coolants (Sharma, Tiwari, & Dixit, 2016b).

Although MQL has been established as an effective clean alternative lubrication technique compared to the conventional cooling, the cutting fluid sprayed by MQL contains air which reduces the thermal conductivity of the fluid. As a consequence, conventional cooling techniques that possess better cooling often outperform MQL. Thus there is a need to improve the thermal conductivity of the cutting fluid for improved MQL performance. A new cutting fluid by incorporating nanoparticle into cutting fluid called nanofluid has been introduced recently. Thermal conductivity is one of the main parameters that effect the tool life and the surface roughness in machining (Wang et al., 2016). The addition of nanoparticles into conventional fluids produces a significant improvement in thermal conductivity due to the increased surface area of nanoparticles being better able to absorb heat (Sharma, Tiwari, & Dixit, 2015). Thus nanoparticles enhance the heat removal rate of cutting fluid as it exhibits good thermal conductivity (Wang et al., 2017). Researchers found that the carbon nanotube (CNT) has the best thermal conductivity compared to other nanoparticles. A summary of nanoparticles enhancing thermal conductivity of cutting fluids are shown in Table 5.3.

Moreover, nanofluids are superior in terms of tribological and thermophysical properties compared to conventional fluids which makes them more effective in reducing cutting forces, surface roughness, tool wear, and cutting temperature. Nanofluids also reduce friction and increase load bearing capacity of friction part. Since nanoparticle-enhanced cutting fluid are highly expensive compared to conventional cutting fluids, nanofluid-enhanced MQL is the preferred cooling technique for clean manufacturing (Sharma et al., 2016).

There are different explanations as to why nanofluids offer greater lubricity over conventional fluids. Nanoparticles have a larger surface area per volume ratio which causes them to be reactive at the tool–workpiece interface resulting in improved machinability. A variety of mechanisms have been proposed such as the ball bearing effect, tribo-film formation, mending effect, and polishing effect (Sharma et al., 2016).

Lee et al. (2009) classifies these mechanisms into two groups. The first group is the direct effect of nanoparticles on lubrication enhancement. The nanoparticles suspended in the cutting fluid function as ball bearings between the friction surfaces. Nanoparticles especially those existed in spherical or spherical-liked shape also have good rolling effect which decreases the sliding friction reducing the cutting forces at the cutting

Table 5.3 Summary of nanoparticles enhance thermal conductivity of cutting fluids.

Researcher	Nanoparticles	Base fluid	Concentration of nanoparticles (wt.%)	Thermal conductivity
Sharma, Tiwari, and Dixit (2016a)	Al_2O_3	Distilled water	6	22.4% improve compare to conventional fluid
ManojKumar and Ghosh (2016)	MWCNT	Sunflower oil	1	47% improve compare to plain sunflower oil
Dambatta et al. (2019)	SiO_2	Water	0.3	More than 3.6%
Sajeevan and Sajith (2016)	ZrO2	Diesel fuel	0.1	10% improve compare to plain diesel fuel

zone (Sharma et al., 2015). Besides, nanoparticles can act as bearings in the tool—workpiece interface forming a protective film covering the rough surfaces. This may enhance the lubricity of cutting fluid resulting a better surface finish. The second group is the secondary effect of the nanoparticle enhancing the surface roughness such as mending and polishing effects.

When the loss of mass at the contact surface is filled up by the deposition of nanoparticles, it is called mending effect, and when the roughness of the contact surface is reduced by nanoparticles abrasion, it is known as the polishing effect. Moreover, nanoparticles can reduce the compressive concentration caused by high contact pressure due to the uniform load distribution (Peng et al., 2009). The lubrication mechanisms of nanoparticle-based fluid are illustrated in Fig. 5.2.

In addition to improved lubricity, nanofluids also have a greater wettability compared to conventional cutting fluids which enhances heat extraction from tool and workpiece surface by spreading the cutting fluid over the entire contact surfaces. The wettability of a fluid is measured by the angle it forms on a flat surface. In other word, nanofluids form smaller droplets compared to normal cutting fluids (Fig. 5.3), thus covering more surface area which leads to better lubricity and thermal conductivity (Ajithkumar & Xavior, 2018).

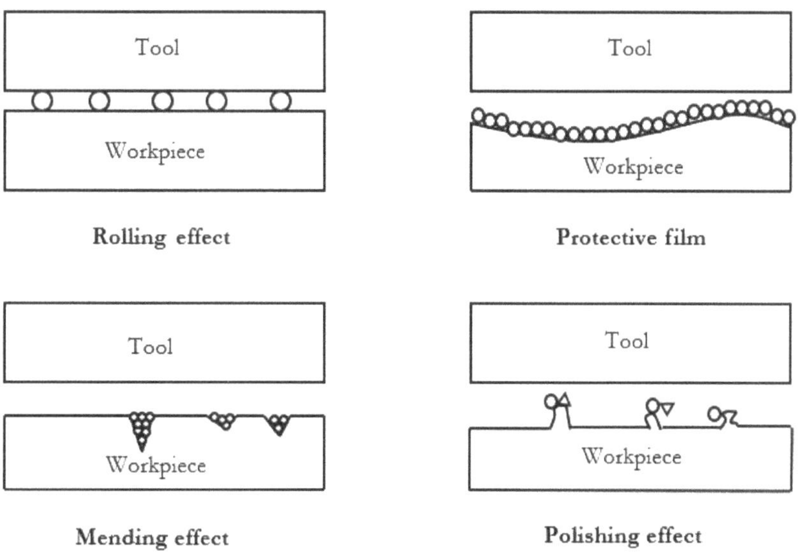

Figure 5.2 Lubrication mechanism of nanoparticles in cutting fluids.

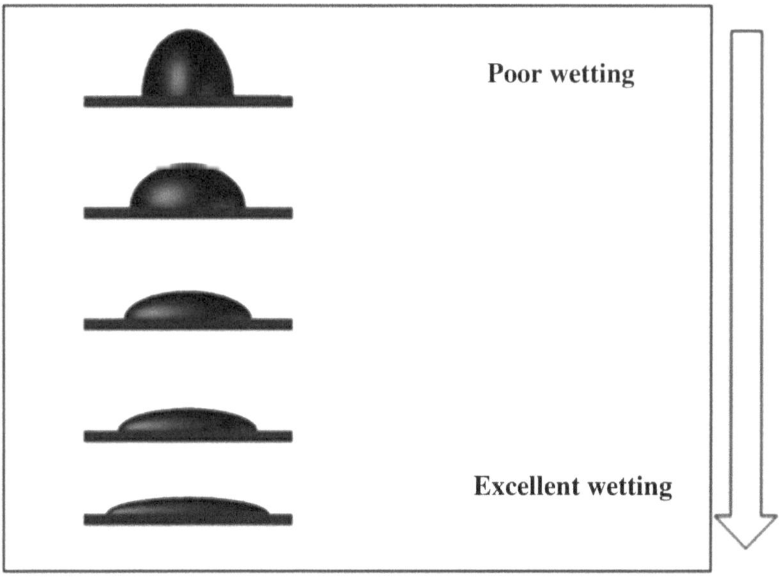

Figure 5.3 Contact angle of nanofluid with cutting surface.

5.3 Critical review of nanofluid-MQL system in clean manufacturing

5.3.1 Nanofluid-minimum quantity lubrication system in turning operation

To investigate the effect of nanofluid on cutting force, numerous studies have been conducted in the recent years. Amrita, Srikant, and Sitaramaraju (2014) investigated the performance of turning AISI 1040 steel under dry, flood, MQL with conventional cutting fluid, and MQL with 0.3 wt.% nanographite-enhanced cutting fluid conditions. It is reported that 0.3 wt.% nanographite-enhanced cutting fluid reduced cutting force by 75%, 54%, and 50%, respectively, when compared to dry, flood, and conventional MQL. In a recent work, researchers incorporated two different nanoparticles namely aluminum oxide (Al_2O_3) and multiwalled carbon nanotube (MWCNT) in rapeseed oil to assess the nano-MQL turning performance of Inconel 718 (Hegab, Darras, and Kishawy, 2018). The results showed that MWCNT performed better than Al_2O_3 in terms of flank wear and surface finish. The experimental results also indicated 4 wt.% of nanofluid for both nanoparticles as the optimal level concentration. The author also concluded that nano-MQL technique offers improved machinability over traditional MQL technique.

Yan, Zhang, and Kuriyagawa (2011) compared the performances of four types of nanoparticles namely molybdenum sulfide (MoS_2), graphite fiber (GF), copper (Cu), and copper oxide (CuO) on reaction-bonded silicon carbide (RB-SiC) ceramic in turning operation. The experimental outcome revealed that copper nanoparticle produced the best surface finish and lowest tool wear and thrust force. Sharma et al. (2016a) experimented the relative performances of three nanoparticles namely aluminum oxide (Al_2O_3), silicon oxide (SiO_2), and titanium oxide (TiO_2). The results revealed that Al_2O_3 performed the best in terms of the highest thermal conductivity and the lowest viscosity compared to SiO_2 and TiO_2, making it a more favorable choice.

AISI 4340 steel was turned under dry, conventional cutting fluid, and Al_2O_3-based nanocutting fluid conditions by Khandekar, Sankar, Agnihotri, and Ramkumar (2012). It was reported that Al_2O_3-based nanocutting fluid showed reduction of 50% and 30% in cutting force when compared to dry and conventional cutting fluid conditions. Sayuti, Sarhan, Tanaka, Hamdi, and Saito (2013) investigated the performance of computer numerical control end-milling duralumin AL-2017-T4 using 0, 0.5, 1.0, and 1.5 wt.% carbon onion nanolubrication system. They reported that cutting force reduced with the increase of carbon onion concentration and 1.5 wt.% carbon onion concentration showed the lowest cutting force by around 22% compared to ordinary lubrication oil. Krishna, Srikant, and Rao (2010) investigated on the performance of SAE-40 and coconut oil suspended with nanoboric acid particles when turning AISI 1040 Steel. They observed that the cutting temperatures, tool flank wear, and surface roughness decreased significantly compared to the base oil due to the lubricity of boric acid. Moreover, as the nanoboric acid percentage increased, the thermal conductivity and heat transfer coefficient also increased and the optimum results were achieve at 0.5 wt.% concentration of nanoboric acid.

Chetan, Ghosh, and Rao (2016) applied alumina- and silver-based nanofluids to investigate the machinability of Nimonic 90 alloy in turning. The experimental results revealed that alumina-based nanofluid greatly reduced tool—workpiece sliding friction forces as well as flank and tool nose wear. They explained that the alumina-based nanofluid produced the lowest range of contact angle and droplet diameter sand surface tension. This small droplet diameter of alumina allows for larger amounts of nanofluid to be dispersed per unit time which eases the ability of the droplets to enter the cutting zone. As a result, chance of droplet rebound is reduced and the lubrication film formation is very likely to occur. Sayuti, Erh, Sarhan, and Hamdi (2014) performed turning on hardened AISI4140 Steel using SiO_2 nanolubrication system under MQL with nozzle angle and stream pressure as parameters. The results showed that the minimum tool wear was obtained with a nanoparticle at concentration of 0.5 wt.%, 2 bar of stream pressure and a nozzle angle of 60 degrees; while the optimum surface roughness was obtained with a nanoparticle concentration of 0.5 wt.%, a reduced stream pressure, and a nozzle angle of 30 degrees.

5.3.2 Nanofluid-minimum quantity lubrication system in milling operation

Uysal et al. (2015) employed molybdenum sulfide (MoS_2) nanoparticle in vegetable cutting fluid for milling of AISI 420 stainless steel. The authors concluded that MoS_2-vegetable oil MQL obtained better tool wear and surface roughness compared to traditional MQL. It was also perceived that flow rate had minor influence in improving surface finish for nano-MQL, whereas an increase in flow rate improved surface finish for normal MQL. This shows that the lubricating effect of the nanoparticle played a significant role in surface finish improvement compared to flow rate.

Uysal (2016) assessed the effect of vegetable cutting fluids on tool wear by preforming milling of AISI 430 ferritic stainless steel using two types of cutting inserts (uncoated WC and TiN coated WC) under different machining conditions namely dry, low feed, and high feed MQL with commercial vegetable cutting fluid and 1 wt.% and 2 wt.% of nanographene-reinforced vegetable cutting fluid. It is reported that MQL with 2 wt.% nanographene-reinforced vegetable cutting fluid showed the lowest flank wear in both cutting inserts when compared to other conditions.

Agapiou (2018) evaluated the performances of three cutting fluids namely semisynthetic coolant, molybdenum sulfide (MoS_2) mixed with the base fluid, and carbon nanoonions (CNO) mixed with the base fluid at 2 wt.% in end milling and drilling on 4120 steel. For the end milling operation, all the three cutting fluids showed similar performance in terms of cutting forces. This can be explained because milling is a discontinuous cutting operation and as such improved cooling and lubrication does not impact significantly as compared to continuous processes like drilling and turning. Nevertheless, CNO produced the best surface finish among the three cutting fluids. In drilling experiments, CNO produced 252 holes drilled showing the best tool life, MoS_2 produced 144 holes, and the base fluid semisynthetic coolant performed the worst with only 11 holes drilled. Based on the above findings, it can be seen that nanofluid does not have substantial impact in end milling but offers a remarkable improvement in terms of tool life in drilling. However, in this research the authors did not investigate tool wear performance in end milling experiments which could have been the most improved aspect.

5.3.3 Nanofluid-minimum quantity lubrication system in drilling and grinding operations

Zhang, Li, Zhang, Jia, and Zhang (2015) experimented grinding of 45 steel using palm oil and paraffin. Four working conditions namely dry, flood, MQL, and MoS_2 nano-MQL were tested. The results showed that the best performance in terms of coefficient of friction and surface finish were obtained by flood cooling closely followed by nano-MQL. Nano-MQL consistently outperformed MQL technique and palm oil performed better than paraffin in terms of surface finish and coefficient of friction. The authors also concluded that the 6 wt.% concentration of nanoparticles provided the best performance. These experimental findings provide strong support that nano-MQL can enhance machinability performance over traditional MQL and that it can obtain results close to that of flood cooling. Furthermore, it emphasizes the viability of vegetable oils as an environmentally friendly alternative to mineral oils.

Vasu and Kumar (2011) performed grinding of EN-31 steel using dry, TRIM E709 emulsifier and TRIM E709 emulsifier with 1% Al_2O_3 nanoparticle to investigate the effect on cutting temperature. It is reported that TRIM E709 emulsifier with 1% Al_2O_3 nanoparticle showed a reduction of 30% in wheel-workpiece temperatures when compared to dry and plain emulsifier cutting conditions. Amrita et al. (2014) found that 0.3 wt.% nanographite-enhanced cutting fluid showed reduction of 70%, 25%, and 20%, respectively, in cutting force temperature when compared to dry, flood, and MQL with conventional cutting fluid conditions. In another study, Li et al. (2017) performed MQL grinding of Ni-based alloy with different nanofluids namely MoS_2, ZrO_2, CNT, Al_2O_3, and SiO_2 mixed individually with palm oil. It was reported that CNT nanofluid showed the lowest grinding temperature followed by Al_2O_3 nanofluid. The researchers explained the reason behind low grinding temperature by CNT nanofluid is because CNT nanofluid has the highest heat transfer coefficient among other nanofluids.

Wang et al. (2017) used different concentration of Al_2O_3 nanofluids to evaluate the tribological performance of MQL grinding process. They concluded that the nanofluid concentration greatly influence machining performance with 2.5 vol.% nanofluid concentration provided lowest tool wear. They explained that unlike pure cutting fluid, nanocutting fluid can form a protective film at the tool—workpiece interface to improve the tribological dynamics and further tool performance.

Wang et al. (2016) utilized six types of nanoparticles (ZrO_2, CNTs, ND, MoS_2, SiO_2, and Al_2O_3) in grinding Nickel Alloy GH4169 under MQL. The experimental finding revealed that aluminum oxide (Al_2O_3) nanoparticles due to their sphere-shaped molecular structures with high viscosity exhibited outstanding lubricity characteristics. It was also observed that nanodiamond cutting fluids enhanced the surface morphology of the machined workpiece due to their polishing effects.

5.3.4 Hybrid nanofluid-minimum quantity lubrication towards clean manufacturing

Hybrid nanofluid, consists of more than one type nanoparticle mixed in cutting fluid, is relatively new development in clean machining. It is hypothesized that the hybrid nanofluid may generate a larger machinability effect by trading off the capabilities of individuals. For instance, nanocopper (a good thermal conductor) and silicon dioxide (SiO_2) (a good lubricant), when combined together in appropriate ratio, may yield much better machinability in terms of surface roughness and temperature reduction from the cutting zone. Nevertheless, appropriate mixing ratio, particle size, concentration, and mixing technique may greatly influence the machining performance for hybrid nanofluid-MQL system. Moreover, the selection of hybrid nanofluid should also consider the type of workpiece and tool. Very few research articles are available in literature which discussed on the effectiveness of hybrid nanofluid in machining.

Singh et al. (2017) utilized alumina-graphene hybrid nanofluid-enhanced MQL during turning of AISI 304 Steel. The experimental results showed that machinability in terms of cutting force and surface roughness reduced significantly. Thus they concluded that hybrid nanofluids tend to exhibit better performance compared to ordinary pure nanofluid. Nevertheless, there is an increased cost associated with synthetization of hybrid nanofluids. Thus the researchers proposed further study to be carried out in order to reduce cost related to the mixing methods of hybrid nanofluids.

Zhang, Li, Jia, Zhang, and Zhang (2015) evaluated the performance of a hybrid nanofluid consisting of MoS_2 nanoparticle and CNT with the aim to obtain an optimum balance by combining the good lubricity of MoS_2 and high heat conductivity coefficient of CNT nanoparticles. The results indicated that the hybrid nanofluid does exhibit better lubricity compared to pure nanoparticle nanofluid. It was hypothesized that the hybrid nanofluid generates a physical synergistic effect that is larger than the simple sum of the effects of which each is individually capable. The researchers proposed to utilize two parts of MoS_2 with one part of CNT at a 6 wt.% with the base fluid for the best outcome.

Sharma, Tiwari, Dixit, and Singh (2017) used Al_2O_3-MoS_2 hybrid nanofluid in turning AISI 304 steel and evaluated their thermophysical properties at different concentrations. A reduction in cutting force, feed force, thrust force, and surface roughness by a fair margin was recorded from the experimental results. It reported that the addition of alumina nanoparticles had improved the thermal conductivity of the cutting fluid by 9.85%. A reduction in viscosity with the increment of temperature was also noticed. Roy and Ghosh (2013) applied hybrid nanofluid consisting of 1 vol.% MWCNT and 3 vol.% alumina utilized twin jet quantity lubrication for turning of AISI steel workpiece. The authors noticed a significant reduction on cutting force and specific energy. In another research, Singh et al. (2017) reported reduction of cutting force when applied graphene-alumina hybrid nanofluid with MQL in turning AISI 304.

5.3.5 Highlights from author's recent work

Authors of this book chapter performed MQL milling of S235JR structural steel plates with different Al_2O_3-palm oil nanofluid concentration and concluded that addition of Al_2O_3 in palm oil improved machinability in MQL milling of S235JR structural steel plates in terms of surface roughness and cutting temperature when compared to traditional plain palm oil MQL milling.

It is evident from Table 5.4 that the surface roughness R_a reduced gradually but significantly as the Al_2O_3-palm oil nanofluid concentration increased from 0 to 2 wt.% for each depth of cut. Based on the literature review stated above, it can be explained that the improvement of surface roughness occurs due to the conversion of sliding friction into rolling friction by the spherical shaped nanoparticles. Other proposition is surface roughness improved due to mending effect which is explained as under

Table 5.4 Effect of depth of cut (DOC) on minimum quantity lubrication (MQL) milling with Al_2O_3-palm oil nanofluid.

Al_2O_3-palm oil nanofluid concentration (wt.%)	DOC = 0.5 mm		DOC = 1.0 mm		DOC = 1.5 mm	
	R_a (μm)	(°C)	R_a (μm)	(°C)	R_a (μm)	(°C)
0	2.31	45.0	1.57	78.0	1.19	71.0
1	1.29	42.7	1.17	73.8	1.13	68.0
2	0.99	41.3	0.88	69.8	0.68	70.9
3	1.09	54.0	1.51	80.0	0.78	83.0

high compressive pressure of cutting process, the nanoparticles experience shearing. Consequently, sheared nanoparticles get embedded into machined surface and compensate for the loss of mass. Furthermore, nanoparticle concentration increases the viscosity of nanofluid which produces viscous films along the workpiece–tool interface, produces polishing effect. Thus both mending and polishing effects produce better surface finish due to reduction of the coefficient of friction between workpiece and tool interfaces.

However, slight increment in surface roughness R_a was noticed when Al_2O_3-palm oil nanofluid concentration increase from 2 to 3 vol.%. Based on the literature review, it can be said that when concentration of nanofluid increases, the thickness of protective layer decreases, which consequently deteriorates surface finish. Moreover, with higher concentration of nanofluid, a large number of nanoparticles are infused through the tiny pores of machined surface which in turn nullify the good work of mending that further result in generation of poor surface quality.

Based on Table 5.4, it was observed that the cutting temperature reduced gradually as the Al_2O_3-palm oil nanofluid concentration increase from 0 to 2 vol.%. This can be attributed due to the addition of metallic oxide nanoparticle increased the thermal conductivity of the base fluid which carried away the heat faster compared to without nanoparticle. Moreover, nanoparticle increases the wettability of the nanofluid which effectively covers the cutting zone, resulting in higher heat dispersion efficiency. The size and shape of nanoparticle may also influence in heat dissipation. Literature indicates that relativity small nanoparticles are mostly spherical in shape, thus possess more surface area, which can perform better in the tiny space at the cutting zone resulting effective heat transfer (Ajithkumar & Xavior, 2018).

However, with the further increase of Al_2O_3-palm oil nanofluid concentration from 2 to 3 vol.%, the cutting temperature increased significantly. This can be explained that high concentration of nanoparticle tends to agglomerate due to strong van der Waals forces among nanoparticle which in turn deteriorates the wettability of nanofluid as shown in Fig. 5.4 (Chinnam, Das, Vajjha, & Satti, 2015), resulting in reversing the positive effects of good wettability of well dispersed nanofluid. Moreover, the high concentration of nanoparticle increases the viscosity of nanofluid which results in high shearing during flow which increases temperature due to high friction among concentrated particles.

In this research, authors also evaluated the surface roughness and temperature for 1 mm depth of cut by changing the oil-based nanofluids (Palm oil) into water-based nanofluids. Fig. 5.5 shows that the water-based nanofluids get the lowest surface roughness at 1 wt.% of nanoparticles concentration, whereas the oil-based nanofluids get the lowest surface roughness at 2 wt.%. It also shows that the overall surface roughness for the oil-based nanofluids are much better than the water-based nanofluids. Besides, in Fig. 5.6, it shows that the best reduction of cutting temperature occur at 1 wt.% for water-based nanofluids and 2 wt.% for oil-based nanofluids. However, for the overall reduction in cutting temperature, water-based nanofluids are much better than the oil-based nanofluids. Thus it indicates that palm oil-based nanofluid performs better compared to water-based nanofluid in terms of surface roughness, however, the better reduction of cutting temperature achieved for water-based nanofluid for all concentrations compared to oil-base.

First, the main difference between water-based and oil-based nanofluids is their chemical properties. From the literature, it can be concluded that an oil-based nanofluid can be considered as a good lubricant but a poor coolant (Kuram et al., 2013). Although, palm oil is a type of vegetable oil that derived from a fruit. It contain a palmitic acid whose chemical formula is $CH_3(CH_2)_{14}COOH$. The COOH group indicates that the palm oil has a good lubricity (Wadumesthrige, Ara, Salley, & Ng, 2009). Therefore the high lubricity of palm oil provides a better reduction in friction force compared to water-based nanofluids resulting better

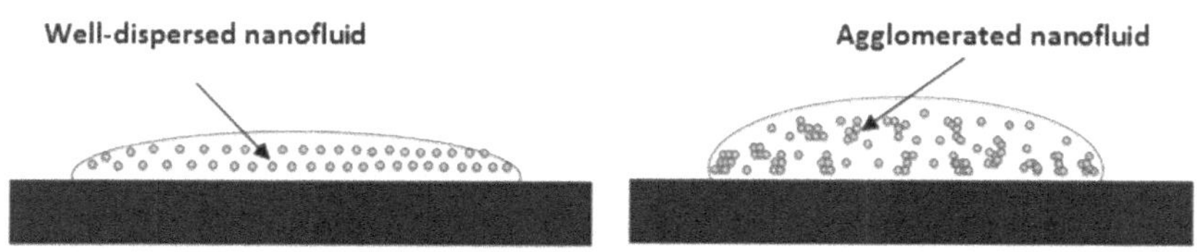

Figure 5.4 Agglomeration effect of nanoparticles on wettability.

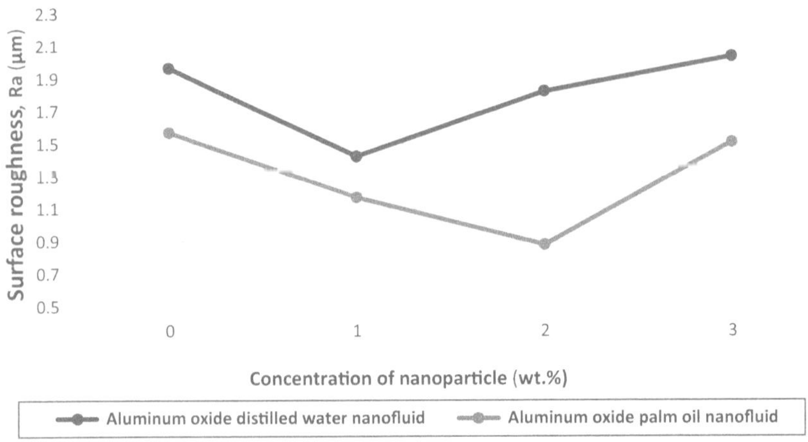

Figure 5.5 Surface roughness when using palm oil- and water-based nanofluids.

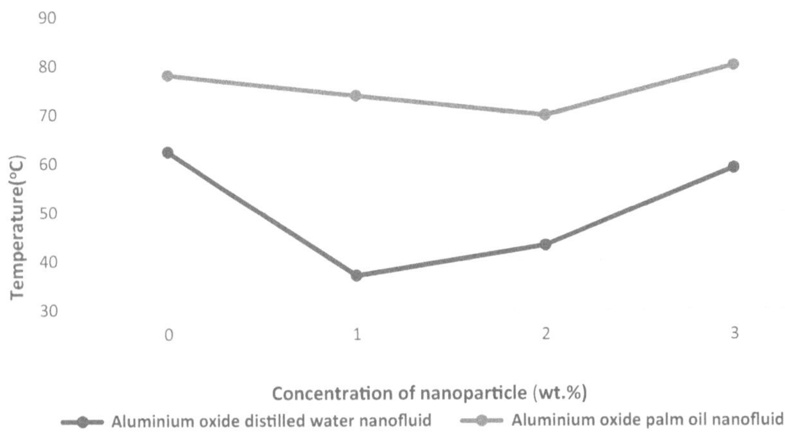

Figure 5.6 Temperature on palm oil- and water-based nanofluids.

surface finish for oil-based nanofluids. However, distilled water has a specific heat capacity of 4.19 kJ/kg/°C, whereas the highest specific heat capacity for palm oil is only 2.857 kJ/kg/°C. This clearly indicates that the water-based coolant provides better cooling compared to oil-based coolant.

5.4 Conclusion

The excessive consumption of mineral oils in machining have caused severe environmental and occupational health hazard. To address this issue, vegetable-based cutting fluids have been employed since they are biodegradable. Nevertheless, vegetable-based cutting fluids suffer from poor machining efficiency. However, MQL technique have been developed as an alternative solution to improve the machining efficiency since it can reduce the amount of cutting fluid required in machining significantly compared to conventional cutting fluid. Nevertheless, MQL technique is still insufficient to meet the requirements of industry when subject to difficult-to-machine high-strength materials namely high-speed steel, nickel, and titanium based alloys. The main challenge for such high-strength materials is the heat removal which leads to premature deformation toward cutting tool as well as substrate surface. One of the emerging solutions for such issue is through incorporating nanoparticles into the base fluid called nanolubricant or nanofluid; hence, the performance of MQL can be improved for such difficult machining materials. Fundamentally, nanoparticles have significantly large surface area compared to its density. Thus well dispersed nanoparticles in base fluid are capable of interacting with the tiny contact surfaces of cutting zone more effectively by dissipating massive heat transfer to cooling media as well as providing better lubricating effect. However, the challenge of identifying dispersion characteristics such as concentration and size of nanoparticle, dispersion mixing process, and parameters of nanoparticle-based cutting fluid are the key factors influencing homogeneity and effectiveness of nanofluid toward desired machinability performance. Furthermore, the hybrid nanofluid-based machining is relatively recent development as an advanced cutting fluid application since the required properties such as lubricity, viscosity, and thermal conductivity can be tailor-made by combining two or more particles of different sizes, shapes, and concentration in the matrix of cutting fluid to maximize machinability performance. Nevertheless, appropriate mixing ratio, particle size, concentration, and mixing technique may greatly influence the machining performance for hybrid nanofluid-MQL system. Future research, development, and innovation are expected to be done in other to establish the field further.

References

Agapiou, J. S. (2018). Performance evaluation of cutting fluids with carbon nano-onions as lubricant additives. *Procedia Manufacturing, 26*, 1429–1440.

Ajithkumar, J. P., & Xavior, M. A. (2018). Influence of nano lubrication in machining operations—A review. *Materials Today: Proceedings, 5*(5), 11185–11192.

Amrita, M., Srikant, R. R., & Sitaramaraju, A. V. (2014). Performance evaluation of nanographite-based cutting fluid in machining process. *Materials and Manufacturing Processes*, *29*(5), 600–605.

Attanasio, A., Gelfi, M., Giardini, C., & Remino, C. (2006). Minimal quantity lubrication in turning: Effect on tool wear. *Wear*, *260*(3), 333–338.

Avila, R. F., & Abrao, A. M. (2001). The effect of cutting fluids on the machining of hardened AISI 1340 steel. *Journal of Materials Processing Technology*, *119*(1–3), 21–26.

Aydin, S. (2017). *Application of quantitative real-time PCR for microbial community analysis in environmental research. Handbook of online and near-real-time methods in microbiology* (pp. 77–86). CRC Press.

Boswell, B., Islam, M. N., Davies, I. J., Ginting, Y. R., & Ong, A. K. (2017). A review identifying the effectiveness of minimum quantity lubrication (MQL) during conventional machining. *The International Journal of Advanced Manufacturing Technology*, *92*(1–4), 321–340.

Brousseau, E. B., Dimov, S. S., & Pham, D. T. (2010). Some recent advances in multi-material micro-and nano-manufacturing. *The International Journal of Advanced Manufacturing Technology*, *47*(1–4), 161–180.

Bruni, C., Forcellese, A., Gabrielli, F., & Simoncini, M. (2006). Effect of the lubrication-cooling technique, insert technology and machine bed material on the workpart surface finish and tool wear in finish turning of AISI 420B. *International Journal of Machine Tools and Manufacture*, *46*(12–13), 1547–1554.

Çakō̄r, O., Yardimeden, A., Ozben, T., & Kilickap, E. (2007). Selection of cutting fluids in machining processes. *Journal of Achievements in Materials and Manufacturing Engineering*, *25*(2), 99–102.

Chetan., Ghosh, S., & Rao, P. V. (2016). Environment friendly machining of Ni–Cr–Co based super alloy using different sustainable techniques. *Materials and Manufacturing Processes*, *31*(7), 852–859.

Chinnam, J., Das, D., Vajjha, R., & Satti, J. (2015). Measurements of the contact angle of nanofluids and development of a new correlation. *International Communications in Heat and Mass Transfer*, *62*, 1–12.

Dambatta, Y. S., Sayuti, M., Sarhan, A. A., Hamdi, M., Manladan, S. M., & Reddy, M. (2019). Tribological performance of SiO2-based nanofluids in minimum quantity lubrication grinding of Si3N4 ceramic. *Journal of Manufacturing Processes*, *41*, 135–147.

Davoodi, B., & Tazehkandi, A. H. (2014). Experimental investigation and optimization of cutting parameters in dry and wet machining of aluminum alloy 5083 in order to remove cutting fluid. *Journal of Cleaner Production*, *68*, 234–242.

Debnath, S., Reddy, M. M., & Yi, Q. S. (2014). Environmental friendly cutting fluids and cooling techniques in machining: A review. *Journal of Cleaner Production*, *83*, 33–47.

El Baradie, M. A. (1996). Cutting fluids: Part I. Characterisation. *Journal of Materials Processing Technology*, *56*(1–4), 786–797.

Gupta, K., Laubscher, R. F., Davim, J. P., & Jain, N. K. (2016). Recent developments in sustainable manufacturing of gears: A review. *Journal of Cleaner Production*, *112*(4), 3320–3330. (Elsevier).

Hegab, H., Darras, B., & Kishawy, H. A. (2018). Sustainability assessment of machining with nano-cutting fluids. *Procedia Manufacturing*, *26*, 245–254.

Kamata, Y., & Obikawa, T. (2007). High speed MQL finish-turning of Inconel 718 with different coated tools. *Journal of Materials Processing Technology*, *192*, 281–286.

Khan, M. M. A., Mithu, M. A. H., & Dhar, N. R. (2009). Effects of minimum quantity lubrication on turning AISI 9310 alloy steel using vegetable oil-based cutting fluid. *Journal of Materials Processing Technology*, *209*(15–16), 5573–5583.

Khandekar, S., Sankar, M. R., Agnihotri, V., & Ramkumar, J. (2012). Nano-cutting fluid for enhancement of metal cutting performance. *Materials and Manufacturing Processes*, *27* (9), 963–967.

Klocke, F., & Eisenblätter, G. (1998). Dry cutting-state of research. *VDI Berichte*, *1399*, 159–188.

Krishna, P. V., Srikant, R. R., & Rao, D. N. (2010). Experimental investigation on the performance of nanoboric acid suspensions in SAE-40 and coconut oil during turning of AISI 1040 steel. *International Journal of Machine Tools and Manufacture*, *50*(10), 911–916.

Kumar Gajrani, K., & Ravi Sankar, M. (2017). Past and current status of eco-friendly vegetable oil based metal cutting fluids. *Materials Today: Proceedings*, *4*(2), 3786–3795.

Kuram, E., Ozcelik, B., & Demirbas, E. (2013). *Environmentally friendly machining: Vegetable based cutting fluids*. *Green manufacturing processes and systems* (pp. 23–47). Berlin, Heidelberg: Springer.

Lawal, S. A., Choudhury, I. A., & Nukman, Y. (2013). Developments in the formulation and application of vegetable oil-based metalworking fluids in turning process. *The International Journal of Advanced Manufacturing Technology*, *67*(5–8), 1765–1776.

Lee, K., Hwang, Y., Cheong, S., Choi, Y., Kwon, L., Lee, J., & Kim, S. H. (2009). Understanding the role of nanoparticles in nano-oil lubrication. *Tribology Letters*, *35* (2), 127–131.

Li, B., Li, C., Zhang, Y., Wang, Y., Jia, D., Yang, M., ... Sun, K. (2017). Heat transfer performance of MQL grinding with different nanofluids for Ni-based alloys using vegetable oil. *Journal of Cleaner Production*, *154*, 1–11.

ManojKumar, K., & Ghosh, A. (2016). Assessment of cooling-lubrication and wettability characteristics of nano-engineered sunflower oil as cutting fluid and its impact on SQCL grinding performance. *Journal of Materials Processing Technology*, *237*, 55–64.

Najiha, M. S., Rahman, M. M., & Yusoff, A. R. (2016). Environmental impacts and hazards associated with metal working fluids and recent advances in the sustainable systems: A review. *Renewable and Sustainable Energy Reviews*, *60*, 1008–1031.

Ni, J., Feng, G., Meng, Z., Hong, T., Chen, Y., & Zheng, X. (2018). Reinforced lubrication of vegetable oils with graphene additive in tapping ADC12 aluminum alloy. *The International Journal of Advanced Manufacturing Technology*, *94*(1–4), 1031–1040.

Ozcelik, B., Kuram, E., Cetin, M. H., & Demirbas, E. (2011). Experimental investigations of vegetable based cutting fluids with extreme pressure during turning of AISI 304L. *Tribology International*, *44*(12), 1864–1871.

Padmini, R., Krishna, P. V., & Rao, G. K. M. (2016). Effectiveness of vegetable oil based nanofluids as potential cutting fluids in turning AISI 1040 steel. *Tribology International*, *94*, 490–501.

Peng, D. X., Kang, Y., Hwang, R. M., Shyr, S. S., & Chang, Y. P. (2009). Tribological properties of diamond and SiO2nanoparticles added in paraffin. *Tribology International*, *42*(6), 911–917.

Pop, L., Puşcaş, C., Bandur, G., Vlase, G., & Nuţiu, R. (2008). Basestock oils for lubricants from mixtures of corn oil and synthetic diesters. *Journal of the American Oil Chemists' Society*, *85*(1), 71–76.

Rahman, M., Kumar, A. S., & Salam, M. U. (2002). Experimental evaluation on the effect of minimal quantities of lubricant in milling. *International Journal of Machine Tools and Manufacture*, *42*(5).

Roy, S., and Ghosh, A. High speed turning of AISI 4140 steel using nanofluid through twin jet SQL system. In International Manufacturing Science and Engineering Conference, Vol. 55461, 2013, V002T04A002. American Society of Mechanical Engineers.

Sajeevan, A. C., & Sajith, V. (2016). Synthesis of stable cerium zirconium oxide nanoparticle—diesel suspension and investigation of its effects on diesel properties and smoke. *Fuel*, *183*, 155—163.

Sayuti, M., Erh, O. M., Sarhan, A. A., & Hamdi, M. (2014). Investigation on the morphology of the machined surface in end milling of aerospace AL6061-T6 for novel uses of SiO2 nanolubrication system. *Journal of Cleaner Production*, *66*, 655—663.

Sayuti, M., Sarhan, A. A., Tanaka, T., Hamdi, M., & Saito, Y. (2013). Cutting force reduction and surface quality improvement in machining of aerospace duralumin AL-2017-T4 using carbon onion nanolubrication system. *The International Journal of Advanced Manufacturing Technology*, *65*(9—12), 1493—1500.

Sharma, A. K., Tiwari, A. K., & Dixit, A. R. (2015). Progress of nanofluid application in machining: A review. *Materials and Manufacturing Processes*, *30*(7), 813—828.

Sharma, A. K., Tiwari, A. K., & Dixit, A. R. (2016a). Characterization of TiO2, Al2O3 and SiO2 nanoparticle based cutting fluids. *Materials Today: Proceedings*, *3*(6), 1890—1898.

Sharma, A. K., Tiwari, A. K., & Dixit, A. R. (2016b). Effects of Minimum Quantity Lubrication (MQL) in machining processes using conventional and nanofluid based cutting fluids: A comprehensive review. *Journal of Cleaner Production*, *127*, 1—18.

Sharma, A. K., Tiwari, A. K., Dixit, A. R., & Singh, R. K. (2017). Investigation into performance of SiO2 nanoparticle based cutting fluid in machining process. *Materials Today: Proceedings*, *4*(2), 133—141.

Sharma, J., & Sidhu, B. S. (2014). Investigation of effects of dry and near dry machining on AISI D2 steel using vegetable oil. *Journal of Cleaner Production*, *66*, 619—623.

Sharma, V. S., Dogra, M., & Suri, N. M. (2009). Cooling techniques for improved productivity in turning. *International Journal of Machine Tools and Manufacture*, *49*(6), 435—453, Yes.

Shashidhara, Y. M., & Jayaram, S. R. (2010). Vegetable oils as a potential cutting fluid—An evolution. *Tribology International*, *43*(5—6), 1073—1081.

Shokrani, A., Dhokia, V., & Newman, S. T. (2012). Environmentally conscious machining of difficult-to-machine materials with regard to cutting fluids. *International Journal of Machine Tools and Manufacture*, *57*, 83—101.

Singh, R. K., Sharma, A., Dixit, A. R., Tiwari, A. K., Pramanik, A., & Mandal, A. (2017). Performance evaluation of alumina-graphene hybrid nano-cutting fluid in hard turning. *Journal of cleaner production*, *162*, 830—845.

Uysal, A. (2016). Investigation of flank wear in MQL milling of ferritic stainless steel by using nano graphene reinforced vegetable cutting fluid. *Industrial Lubrication and Tribology*, *68*(4), 446—451.

Uysal, A., Demiren, F., & Altan, E. (2015). Applying minimum quantity lubrication (MQL) method on milling of martensitic stainless steel by using nano MoS2 reinforced vegetable cutting fluid. *Procedia-Social and Behavioral Sciences*, *195*, 2742—2747.

Vasu, V., & Kumar, K. M. (2011). Analysis of nanofluids as cutting fluid in grinding EN-31 steel. *Nano-Micro Letters*, *3*(4), 209—214.

Vieira, J. M., Machado, A. R., & Ezugwu, E. O. (2001). Performance of cutting fluids during face milling of steels. *Journal of Materials Processing Technology*, *116*(2—3), 244—251.

Wadumesthrige, K., Ara, M., Salley, S. O., & Ng, K. S. (2009). Investigation of lubricity characteristics of biodiesel in petroleum and synthetic fuel. *Energy & Fuels*, *23*(4), 2229—2234.

Wang, Y., Li, C., Zhang, Y., Yang, M., Li, B., Jia, D., ... Mao, C. (2016). Experimental evaluation of the lubrication properties of the wheel/workpiece interface in minimum quantity lubrication (MQL) grinding using different types of vegetable oils. *Journal of Cleaner Production*, *127*, 487—499.

Wang, Y., Li, C., Zhang, Y., Yang, M., Zhang, X., Zhang, N., & Dai, J. (2017). Experimental evaluation on tribological performance of the wheel/workpiece interface in minimum quantity lubrication grinding with different concentrations of Al2O3 nanofluids. *Journal of Cleaner Production*, *142*, 3571–3583.

Weinert, K., Inasaki, I., Sutherland, J. W., & Wakabayashi, T. (2004). Dry machining and minimum quantity lubrication. *CIRP Annals*, *53*(2), 511–537.

Yan, J., Zhang, Z., & Kuriyagawa, T. (2011). Effect of nanoparticle lubrication in diamond turning of reaction-bonded SiC. *International Journal of Automotive Technology*, *5*(3), 307–312.

Zhang, D., Li, C., Zhang, Y., Jia, D., & Zhang, X. (2015). Experimental research on the energy ratio coefficient and specific grinding energy in nanoparticle jet MQL grinding. *The International Journal of Advanced Manufacturing Technology*, *78*(5–8), 1275–1288.

Zhang, Y., Li, C., Jia, D., Zhang, D., & Zhang, X. (2015). Experimental evaluation of MoS2 nanoparticles in jet MQL grinding with different types of vegetable oil as base oil. *Journal of Cleaner Production*, *87*, 930–940.

CHAPTER 6

Re-envisioning sustainability: circular economy and flourishing as promising paths

Xinyi Kuang[1], Miying Yang[2] and Steve Evans[1]
[1]Institution for Manufacturing, University of Cambridge, United Kingdom
[2]College of Engineering, Mathematics and Physical Sciences, University of Exeter, United Kingdom

6.1 Introduction

Circular economy (CE) has recently gained traction as a pathway towards a more promising sustainable future. Firstly, introduced by the environmental economist Pearce and Turner in 1990, circular economy challenges the dominant linear economy represented by the "take-make-disposal" model. The concept of circular economy absorbs the idea of biomimicry to mirror the human system through natural processes. Without negative impact on the environment, biomimicry inspires the regenerative design by creating products and processes that function similar to natural components of the ecosystem (Benyus, 1997).

The perspective of flourishing, on the other hand, proposed by Ehrenfeld as an ideal state of sustainability, refers to "the possibility that humans and other life will flourish on the Earth forever." (Ehrenfeld & Hoffman, 2013). Flourishing contradicts the modern dominating GDP-driven economics model. In addition to the technological and material development, flourishing demands a radical shift of the behavioral, cultural, and institutional underpinnings of the society (Ehrenfeld & Hoffman, 2013). Flourishing raises consciousness of interconnectedness between human and nature, and arouses caring for not only ourselves, but also the entire environment of which we are a part.

Given resemblances of the sustainability vision presented by circular economy and flourishing, this chapter aims at addressing two critical questions: (1) How has circular economy been conceptualized and operationalized as a promising sustainability solution? (2) How well can circular economy contribute to the state of sustainability-as-flourishing?

Sustainable Manufacturing
DOI: https://doi.org/10.1016/B978-0-12-818115-7.00015-8

To answer these questions, the remainder of this chapter proceeds as follows. The second section systematically reviews various school of thoughts, definitions, and underpinning principles that circular economy takes root in, followed by the circular activities taken by policymakers and practitioners as operationalizations of circular economy understandings. The third section firstly addresses the notion of flourishing as the path towards sustainability, and then illustrates the pros and cons of circular economy when it comes to sustainability-as-flourishing. In the end we develop a framework to view the complex world and place circular economy and flourishing on it to indicate their positions. By using the conceptual model as a lens to re-envision sustainability, we further explore how circular economy and flourishing can jointly move the path towards sustainability.

6.2 Circular economy as context

6.2.1 School of thoughts, definitions, and underpinning principles of circular economy

Since the notion of Circular Economy was first introduced by the environmental economists Pearce and Turner (1990) in their book 'Economics of Natural Resources and the Environment', it has taken root in various school of thoughts to complete its grounded theories: cradle-to-cradle design (McDonough & Braungart, 2002), general systems theory (Bertalanffy, 1950), industrial ecology (Graedel & Allenby, 1995), natural capitalism (Lovins, Lovins, & Hawken, 1999), performance economy (Stahel, 2010), blue economy (Pauli, 2010), laws of ecology (Commoner, 1971), biomimicry (Benyus, 1997), and regenerative design (Lyle, 1994).

Scientific endeavors of scholars and practitioners built upon these school of thoughts, therefore, have been made to unlock its underpinning principles and conceptualizations. However, there is no consensus being reached regarding the definition of Circular Economy due to the difficulty of describing or specifying it (de Jesus & Mendonça, 2018). This opinion was echoed in literature that '. . .there is no one single definition of CE.' (Matus, Xiao, & Zimmerman, 2012) and that 'there are various possibilities for defining Circular Economy' (Lieder & Rashid, 2016). Many scholars such as Ghisellini, Cialani, and Ulgiati (2016), Lieder and Rashid (2016), Blomsma and Brennan (2017), Murray, Skene, and Haynes (2017), Geissdoerfer, Savaget, Bocken, and Hultink (2017), and Kirchherr, Reike, and Hekkert (2017) attempted to conceptualize circular economy. Among all these definitions, the most renowned definition came from Ellen MacArthur Foundation (2013), defining the Circular Economy as "an industrial system that is restorative or regenerative by intention and design. It replaces the 'end-of-life' concept with restoration, shifts towards the use of renewable energy, eliminates the use of toxic chemicals, which impair reuse, and aims for the elimination of waste through the superior design of materials, products, systems, and, within this, business models." (Geissdoerfer et al., 2017). This definition implies many different ways to shift towards circular economy including the design of materials, products, systems, and business models as a whole.

These different factors (materials, products, systems, business models, etc.) involved in the shift also distinct different focuses of circular economy conceptualization. One stream of circular economy concept is with an emphasis of systems perspective and business models. In light of Brennan, Tennant, and Blomsma (2015) and Ellen MacArthur Foundation (2013), innovating novel business models play a significant role in the transition towards Circular Economy. van Eijk (2015) also indicated that one advantage of the circular economy is its capability of decoupling economic growth through new business models based on services rather than natural resources. Kirchherr et al. (2017) comprehensively and systematically investigated circular economy definitions based on an analysis of 114 definitions, stating that the systems perspective of circular economy concepts calls for a fundamental shift rather than the incremental optimization of the current linear 'take-make-disposal' economy model. The definition of Circular Economy proposed by Kirchherr et al. (2017) reads: "A circular economy describes an economic system that is based on business models which replace the 'end-of-life' concept with reducing, alternatively reusing, recycling and recovering materials in production/distribution and consumption processes, thus operating at the micro level (products, companies, consumers), meso level (eco-industrial parks), and macro level (city, region, nation and beyond), with the aim to accomplish sustainable development, which implies creating environmental quality, economic prosperity, and social equity, to the benefit of current and future generations."

Whereas some definitions of circular economy emphasize the perspectives of material flows. In the view of material flow, circular economy has been regarded as a 'self-sufficient' industrial system that conducts through closing the loop of material and energy flows, in order to reduce the input of virgin materials and output of wastes (Geng, Sarkis, Ulgiati, & Zhang, 2013; Haas, Krausmann, Wiedenhofer, & Heinz, 2015; Kama, 2015;

Yuan, Bi, & Moriguichi, 2006). The cradle-to-cradle thinking proposed by Braungart et al. (2007) further distinguishes material flows between biological cycle and technical cycle respectively. This separation of material flows was later visualized in the Butterfly diagram by Ellen MacArthur Foundation (2013), elaborating two flows of circulation of technical and biological materials. In the biological cycle, biological (renewable) materials are utilized to design products for reuse purpose and will ultimately return to the biosphere (Smol, Kulczycka, Henclik, Gorazda, & Wzorek, 2015). Whilst technical (non-renewable) materials are mainly recovered and restored in cascading loops (Govindan & Hasanagic, 2018). The tighter the circles in the butterfly diagram, the more complete the circular economy strategy can be because products are retained in the loops at their highest value (Bastein, Roelofs, Rietveld, & Hoogendoorn, 2013; Ellen MacArthur Foundation, 2015a, 2015b).

The various schools of thoughts and different focuses of definitions shed the light on investigating the principles underpinning Circular Economy. Likewise, no common version of the principles has been agreed in this regard. Collectively, at least twelve different principles can be directly extracted from the published reports by Ellen MacArthur foundation (EMF), McKinsey and Company, and other researchers (Ellen MacArthur Foundation, 2013; Ellen MacArthur Foundation, 2017; Ellen MacArthur Foundation and McKinsey, 2014; Ellen MacArthur Foundation and McKinsey Center for Business and Environment, 2015; McKinsey and Company, 2016; Esposito, Tse, & Soufani, 2018), illustrated as below:

1. Design out waste and pollution
2. Keep products and materials in use
3. Regenerate natural systems
4. Preserve and enhance natural capital by controlling finite stocks and balancing the flow of renewable resources
5. Optimize resource yields by circulating products, components, and materials in use at the highest possible levels at all times
6. Make the system more effective by eliminating negative externalities
7. Circularity introduces a strict differentiation between consumable and durable components of a product
8. The energy required to fuel this cycle should be renewable by nature
9. Build resilience through diversity
10. Waste is food
11. Think in systems
12. Think in cascades

6.2.2 Policy making and industrial activities as an operationalization of circular economy understandings

Moving beyond the conceptual definitions and principles, circular economy has increasingly gained importance on the agenda of policymakers (Brennan et al., 2015). Germany started implementing circular economy as a pioneer in 1996 through enactment of its national law "Closed Substance Cycle and Waste Management Act" (Su, Heshmati, Geng, & Yu, 2013). Subsequently, Japan accelerated the transition towards a recycling based society with quantitative targets by issuing a comprehensive legal framework, namely 'Japan's 2002 Basic Law for Promoting the Creation of a Recycling-Oriented Society' (METI, 2004). In 2009, China announced the "Circular Economy Promotion Law" which defines Circular Economy as "a general term for the activity of reducing, reusing, and recycling in production, circulation, and consumption" (Fischer & Pascucci, 2017). In 2015 European Commission declared the policies of "Towards a circular economy: A zero waste program for Europe" (European Commission, 2015a, 2015b) and European Union's 2015 Circular Economy Package (European Commission, 2015c), in which CE has been defined as an economy "where the value of products, materials and resources is maintained in the economy for as long as possible, and the generation of waste minimized". Compared to China where a top-down approach has been mainly adopted for implementing circular economy, other countries such as Brazil, India, and the United States advocate the development of CE with bottom-up policies, whilst the EU combines the strength of both (Geng, Sarkis, & Bleischwitz, 2019; Ghisellini et al., 2016). Yet Geng et al. (2019) specified there still are many gaps existing in the current economic model for implementing the circular economy policy. For instance, the costs of adjusting production process and organizational routine for circular economy are often overlooked or assumed to be very low.

Simultaneously, industrial activities have been taken in support of CE policies as an operationalization of CE understandings. Various 'R' activities are viewed as the 'how-to' of Circular Economy (Reh, 2013; Zhu, Geng, & Lai, 2010). Chinese Circular Economy Promotion Law initially set the 3R framework at the core of CE implementation that consists of reduce, reuse and recycling (Kirchherr et al., 2017). This framework was later supplement with Recover as the fourth R, according to the 'European Union (EU) Waste Framework Directive' (European Commission, 2008).

Lacy and Rutqvist (2015) represents 5Rs as reduce, reuse, refurbish, repair, and recycle. VanBuren, Demmers, Heijden, and Witlox (2016) further expand the scope to 9Rs refuse, rethink, redefine, reuse, reform, remanufacture, redefine, recycle, and recover. The most comprehensive framework so far has outlined 10R activities as refuse, rethink, reduce, reuse, repair, refurbish, remanufacture, repurpose, recycle, and recover (Potting, Hekkert, Worrell, & Hanemaaijer, 2017). Though the number of Rs varies from list to list, many R-lists have established a priority order for implementing R activities for circularity (Henry, Bauwens, Hekkert, & Kirchherr, 2020). Sihvonen and Ritola (2015) set the 4R following the priority order of reduce, reuse, recycle, and recover for circular economy implementation, with specific circular activities to be addressed in each R. For instance, specific R activities such as resale, repurpose, repair, refurbish, remanufacture, and resynthesize are considered to support the 'Reuse' objective as the second priority to enable circularity (Sihvonen & Ritola, 2015).

In this section, we initially review the various schools of thoughts behind the formation of Circular Economy. Then we explore the conceptualization and definition of Circular Economy with different focuses of either the business model perspective or the material flow perspective. Investigating the core principles underpinning circular economy, we have identified at least 12 different principles suggested by Ellen MacArthur foundation (EMF), McKinsey and Company, and other scholars. Subsequently, a further review was provided regarding how policymakers have interpreted these circular economy concepts and principles in the policy making. In accordance with these policies, ten different types of 'R' activities have been specifically outlined to demonstrate the industrial activities as an operationalization of CE understandings. Bearing all these understanding of circular economy in mind, the next section explores how circular economy contributes to the realm of sustainability-as-flourishing.

6.3 Could circular economy be a promising solution for sustainability-as-flourishing?

6.3.1 Sustainability-as-flourishing as a path towards a truly sustainable world

Ehrenfeld and Hoffman (2013) defined sustainability-as-flourishing which is about "the possibility that human and other life will flourish on Earth forever." It delineates a story radically different from Brundtland

Commission's definition of sustainable development. Ehrenfeld and Hoffman (2013) criticized the Brundtland Commission's definition as it failed to shake the current pattern of the world but merely reinforce it by doing things better. However, merely doing things better will not lead us towards sustainability. Ehrenfeld and Hoffman (2013) argued that the world adhering to the Brundtland Commission's sustainable development still rests on the same material-oriented paradigm and goes on the same path of economic development on growth and efficiency.

Different from Brundtland Commission's definition of sustainable development, flourishing attempts to reformulate the metaphor of sustainability and refines our fundamental way of viewing ourselves and the world around us. Ehrenfeld explains that 'flourishing' means 'not only to grow, but to grow well, to prosper, to thrive, to live to the fullest.' In line with this positive vision, a radical shift is called for to establish a new world on cultural, personal, and spiritual growth on top of the technological and material development (Ehrenfeld & Hoffman, 2013).

Sustainability-as-flourishing thinking has laid the foundation for researchers to develop solutions for sustainability. Evans et al. (2017) proposed that new sustainable business models can be enabled by systematically reconsidering the purpose, design, and governance of the value network, and if the planet system flourishes as described by Ehrenfeld and Hoffman (2013), each aspect of sustainability—social, economic, and ecological—should stay healthy and well-balanced. Likewise, Breuer and Lüdeke-Freund (2014) defined sustainable business model based on 'sustainability-as-flourishing' (Ehrenfeld & Hoffman, 2013) and 'non-declining capital rule' (Dresner, 2008), as a business model to create, deliver, and capture value for all its stakeholders without depleting the natural, economic, and social capital it relies on.

In view of circular business model as one type of sustainable business models (Geissdoerfer, Morioka, de Carvalho, & Evans, 2018), and of circular economy being acknowledged as an emerging solution for achieving sustainability, we are interested to investigate whether circular economy could be a promising solution for achieving sustainability-as-flourishing in future. In the preceding sections, we have reviewed where circular economy takes roots in, and what circular economy means conceptually, and how circular economy can be operationalized by policymakers and practitioners. Though emerged and implemented at early stages with significant impediments, circular economy still has gained many attentions and achieved much progression. Thereby we use flourishing as the lens to view if circular economy is moving us through the path towards sustainability-as-flourishing, which has depicted a desirable and authentic vision of a sustainable future.

6.3.2 What circular economy share in common with sustainability-as-flourishing

Ehrenfeld related sustainability to flourishing over fifteen years ago, and lately he published the book 'The right way to flourish: Reconnecting to the real world' (Ehrenfeld, 2019) to continue the argument that there are two main historic facts hindering flourishing: (1) the depictions of how the natural world works and (2) the depictions of how the human species work. We observe that circular economy positively contributes to the first part considering circular economy is derived from biomimicry and industrial ecology, thus to a certain extent moving the world towards sustainability-as-flourishing. However, due to the scarcity of social dimensions as the inherent defects of circular economy conceptualization, flourishing, as the ultimate ideal state for sustainability, therefore, still remains distant from where we are now. In this section, we state there are mainly four positive aspects that circular economy shares in common with sustainability-as-flourishing.

6.3.2.1 First, both shifts for flourishing and circular economy contradict the assertion that more eco-efficiency necessarily equates to sustainability

Ehrenfeld and Hoffman (2013) argued most efforts adhere to Brundtland Commission's definition of sustainable development are built upon eco-efficiency as the core principle. However, the efforts trying to eliminate unsustainability is not necessarily to generate sustainability. Trapped into the same wrong path, the world addicts to a technology fix to create more eco-efficiency and remains at a certain level of unsustainability. Sustainability-as-flourishing challenges the old fallacious path and depicts a new paradigm, calling for a fundamental shift in the ways for us to rethink and to reorganize the society.

Similarly, circular economy challenges the conventional, linear take-make-disposal path by distinguishing eco-efficiency from eco-effectiveness, the latter notion being used to sort out a feasible relationship between ecological systems and economic growth. The dilemma of eco-efficiency is that it failed to unify with the economic goal of eternal growth (Braungart, Mcdonough, & Bollinger, 2007). According to Braungart's cradle-to-cradle philosophy, 'instead of reducing or delaying the cradle-to-grave flow of materials, eco-effectiveness creates metabolisms where materials are used over and over again at a high level of quality' (Ellen MacArthur Foundation, 2013).

6.3.2.2 Second, both shifts for flourishing and circular economy look at the concept of value rather than purely economic terms

Sustainability-as-flourishing requires a deep shift in values. The issue of our entrenched culture is that GDP becomes the only measure of the value of goods and services traded in the markets, so that relationships among individuals become primarily transactional and maximization of utility becomes the eventual goal (Ehrenfeld & Hoffman, 2013). However, what is hidden behind the economic measures of material success is a much-diminished set of values. Though economic value has trumped a majority of other dimensions of value, sustainability-as-flourishing calls for a better value metric beyond the extrinsic acquisitions and accomplishments, but to deal with well-being and to measure the value of a quality life.

Circular economy also aims at replacing the linear model and delivering sustainable value. When it comes to a firm's creation for sustainable value, Evans et al. (2017) suggested that a more holistic meaning of value to integrate social, economic, and environmental goals should be considered. In search of different goals, different organizations engage in value networks for exchanging tangible and intangible value, requiring specific sustainable value propositions to be created for all roles involved in the network (Allee, 2008; Ouden, 2012; Evans et al., 2017). Conventional business model research built on the linear model generally divides value system into three core elements: value proposition, value creation and delivery, and value captured. In the context of sustainability, Yang, Evans, Vladimirova, and Rana (2017) extensively deem sustainable business model as a wider scope of value system to incorporate value surplus, value absence, value missed and value destroyed underpinned by the triple-bottom line perspective of sustainable value. For instance, Yang and Evans (2019) explored the sustainable value creation in various archetypes of product-service system business model, which has been regarded as a typical and successful circular business model. According to their research, social value delivered by PSS models include increased jobs and improved employee salary and satisfaction, whilst environmental value comprises longer product life, freedom to design for sustainability, increased product usage, increased recycling, remanufacturing and reuse, increased resource and energy efficiency, reduced carbon emission, etc.

6.3.2.3 Third, both shifts for flourishing and circular economy address ecosystem resilience

Ehrenfeld (2019) discussed the biological aspect of flourishing and stated the fundamental purpose of life is to stay alive. He cited Maturana and Varela (1980)'s opinion that the primary but unique process of living systems indicates continuous self-creating or self-reproducing. In this way, the continuous rise and survival of every new species contribute to species variation and reaches the state of flourishing.

One of the key principles behind circular economy, though relatively less discussed and researched in literature, is to build resilience through diversity (Ellen MacArthur Foundation, 2013). As confirmed by Braungart, 'Natural systems support resilient abundance by adapting to their environments with an infinite mix of diversity, uniformity and complexity.' (Ellen MacArthur Foundation, 2013). The main difference here is that flourishing is only used in reference to living beings and not supposed to be applied on business. This means a "flourishing business" ought not to exist (Ehrenfeld, 2019). While circular economy encourages industries to build resilience with many connections and scales by mimicking natural systems as successful models (Ellen MacArthur Foundation, 2013).

6.3.2.4 Fourth, both shifts for flourishing and circular economy are concerned with natural system

Flourishing pertains to all natural systems including not only human beings but also other lives. However, the modern society has set physical and mental barriers between us and the world out there, thus we are suffering the consequences of this disconnection such as species extinction and natural disasters (Ehrenfeld, 2019). Only when we authentically perceive who we are, how we connect with the environment, and what we should do to care for the world, can we remove these barriers and enable sustainability as flourishing.

In circular economy the careful management of biological nutrients also demonstrates great care for the biosphere. Cradle-to-cradle thinking separates materials flow into biological cycles and technical cycles. Biological cycle is designed for the non-toxic and biodegradable ingredients to re-enter the biosphere safely, while technical cycle is designed to circulate durable, toxic, and non-degradable components at high quality without entering the biosphere (McDonough & Braungart, 2002). The butterfly diagram depicting the relationship between biosphere and technical sphere (Ellen MacArthur Foundation, 2013) indeed reflects the connectedness of human life to the rest of the world.

6.3.3 What are the challenges of circular economy to approach sustainability-as-flourishing

As stated above, circular economy significantly overlaps many far-sighted visions of flourishing and attempts to move the world towards the next level of sustainability. Circular economy looks beyond eco-efficiency, delivers the sustainable value, addresses ecosystem resilience, and connects with the natural system. Despite the merits of circular economy, there are some inherent problems of circular economy that hinder it from reaching the state of sustainability-as-flourishing.

There are two most commonly discussed imperfections of circular economy regarding sustainability: the lack of social dimension in general and the short of energy efficiency in specific cases. Circular economy has addressed the interconnections between the environmental and economic activities (Andersen, 2007) but undervalue social issues (Murray et al., 2017). Kirchherr et al. (2017) confirmed this point in the report which comprehensively reviewed 114 extant definitions of circular economy, only to find 18%–20% CE definitions consider the social dimension of sustainability. Geissdoerfer et al. (2017) stated that the notion of circular economy focuses on economic benefits but neglects large parts of the social dimension and simplifies the environmental perspective. Even in limited number of researches the social dimension of circular economy under addressed, scholars mostly look at the influence of circular economy on job creation without a clear understanding how it can contribute to social wellbeing (Beuren, Gomes Ferreira, & Cauchick Miguel, 2013; Frey, Stutzer, & MacCulloch, 2003). Another drawback is that in some circumstances circular economy may lack energy efficiency and harm sustainability (Geissdoerfer et al., 2017). In some cases, the cost paid for energy consumption and the overall impact on sustainability cannot offset the total effect of obtaining material for circularity. Especially with the massive growth of demand, there is a technical impossibility for meeting the required energy to circulate materials in cascaded loops (Allwood, 2014; Andersen, 2007). The similar criticism goes to circular business models. Despite circular business model having been manifested as one type of sustainable business model (Geissdoerfer et al., 2018), scholars have argued that CE-oriented business models fail to always capture the full potential of sustainability or to realize the ultimate target of sustainability even when fully circular (Pieroni, McAloone, & Pigosso, 2019).

The inherent scarcity of social dimension of circular economy poses significant challenges for achieving sustainability-as-flourishing. Lahane, Kant, and Shankar (2020) referred circular economy as an emerging sustainability strategy which requires a thorough reform for the entire system of human activities. We argue that circular economy in practice changes only limited ways of people to live or consume. By mirroring natural system as successful models to the business world, circular economy benefits environmental sustainability without diminishing the economic growth and prosperity. The purpose of balancing environmental and economic dimensions therefore can be fulfilled in this way, but as outlined above, we observe that circular economy does not really provide the depictions of how the human species work.

Different from circular economy, flourishing begins with the social metrics by asking two crucial questions (Ehrenfeld, 2008; Ehrenfeld, 2019; Ehrenfeld & Hoffman, 2013): '(1) Is there an intrinsic purpose to all life, and, if so, is human life different from other forms? (2) How are we (as a species) doing to realize our purpose, and if we are not doing so well, how can we do better?' Based on the conversation between Ehrenfeld and Hoffman (2013), sustainability-as-flourishing suggests a twofold shift. At the individual level, sustainability-as-flourishing requires everyone to shift from Having mode (defining ourselves by the materials we possess) to Being mode (defining ourselves by the extent to which we act authentically). At the systemic level, sustainability-as-flourishing needs the society to shift from the pure rationalism and enlightenment culture as its underpinnings towards a more pragmatic thinking. Without giving much attention to the economic benefit, sustainability-as-flourishing emphasizes on the social dimension of human beings beyond the economic or technological aspects of sustainability, by looking at the cultural behavioral, and institutional underpinnings. In this way, the natural system of which we are a part will be ultimately cared.

6.4 Circular economy and flourishing shifts as a joint path towards sustainability: a conceptual model

6.4.1 A conceptual model to re-envision sustainability

In last section, we have addressed how circular economy aims at challenging the traditional linear model, and how sustainability-as-flourishing could tackle the deeply embedded structural root of the dominating

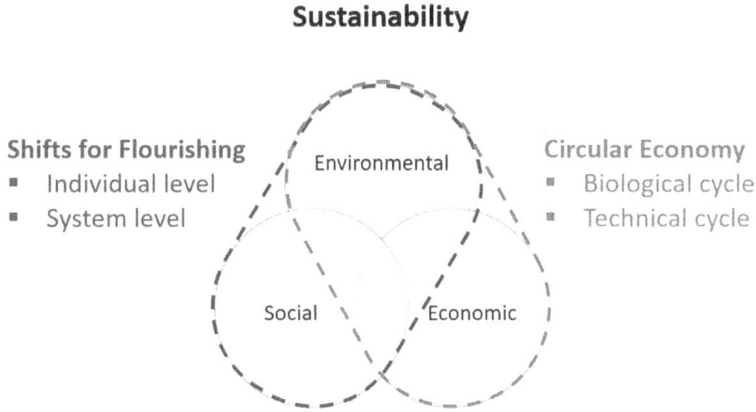

Figure 6.1 Circular economy and shifts for flourishing mutually complement triple-bottom line of sustainability.

GDP-driven economics paradigm. Fig. 6.1 illustrates how circular economy and shifts for flourishing address Elkington's (1997) the triple-bottom line perspectives of sustainability. Circular economy advocates a much healthy economic growth pattern as opposed to the linear economy, simplifies the environmental aspect into to resource input, waste and emission output, and neglects large parts of the social dimension (Geissdoerfer et al., 2017), while the shifts towards sustainability-as-flourishing are more concerned with the human core and the environment around us, expressing the disapproval of the modern economics-based paradigm (Ehrenfeld & Hoffman, 2013). Hardly can shifts for sustainability-as-flourishing or circular economy be able to deliver the three-dimensional sustainable value.

To flourish, a new model of what is authentic and genuine awaits being created (Ehrenfeld & Hoffman, 2013), but this model sounds difficult to visualize and even harder to implement. To deepen the understanding of the gap between circular economy and sustainability-as-flourishing, in this section we develop a conceptual model (shown in Figs. 6.2 and 6.3) to seek the underlying solution for a sustainable path. This conceptual model provides a new lens to better indicate the position of circular economy and flourishing shifts regarding the path towards sustainability. By pointing out where we are now and where we are heading for, some crucial aspects to re-envision sustainability have been discussed under the view of the conceptual model.

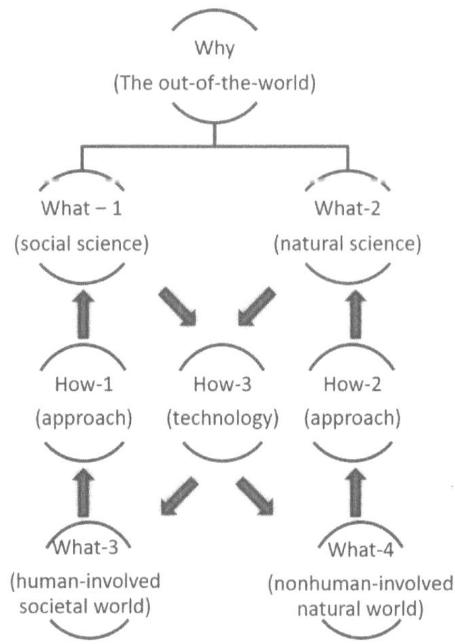

Figure 6.2 A conceptual model to re-envision sustainability (1st and 2nd dimensions).

This conceptual model was originally inspired by the renowned Golden Circle proposed by Simon Sinek (2011). In addition to the conventional why-how-what pattern, this conceptual model is four-dimensional, and acquires a deeper insight into the essence of the world. Fig. 6.2 depicts the first two dimensions and Fig. 6.3 the other two dimensions. There are five fundamental elements contained in this conceptual model: 'why', 'how', 'what', 'momentum', and 'time'. In the following we firstly address how each element supports or holds back to way towards sustainability as flourishing.

6.4.1.1 'Why' as the first element

'Why' is positioned at the top level and considers the world from the inside. There are four reasons why the 'Why' plays the key role to permitting flourishing. To start, as brought about by Ehrenfeld and Hoffman (2013), sustainability lies in a shift of everyone's consciousness about who we are and how we govern, respect, and become a part of the world. However, many modern thinkers are obsessed in answering 'how we should live' but outsource the 'Why' questions to theologians (Ehrenfeld, 2019). A positive example is Maslow's (1943) five-tier hierarchy of needs

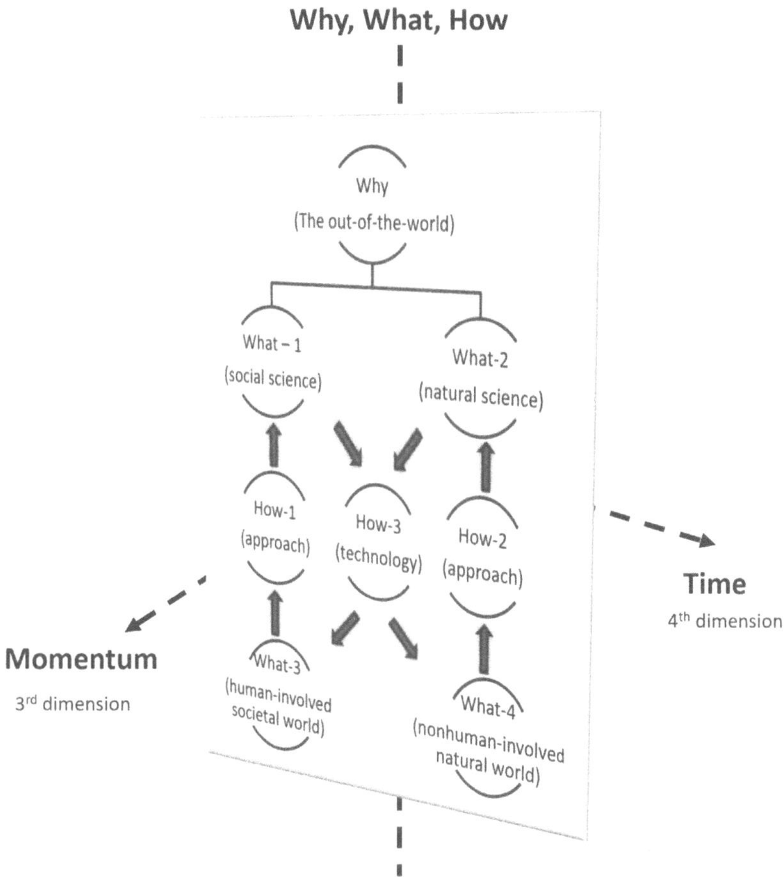

Figure 6.3 A conceptual model to re-envision sustainability (3rd and 4th dimensions).

springs from an inside look at human needs, which sets self-actualization at the highest existential level.

Besides, 'Why' represents where the spiritual domain places, which refers to an ultimate or immaterial reality. Ehrenfeld and Hoffman (2013) believes putting spiritual domain back to the place is particularly crucial to sustainability, as it invokes our deepest values and meanings by which we live. Furthermore, in contrast to the material world, the out-of-the-world domain of spirituality recover our sense of reconnecting to the other human community and nature (Ehrenfeld & Hoffman, 2013). This is echoed by what Schopenhauer (1883) wrote in the book 'The World as Will and Representation' that the best part of life can be gained through a true delight in art and a pure joy in beauty which lift us out of real existences, in order to reach a pure will-less, painless, and timeless subject of knowledge.

Moreover, Ehrenfeld and Hoffman (2013) stated that 'Sustainability-as-flourishing' refers to a state of Being in which the individual realizes a sense of wholeness, completion, or perfection. 'Flow', a psychological concept originated by Csikszentmihalyi (1991), was related to flourishing as it describes a state where our whole beings are completely involved in an activity with full concentration paid and utmost skills used (Ehrenfeld, 2019).

Last but not least, 'Why' is where 'love' resides. Sustainability is not going to present itself until our culture starts transforming to foster love that is at the center of caring and interconnections (Ehrenfeld & Hoffman, 2013). Schopenhauer (1883) cited Hesiod and Parmenides' words stating that 'Eros was the first, the creator, the principle out of which all things emerged' (Eros is Greek word meaning different kinds of love).

6.4.1.2 'What' as the second element

In this conceptual model for viewing the complex world, 'What' can be represented with two different types or layers. First, at the bottom of the graph 'what' refers to the representation of the extrinsic world, separated by human involved (Marked as What-3) part and non-human involved part (Marked as What-4). In the context of the Butterfly diagram of circular economy, the left side refers to the technical sphere and the right side refers to the biosphere.

Second, the upper layer of 'What' refers to scientific knowledge on what we have known about the social (human involved as What-1) world and natural (non-human involved as What-2) world. That's the 'What' type of questions that science generally answers. The way to look at the conceptual mode from either bottom-up or top-up direction prepare the ground for the two reasoning approaches – Induction and Deduction.

6.4.1.3 'How' as the third element

'How' locates in between the two types of 'What'. From 'What' represents the world to 'What' science to learn, 'How' stands for the ways and methods we adopted to study and understand the representation of the world (Marked as How-1 and How-2). On the other way around, 'How' indicates how the technologies and tools we created from social and natural science (Marked as How-3), to reinforce, reshape, and reconstruct the world we are living in. However, to flourish, obstacles reside in both method-based 'How' and technology-based 'How'.

With regard to the former type of 'How' as approaches and methods, one problem is that despite scientific methodology has dominated our culture to create truth, there are limits of scientific methodology to perceive the system complexity of highly interconnected objects or capture the holistic qualities of life (Ehrenfeld & Hoffman, 2013). In fact, to learn sustainability as a complex system requires understanding beyond the Cartesian mechanical model of the natural world and the Enlightenment concept of human nature as embedded in the modern world (Ehrenfeld, 2019). Another issue is that current scientific methodology is quantitative-dominated and data-driven. The drawback of scientific method's domination is that it inevitably tears the whole system apart, leaving only part of the pieces to be quantified and controlled which can further reproduce and reconstruct the whole system. Ehrenfeld and Hoffman (2013) argued even with the help of supercomputer models and algorithms, it is a mistake to frame everything through the scientific method and objective reality, because merely objective, quantitative, disparate metrics, rational reasoning cannot capture the full picture of the complex world. The third observation goes to the defects of scientific methodology for social science. Soros (1994) stated that there ought to be differences in the method of studying natural science and social science. It is a fallacy to employ the entire same logic and approaches of studying natural science as the lens for social science, partly due to reflexivity effect on the latter one. The only solution as he mentioned, is to 'deprive social science of the status it enjoys on account of natural science (Soros, 1994)'. The quantitate-oriented culture further promotes the idea of ranking things and reinforces the materialist vision of the standard good life but hampers the critical cultural to shift from quantity to quality. Different from other measure, flourishing is qualitative and cannot fully quantified without harming its basic meaning (Ehrenfeld, 2019).

In terms of the latter type of 'How' as technology and tools, some technologies have shaped our cognition and cultural surrounding and perpetuated the world at a state of unsustainability. Postman (2006) highlighted the negative effect of television and other media as the new medium for knowledge transmission, harms the seriousness of thoughts and cognition which are used to be conveys by printed books. In addition, Harari (2016) recognized that in future everything framed by digitalization may become our new belief and thus may negatively affect the humanity. Moving towards a sustainable society requires us to cease the habitual ways of acting individually and societally (Ehrenfeld & Hoffman, 2013). Some of these technology developments, on the contrary, reinforce our cultural surroundings to feed our habitual behavior.

6.4.1.4 'Momentum' as the fourth element

Fig. 6.2 reveals the first two dimensions of the conceptual model, and Fig. 6.3 further delineates the third dimension as 'Momentum' which implies the impact of each element at a specific time, as well as the fourth dimension as "time" to represent its dynamic essence. The element of 'momentum' reveals the essence of the system which is not static but dynamic. For each point on the graph, regardless of 'What', 'How' or 'Why', its momentum is ever changing and can be transferred from one point to the other. An example is the current cutting-edge technologies such as artificial intelligence and internet have gained the powerful momentum to shape the system, while the obsolete technology has lost momentum and disappeared in the extrinsic world.

6.4.1.5 'Time' as the fifth element

Sustainability must be a dynamic concept and flourishing itself represents change and striving (Ehrenfeld & Hoffman, 2013). The conceptual model consolidates the role of time as the fourth dimension to reveal its dynamic and ever-changing essence. This factor also can be named as 'When' in parallel with the 'Why', 'How', and 'What' factors. The momentum accumulated at each point, is closely related to the 'Time' dimension. As stated above, up and fall of momentum at each point can be transferred from one to the other over time. A most typical example in modern world is the monetary value which can be easily transferred to technology investment and scientific funding. Affluent nations take advantage of money, technology, and science as means to gain environmental superiority.

Measured by time, the bottom 'What' (What-3,4) represents our most changeful extrinsic world; the 'How' (How-1,2,3) represents the update of technologies and tools changing still fast but slight slower than the extrinsic world; the upper 'What' represents (What-1,2) that the scientific knowledge develops much slower where the knowledge boundary is inch by inch pushed forward by researchers; the top 'Why' almost does not evolve much over time as tracing back to Aristotle and Plato's era, their philosophical thinking are still deemed as truth today.

6.4.2 Moving towards a sustainable future

Having described each element in above four-dimensional conceptual model to unlock the understanding towards sustainability-as-flourishing, in the following we are going to dig up deeply from several important aspects with the conceptual model as a lens, to seek the underlying solutions for the path towards a sustainable future.

The first angle is to understand that the sequence for implementing this model is critical for sustainability. Flourishing suggests the shifts should start with the top 'Why', by asking the core question that who we are as human beings. Instead of defining ourselves by the bottom 'What' – the material goods we possess in the extrinsic world due to the overwhelming influence of the modern culture, flourishing requires us to authentically answer the 'Why' question in order to redefine our value, culture, beliefs, and behaviors (Ehrenfeld & Hoffman, 2013). Likewise, Sinek (2011) in his book 'Starting from why' indicates the right sequence of implementing the Golden Circle is the key to success, since 'The power of why is not opinion, it is biology'. Another book 'Tao Te Ching', an ancient Chinese philosophy text written around 600 BC, expressed the similar wisdom: 'All things are produced by the Tao, and nourished by its outflowing operation. They receive their forms according to the nature of each and are completed according to the circumstances of their condition. Therefore, all things without exception honor the Tao, and exalt its outflowing operation. This honoring of the Tao and exalting of its operation is not the result of any ordination, but always a spontaneous tribute. Thus, it is that the Tao produces (all things), nourishes them, brings them to their full growth, nurses them, completes them, matures them, maintains them, and overspreads them.' (Tse & Legge, 2009). Here Tao in ancient Chinese refers to 'the Why'. The conceptual model interprets the first sentence of this paragraph as: 'Why' produces everything, penetrates all the way down through to generate 'How' and 'What' into forms, and cultivates momentum along with the circumstances of each condition. Going back to circular economy, we argue that one important reason keeping circular economy away from the state of sustainability-as-flourishing, is that circular economy starts changing the unsustainable system from 'What', with only little 'Why' aspect to be addressed. By imitating the bottom right 'What' of ecosystem (What-4), to understand 'How' the nature model works, circular economy develops its concepts and knowledge into the scientific 'What', and attempts to apply an analogical model to the bottom left 'What' (What-3), i.e. the modern industrial society. While sustainability demands us to cease consuming the way that we do today because 'it is not that we consume, but the way we consume that matters (Ehrenfeld & Hoffman, 2013).' Without fundamentally reconsidering the 'Why' question, circular economy challenges the underpinnings of the framework in a mild way and has limited impacts on the economic system and consuming behavior.

The second angle is to understand how this conceptual model can be implemented at macro, meso, and micro levels. Sustainability-as-flourishing demands a shift individually and societally. In this conceptual model, at the individual level, 'Why' question asked at the individual level indicates the purpose and meaning of life derived from heart, as the first but fundamental step towards flourishing; the upper 'What' (What-1,2) refers to left brain and right brain controlling one's thinking pattern; 'How' (How-1,2,3) implies one's skillset to fulfill sustainability tasks; the bottom 'What' (What-3) can be the appearance and actions of this person. At the corporation's level, 'Why' refers to a firm's vision, for instance, demonstrating the core value by caring for the world; the upper 'What' (What-1,2) is the firm's sustainability strategy in line with its vision; 'How' (How-1,2,3) can be the tactics adopted in the daily operation, like lean manufacturing, in support of the vision and strategy; The bottom 'What' (What-3,4)is the company's actual performance that are generally measured by a set of economic, environmental, and social performance indicators. At the society level, 'Why' incorporates to the philosophical and spiritual source; The upper 'What' (What-1,2) refers social and natural science developed in the process the knowledge evolvement; 'How' (How-1,2,3) refers to approaches, methods, and technologies used to understand the two-way world; The bottom 'What' (What-3,4) represents the extrinsic world consisting of human, other animals, and the surrounding environment.

The third angle under exploration is regarding 'bring back to future'. In the contemporary world, each aspect of the society has been shaped by technology, technocracy, and scientific principles. 'Bring back to future' suggests we should return to the mind-set containing elements used to be treated as vital in the past world (Ehrenfeld & Hoffman, 2013). The term of 'bring back to future' recalls me an important debate in 2019 regarding restorative economy, among the sustainability cohort organized by Institute for Manufacturing (IfM), University of Cambridge: If being sustainable means retaining the current economic, environment, and social statures in future, what should we make the world restorative and return it to a certain point in the past? Indeed, sustainability implies a comparison between the systemic future condition and the present state (Ehrenfeld, 2019). And this insight brings significance to the 'restorative'. Here the meaning of 'restorative' is beyond the EMF's definition of restorative and regenerative cycles that correlate to the biological and technical cycle of circular economy. With the lens of the conceptual mode, we need to carefully decide 'What are the specific sustainable value to be selected for restoration in future, who should make the decision, and how to ensure the decision is ethically acceptable?'. While 'When' asks another crucial aspect - Back to which specific point in the past time, are we going to restore and recover the sustainability statues for a sustainable future? 'How' and 'What' are relatively easier to be answered especially with the certain sustainability knowledge and technology in hand. Overall, the sustainable value at each dimension of the conceptual model should be carefully considered for restoration.

The fourth angle is that we conclude product-service systems (PSS) as a powerful 'How' can enhance both flourishing and circular economy. For one thing, PSS has been deemed as a pioneering sustainable business models that shifts the manufacturing and consumption from the linear economy towards circular economy (Spring & Araujo, 2017; Yang, Smart, Kumar, Jolly, & Evans, 2018). The essence of PSS is to offer functionalities as opposed to products to customers. In contrast to externalities of the product-based business models, PSS enables sustainability by internalizing externalities (Evans et al., 2017). PSS also allows firms to be responsible for any economic, environmental and social issues arise during the product's life cycle (Evans et al., 2017). Yang and Evans (2019) specifically develops a novel framework of PSS business model archetypes to systematically understand how each dimension of sustainable value can be delivered by different PSS archetypes. For another, Ehrenfeld (2019) also recognized that in the context of capitalism, businesses can view themselves as services providers to enable the consumers to act more authentically. This view is in accordance with the nature of PSS and contributes to the authenticity of Being for flourishing.

The last angle is to seek an underlying solution assisted by the conceptual model. We have illustrated how the shifts for sustainability-as-flourishing start from the 'Why' domain at the top of the conceptual model, following a top-down path to influence all other elements. As for circular economy, the reform comes from bottom 'What' domain and follows the bottom-up direction, leaving 'Why' for sustainability and 'What' for the social dimension as blank. We propose that to reach the state of sustainability-as-flourishing, the 'Why'-oriented question should be prioritized with the upmost weight followed by the answers of 'How-to' and 'What'. Flourishing for sustainability argues that we ought not to be defined by 'Economic Man' through the material we possess but should stay as the authentic Being. For the long run, nurturing organization culture change, cultivating leadership for sustainability, and enhancing flourishing for engineering education (Fitzpatrick, 2017; Nicholson & Kurucz, 2019; Vodonick, 2018) could be more radical and effective ways moving us towards the sustainability-as-flourishing. Regarding 'How-to' and 'What' type of questions, Ehrenfeld (2019) further emphasized that systems thinking can positively contribute to the sustainability-as-flourishing since it is a cultural attribute. Following this point, means embedded with system thinking such as lean thinking, PSS system, and circular economy, can jointly promote the progression towards a sustainable future.

6.5 Conclusion

This chapter has explored the underlying drivers and impediments for circular economy to achieve sustainability-as-flourishing, and proposes a conceptual model giving a new way to visualize flourishing and to re-envision sustainability. The conceptual model provides a useful lens to view the deepest essence of sustainability-related issues and to discuss the solutions from several critical aspects.

The research contributes to knowledge by proposing the conceptual model that expands the Golden Circle introduced by Simon Sinek (2011) and firstly relates the model to sustainability from the flourishing perspective. Considering the notion of flourishing is difficult to visualize and even harder to implement, the conceptual model as an interpretation for the complex world attempts to put all the essential domains together for understanding the complexity of sustainability. This allows for a systematic approach to examine the nature of the complex word and to move the way towards a sustainable future. Upon reviewing the conceptualization and operationalization of circular economy, another contribution lies in the analysis of the pros and cons of circular economy when it comes to the fundamental shifts required by sustainability-as-flourishing. This provides a deep insight for understanding how far circular economy is from the sense of flourishing, and what are the fundamental reasons behind.

Based on the conceptual model developed, the last section further initiates discussion on several crucial aspects of sustainability, and understands why flourishing should start from 'Why', what might be the sustainable value to be restored, how PSS can enhance circular economy and contribute to sustainability-as-flourishing, what a shift from quantity to quality is essential for flourishing, and what can be sustainability solutions for the long run, etc. All these discussions support the decision-makings for researchers, policy-makers, and practitioners in the field of sustainability.

References

Allee, V. (2008). Value network analysis and value conversion of tangible and intangible assets. *Journal of Intellectual Capital*, *9*(1), 5–24.

Allwood, J. (2014). *Squaring the circular economy: the role of recycling within a hierarchy of material management strategies. Handbook of Recycling* (pp. 445–477). Elsevier.

Andersen, M. (2007). An introductory note on the environmental economics of the circular economy. *Sustainability Science*, *2*(1), 133–140.

Bastein, T., E. Roelofs, E. Rietveld, and A. Hoogendoorn. (2013). Opportunities for a Circular Economy in the Netherlands. TNO, Report commissioned by the Netherlands Ministry of Infrastructure and Environment.

Benyus, J. M. (1997). *Biomimicry*. New York, NY: William Morrow.

Bertalanffy, L. V. (1950). An outline of general system theory. *The British Journal for the Philosophy of Science*, *1*(2), 134–165.

Beuren, F., Gomes Ferreira, M., & Cauchick Miguel, P. (2013). Product-service systems: a literature review on integrated products and services. *Journal of Cleaner Production*, *47* (C), 222–231.

Blomsma, F., & Brennan, G. (2017). The emergence of circular economy: a new framing around prolonging resource productivity. *Journal of Industrial Ecology*, *21*(3), 603–614. Available from https://doi.org/10.1111/jiec.12603.

Braungart, M., Mcdonough, W., & Bollinger, A. (2007). Cradle-to-cradle design: creating healthy emissions – a strategy for eco-effective product and system design. *Journal of Cleaner Production*, *15*(13), 1337–1348.

Brennan, G., Tennant, M., & Blomsma, F. (2015). Business and production solutions: closing loops and the circular economy. In H. Kopnina, & J. Blewitt (Eds.), *Sustainability: key issues*. London, United Kingdom: Routledge.

Breuer, H., & Lüdeke-Freund, F. (2014). Normative Innovation for Sustainable Business Models in Value Networks. The Proceedings of XXV ISPIM Conference-Innovation for Sustainable Economy and Society, June, 17.

Commoner, B. (1971). *The closing circle: nature, man, and technology*. New York: Random House.

Csikszentmihalyi, M. (1991). *Flow: the psychology of optimal experience*. New York: Harper Perennial.

de Jesus, A., & Mendonça, S. (2018). Lost in transition? drivers and barriers in the Eco-innovation road to the circular economy. *Ecological Economics*, *145*(July 2017), 75–89. Available from https://doi.org/10.1016/j.ecolecon.2017.08.001.

Dresner, S. (2008). *The Principles of Sustainability. Business & Economics*. Earthscan publishers.
Ehrenfeld, J. (2019). *The right way to flourish: reconnecting to the real world* (1st. ed.)., Routledge.
Ehrenfeld, J., & Hoffman, A. (2013). Flourishing: A frank conversation about sustainability.
Ehrenfeld, J. R. (2008). *Sustainability by design: a subversive strategy for transforming our consumer culture*. New Haven, CT: Yale University Press.
Elkington, J. (1997). *Cannibals with forks: the triple bottom line of 21st century business*. Oxford: Capstone.
Ellen MacArthur Foundation. (2013). Towards the Circular Economy. Ellen MacArthur Foundation, 1, 1–96. https://doi.org/10.1162/108819806775545321.
Ellen MacArthur Foundation (2015a) Cradle to Cradle in a Circular Economy - Products and Systems. Available at: https://www.ellenmacarthurfoundation.org/circulareconomy/schools-ofthought/cradle2cradle (Accessed: 21 March 2017).
Ellen MacArthur Foundation. (2015b). Delivering the Circular Economy: A Toolkit for Policymakers. 19–32.
Ellen MacArthur Foundation. (2017). A NEW TEXTILES ECONOMY:Redesigning fashion's future. https://www.ellenmacarthurfoundation.org/assets/downloads/publications/A-NewTextiles-Economy_Full-Report.pdf.
Ellen MacArthur Foundation and McKinsey & Company, (2014). Towards the Circular Economy:Accelerating the Scale-up across Global Supply Chains, Rapport. World Economic Forum, Switzerland.
Ellen MacArthur Foundation and McKinsey & Company Center for Business and Environment (2015), GROWTH WITHIN: A CIRCULAR ECONOMY VISION FOR A COMPETITIVE EUROPE. (2015). Available at: https://www.ellenmacarthurfoundation.org/assets/downloads/publications/EllenMacArthurFoundation_Growth-Within_July15.pdf.
Esposito, M., Tse, T., & Soufani, K. (2018). Introducing a circular economy: new thinking with new managerial and policy implications. *California Management Review*, *60*(3), 5–19. Available from 10.1177/0008125618764691.
European Commission, (2008). Directive 2008/98/EC of the European Parliament and of the Council of 19 November 2008 on Waste and Repealing Certain Directives. Available at. http://eur-lex.europa.eu/legal-content/EN/TXT/PDF/?uri = CELEX:32008L0098&from = EN.
European Commission, (2015a). An EU action plan for the circular economy. COM 614,21. https://doi.org/10.1017/CBO9781107415324.004.
European Commission, (2015b). Closing the Loop - an EU Action Plan for the Circular Economy (COM(2015) 614 Final).
European Commission, (2015c). Communication from the Commission to the European Parliament, the Council, the European Economic and Social Committee and the Committee of the Regions. Closing the Loop - an EU Action Plan for the Circular Economy (Brussels, Belgium).
Evans, S., Vladimirova, D., Holgado, M., Van Fossen, K., Yang, M., Silva, E. A., & Barlow, C. Y. (2017). Business model innovation for sustainability: towards a unified perspective for creation of sustainable business models. *Business Strategy and the Environment*, *26*(5), 597–608. Available from https://doi.org/10.1002/bse.1939.
Fischer and Pascucci, (2017). Chinese National People's Congress, 2008. Regulation on Promotion of Circularly Economy.
Fitzpatrick, J. (2017). Does engineering education need to engage more with the economic and social aspects of sustainability? *European Journal of Engineering Education*, *42*(6), 916–926.

Frey, B., Stutzer, A., & MacCulloch, R. (2003). Happiness and economics; how the economy and institutions affect human well-being. *Economic Journal, 113*(488), 409–411.

Geissdoerfer, M., Morioka, S. N., de Carvalho, M. M., & Evans, S. (2018). Business models and supply chains for the circular economy. *Journal of Cleaner Production, 190*, 712–721. Available from https://doi.org/10.1016/j.jclepro.2018.04.159.

Geissdoerfer, M., Savaget, P., Bocken, N. M. P., & Hultink, E. J. (2017). The circular economy – a new sustainability paradigm? *Journal of Cleaner Production, 143*, 757–768. Available from https://doi.org/10.1016/j.jclepro.2016.12.048.

Geng, Y., Sarkis, J., & Bleischwitz, R. (2019). How to globalize the circular economy. *Nature, Nature, 565*(7738), 153–155.

Geng, Y., Sarkis, J., Ulgiati, S., & Zhang, P. (2013). Measuring China's circular economy. *Science (New York, N.Y.), 339*(6127), 1526–1527.

Ghisellini., Cialani., & Ulgiati. (2016). A review on circular economy: the expected transition to a balanced interplay of environmental and economic systems. *Journal of Cleaner Production, 114*(C), 11–32.

Govindan, K., & Hasanagic, M. (2018). A systematic review on drivers, barriers, and practices towards circular economy: a supply chain perspective. *International Journal of Production Research, 56*(1–2), 278–311.

Graedel, T. E., & Allenby, B. R. (1995). *Industrial ecology*. Englewood Cliffs, N.J.: Prentice Hall.

Haas, W., Krausmann, F., Wiedenhofer, D., & Heinz, M. (2015). How circular is the global economy? an assessment of material flows, waste production, and recycling in the European Union and the world in 2005. *Journal of Industrial Ecology, 19*(5), 765–777.

Harari, Y. (2016). *Homo deus: a brief history of tomorrow*. Yuval Noah Harari.

Henry, M., Bauwens, T., Hekkert, M., & Kirchherr, J. (2020). A typology of circular start-ups: analysis of 128 circular business models. *Journal of Cleaner Production, 245*, 118528. Available from https://doi.org/10.1016/j.jclepro.2019.118528.

Kama, K. (2015). Circling the economy: resource-making and marketization in EU electronic waste policy. *Area, 47*(1), 16–23.

Kirchherr, J., Reike, D., & Hekkert, M. (2017). Conceptualizing the circular economy: an analysis of 114 definitions. *Resources, Conservation and Recycling, 127*(September), 221–232. Available from https://doi.org/10.1016/j.resconrec.2017.09.005.

Lacy, P., & Rutqvist, J. (2015). *Waste to wealth: the circular economy advantage*. Peter Lacy and Jakob Rutqvist.

Lahane, S., Kant, R., & Shankar, R. (2020). Circular supply chain management: a state-of-art review and future opportunities. *Journal of Cleaner Production, 258*, 120859. Available from https://doi.org/10.1016/J.JCLEPRO.2020.120859.

Lieder, M., & Rashid, A. (2016). Towards circular economy implementation: a comprehensive review in context of manufacturing industry. *Journal of Cleaner Production, 115*, 36–51.

Lovins, A. B., Lovins, L. H., & Hawken, P. (1999). *Natural capitalism: the next industrial revolution*. Earthscan.

Lyle, J. T. (1994). *Regenerative design for sustainable development*. New York: John Wiley & Sons.

Maslow, A. H. (1943). A theory of human motivation. *Psychological Review, 50*(4), 370–396.

Maturana, H. R., & Varela, F. J. (1980). *Autopoiesis and cognition: the organization of the living*. Boston: Reidel.

Matus, K., Xiao, X., & Zimmerman, J. (2012). Green chemistry and green engineering in China: drivers, policies and barriers to innovation. *Journal of Cleaner Production, 32*(C), 193–203.

McDonough, W., & Braungart, M. (2002). *Cradle-to-Cradle: remaking the way we make things*. New York, NY: North Point Press.
McKinsey and Company (2016). The circular economy: Moving from theory to practice. https://www.mckinsey.com/business-functions/sustainability-and-resourceproductivity/our-insights/the-circular-economy-moving-from-theory-to-practice.
METI. (2004). *Handbook on Resource Recycling Legislation and 3R Trends in 2003*. Tokyo: Japanese Ministry of Economy. Trade and Industry.
Murray, A., Skene, K., & Haynes, K. (2017). The circular economy: an interdisciplinary exploration of the concept and application in a global context. *Journal of Business Ethics, 140*(3), 369–380.
Nicholson, J., & Kurucz, E. (2019). Relational leadership for sustainability: building an ethical framework from the moral theory of 'ethics of care'. *Journal of Business Ethics, 156*(1), 25–43.
Ouden, E. (2012). *Innovation design: creating value for people, organizations and society*. London; New York: Springer.
Pauli, G. A. (2010). *The blue economy: 10 years, 100 innovations, 100 million jobs*. Taos, NM: Paradigm Publications.
Pearce, D. W., & Turner, R. K. (1990). *Economics of natural resources and the environment*. London: Harvester Wheatsheaf.
Pieroni, M. P. P., McAloone, T. C., & Pigosso, D. C. A. (2019). Business model innovation for circular economy and sustainability: a review of approaches. *Journal of Cleaner Production, 215*, 198–216. Available from https://doi.org/10.1016/j.jclepro.2019.01.036.
Postman, N. (2006). *Amusing ourselves to death: public discourse in the age of show business*. London: Penguin.
Potting, J., Hekkert, M.P., Worrell, E., Hanemaaijer, A. (2017). Circular Economy: Measuring Innovation in the Product Chain. Available at. file:///C:/Users/44742/Downloads/Circular.pdf.
Reh, L. (2013). Process engineering in circular economy. *Particuology, 11*(2), 119–133.
Schopenhauer, A. (1883). *The world as will and idea* (pp. 22–24). London: English and foreign philosophical library, Trubner.
Sihvonen, S., & Ritola, T. (2015). Conceptualizing ReX for aggregating end-of-life strategies in product development. *Procedia CIRP, 29*, 639–644.
Sinek, S. (2011). *Start with why: how great leaders inspire everyone to take action* (New ed.). New York; London: Portfolio/Penguin, Simon Sinek.
Smol, M., Kulczycka, J., Henclik, A., Gorazda, K., & Wzorek, Z. (2015). The possible use of sewage sludge ash (SSA) in the construction industry as a way towards a circular economy. *J. Clean. Prod., 95*, 45–54.
Soros, G. (1994). *The alchemy of finance: reading the mind of the market*. New York; Chichester: J. Wiley.
Spring, M., & Araujo, L. (2017). Product biographies in servitization and the circular economy. *Industrial Marketing Management, 60*, 126–137.
Stahel, W. R. (2010). *The performance economy* (2nd ed.). Hampshire, UK: Palgrave Macmillan UK.
Su., Heshmati., Geng., & Yu. (2013). A review of the circular economy in China: moving from rhetoric to implementation. *Journal of Cleaner Production, 42*(C), 215–227.
Tse, L., & Legge, J. (2009). *Tao te ching or the tao and its characteristics*. S.l.: The Floating Press.
van Eijk, N. (2015). *Barriers & Drivers Towards a Circular Economy, 8*(7). Available from http://www.circulairondernemen.nl/uploads/e00e8643951aef8adde612123e824493.pdf.
VanBuren, N., Demmers, M., Heijden, R., & Witlox, F. (2016). Towards a circular economy: the role of dutch logistics industries and governments. *Sustainability, 8*(7), 647.

Vodonick, J. (2018). The key to organizational sustainability: nurturing a culture of change. *Systems Research and Behavioral Science*, *35*(4), 458–468.

Yang, M., & Evans, S. (2019). Product-service system business model archetypes and sustainability. *Journal of Cleaner Production*, *220*, 1156–1166. Available from https://doi.org/10.1016/j.jclepro.2019.02.067.

Yang, M., Evans, S., Vladimirova, D., & Rana, P. (2017). Value uncaptured perspective for sustainable business model innovation. *Journal of Cleaner Production*, *140*, 1794–1804.

Yang, M., Smart, P., Kumar, M., Jolly, M., & Evans, S. (2018). Product-service systems business models for circular supply chains. *Production Planning & Control*, *29*(6), 498–508.

Yuan, Z., Bi, J., & Moriguichi, Y. (2006). The circular economy: a new development strategy in China. *Journal of Industrial Ecology*, *10*(1–2), 4–8.

Zhu, Q., Geng, Y., & Lai, K. (2010). Circular economy practices among Chinese manufacturers varying in environmental-oriented supply chain cooperation and the performance implications. *Journal of Environmental Management*, *91*(6), 1324–1331.

CHAPTER 7

Sustainable casting processes through simulation-driven optimization

Michail Papanikolaou and Prateek Saxena
Sustainable Manufacturing Systems Centre, Manufacturing Theme, Cranfield University, Cranfield, United Kingdom

7.1 Introduction

Shape casting is a metal forming process in which the molten metal flows into a mold cavity (Jolly, 2003). The molten metal is solidified to replicate the shape of the mold. The mold consists of a cavity that determines the shape and size of the final part. The dimensions of the cavity are kept slightly larger than the intended part in order to take into account shrinkage that might occur during solidification. The final part obtained by this manufacturing process is known as casting. Casting is very useful in manufacturing complex parts to net shape or near-net shape, depending upon the type of the process used. In principle, any metal that can be melted can be used for producing castings. The process is suitable for mass production and is not limited by the size of the part. However, there remain some concerns, namely, mechanical properties, surface roughness, porosity, dimensional accuracy, and high energy consumption.

Based on the filling of the mold, casting processes can be classified as *open* and *closed mold* as illustrated in Fig. 7.1. In an open mold system, molten metal is poured directly into the mold cavity whereas, in a closed mold system, a gating system is utilized to fill the cavity. The molten metal is made to flow through a passageway (gating system) connecting the cavity and the pouring cup. As the molten metal enters the cavity, the cooling process starts. This process involves phase change and is called solidification. This step is critical as most of the casting properties are established during solidification. Once the part has sufficiently cooled down, it is extracted from the mold cavity. Based on the type of casting operation, the part is sometimes sent for postprocessing operation, which

Sustainable Manufacturing
DOI: https://doi.org/10.1016/B978-0-12-818115-7.00003-1

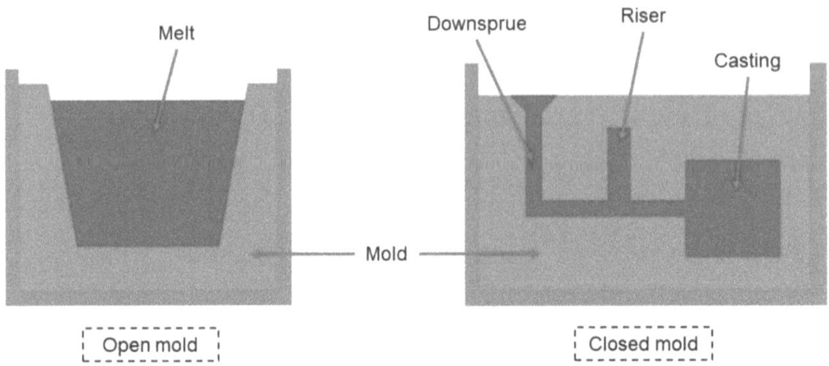

Figure 7.1 Types of mold: open mold system (left) and closed mold system (right) for casting.

involves cleaning the part, excess metal removal, and occasionally heat treatment.

On the basis of the mold type, there are two categories, that is, expendable and permanent mold casting processes. In expendable mold casting, the mold is destroyed to remove the casting. Thus such molds are prepared using plaster or sand, coupled together with a binder to maintain the strength of the mold. Recently, several researchers have commented on utilizing the capabilities of additive manufacturing methods to print the sand mold using binder jetting (Saxena, Papanikolaou, Pagone, Salonitis, & Jolly, 2020). Sand casting is a typical example of the casting process that utilizes expendable molds. As the mold needs to be destroyed and cannot be reused for manufacturing another part, such a process is not suitable for mass production. In permanent mold casting, the mold is made to withstand multiple cycles; consequently, mass production is feasible. The mold is a split cavity that opens and allows removal of the casting. It is usually made of metals. Die casting is a characteristic example of permanent mold casting.

7.1.1 Expendable mold casting processes

7.1.1.1 Sand casting

Sand casting is one of the most common expendable mold casting processes. The versatility of this process allows the majority of metals (with a high melting point) and casting alloys to be cast without having any limitation on the size of the part. The molten metal is poured into the mold cavity and is left for solidification. Once solidified, the casting is removed by breaking the mold and cleaned. A subsequent heat treatment operation is sometimes carried out to improve the material properties. A pattern is required in order to prepare the mold cavity. A pattern is an exact replica of the casting to be produced. Casting sand, mixed with a binder is packed around the pattern, which is subsequently removed by splitting the mold into two halves and the cavity is formed. A pattern can be made out of wood, metal, or plastic depending upon the number of castings required to be produced using the same pattern. A core is used to determine the intricate features of the part. The core is a replica of the internal features of the part and is added to the mold cavity. Thus when the molten metal is poured into the cavity, it flows in the free space between core and the mold cavity, and thus both internal and external features can be conveniently fabricated.

7.1.1.2 Shell molding

This casting process is similar to sand casting while a thin shell with a thickness of about 9 mm is utilized as a mold (Groover, 2010). Shells are manufactured by applying a sand–resin mixture on a metal pattern as illustrated in Fig. 7.2. More specifically, sand is mixed with resin and collected in a box. The metal pattern is subsequently heated up and fixed on top of the box. The box is then inverted and the sand falls-off over the heated pattern. This causes curing of the thin layer of sand and a thin shell is formed. The box is brought back to the original position so that the

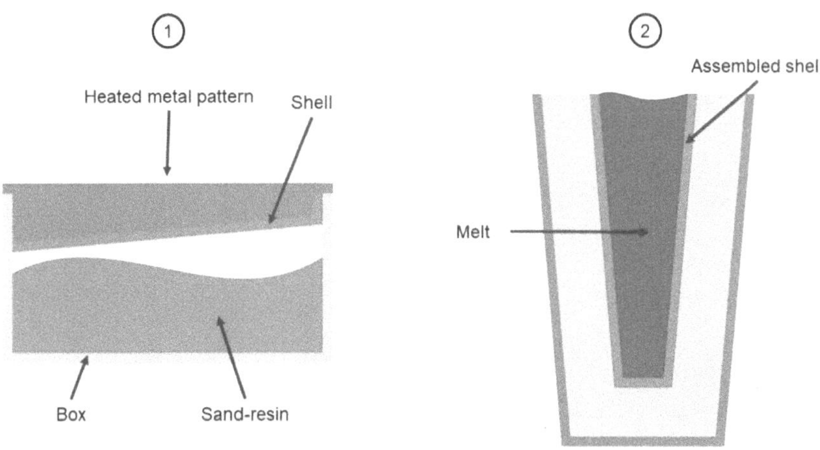

Figure 7.2 Shell mold casting process: (1) shell manufacturing and (2) casting.

loose sand gets collected at the bottom and subsequently the curing process of the shell is completed using an oven. The shell is then separated from the pattern and the two halves of the shell are assembled inside a box, forming a cavity. Metal shot or sand is used to fill in the space between the shell and the box. Molten metal is poured and allowed to solidify. The finished part is then removed from the mold. The shell molding process is suitable for mass production of gears, camshafts, bushings, and many other parts.

7.1.1.3 Investment casting

Another expendable mold casting process is investment casting which is also referred to as lost-wax casting (Groover, 2010). The investment casting process workflow is shown in Fig. 7.3. This process makes use of wax patterns, which are usually manufactured by injection molding. These

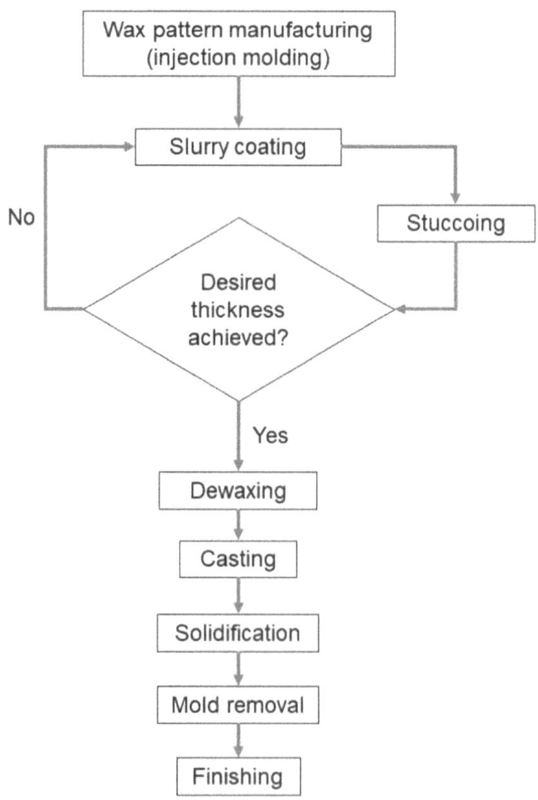

Figure 7.3 Process workflow of investment casting.

patterns include the sprue, risers, running, and gating systems. Following their manufacturing, wax patterns are dipped into a slurry containing ceramic materials; the wet slurry is subsequently stuccoed with sand particles while the aforementioned loop might be repeated several times until the desired shell thickness is achieved. After the ceramic shell (mold) has been manufactured, the wax inside the shell is melted and removed using a stem-dewaxing autoclave or a flash fire furnace. This process is also called dewaxing. The ceramic mold is preheated before pouring in the molten metal and later broken down to remove the casting. The investment casting process may be complemented by typical finishing processes such as grinding or sand blasting.

7.1.2 Permanent mold casting processes

The main disadvantage of the expendable mold casting process is that the mold can only be used once (Groover, 2010). Permanent molds, on the other hand, can be reused and are thus more cost-effective compared to the expendable ones. The permanent mold process uses two split cavity metal molds. These molds are typically made from cast iron or steel. Permanent molds are machined and the gating system design is incorporated in the mold. Two main types of the permanent mold casting process will be discussed in the following subsections, namely, (1) die casting and (2) centrifugal casting.

7.1.2.1 Die casting

In the die casting process, the molds are referred to as dies. In high-pressure die casting, molten metal is injected into the mold cavity at high pressure (~7–350 MPa) (Groover, 2010). The molten metal is left to solidify while maintaining high pressure at the same time. The mold is then opened up and parts are removed. The casting process is carried out in die casting machines. Based on how the molten metal is injected, two main types of die casting process can be identified: (1) hot chamber and (2) cold chamber die casting. In hot chamber die casting, the metal is melted in a container that is attached to the casting machine followed by injection of the molten metal at high pressure. Production rates of about 500 parts per hour can be achieved with this method. In cold chamber die casting, the metal is melted externally and subsequently poured into the chamber from which it is then injected into the mold cavity with the help of a piston. The die can be made either by a single cavity or by

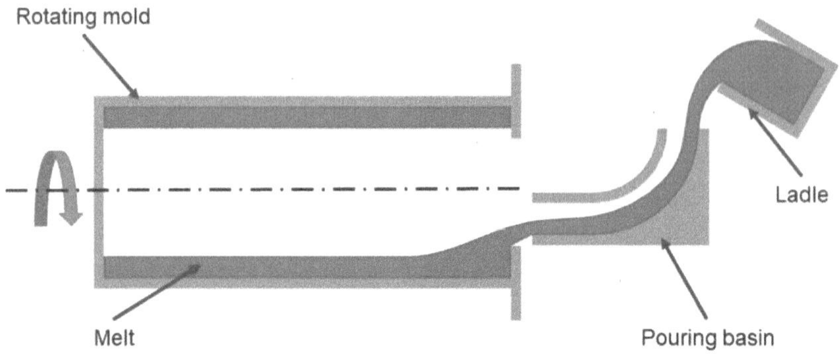

Figure 7.4 True centrifugal casting process.

multiple cavities allowing the production of many parts in a single run of the process.

7.1.2.2 Centrifugal casting

In this casting process, the mold is rotated at high rotational speeds. The centrifugal force pushes the molten metal toward the boundaries of the mold. There are three types of centrifugal casting processes, namely, (1) true centrifugal, (2) semicentrifugal, and (3) centrifuge casting. True centrifugal casting is used to produce tubular parts such as tubes, pipes, rings, as illustrated in Fig. 7.4. The mold can be held stationary or be rotated during the filling process. The axis orientation can be either horizontal or vertical. Due to the high rotational speeds imposed, the molten metal obtains the shape of the mold.

Semicentrifugal casting differs from the true centrifugal casting with regard to the characteristics of the final part. This casting process is used to manufacture solid castings such as wheels and pulleys. Finally, in the centrifuge casting process, the part cavity is located away from the axis of rotation and the centrifugal force is exploited in order to fill in the cavity. Unlike the two aforementioned casting processes, radial symmetry is not a requirement in this process, which is suitable for producing small parts.

7.1.3 Casting quality

In this section, common defects encountered in casting processes are discussed. The focus is shifted to two main types of defects: (1) defects being common in all casting operations and (2) defects related to sand casting.

7.1.3.1 Common casting defects

Some of the common casting defects are listed below (Groover, 2010):

- Misruns: this defect is caused when the molten metal solidifies before filling the mold cavity completely and can be caused due to: (1) low pouring temperatures, (2) improper fluidity, (3) slow pouring, and (4) thin cross-section of the mold cavity.
- Cold shuts: they arise when two parts of the molten metal fail to diffuse properly due to premature freezing.
- Cold shots: this defect results in the formation of small globules of metal that get trapped inside the casting. It can be avoided by improving the gating design.
- Shrinkage cavity: this is caused due to solidification shrinkage resulting in a depression on top of the casting. It can be avoided by improving the riser design.
- Microporosity: it is found mainly in alloys and results in small voids within the casting. Microporosity is caused due to localized solidification shrinkage.
- Hot tearing: this defect is caused due to restraining from a contraction by an unyielding mold. It can be avoided by immediate removal of the part after solidification.

7.1.3.2 Sand casting defects

Defects specific to sand castings are listed below (Groover, 2010):

- Sand blow: this is caused due to the release of gases while the molten metal is poured. This defect type results in small cavities in the casting.
- Pinholes: they are also caused by the gas released during the pouring of molten metal. However, they result in multiple cavities below the casting surface.
- Sand wash: this is caused due to erosion of the mold during pouring.
- Scabs: these defects occur on the surface of the casting and are caused due to mixed layers of sand and metal. This happens when the mold flakes get detached and mixed with the molten metal surface.
- Penetration: in situations when the fluidity of molten metal is higher than the desired range, the molten metal penetrates into the mold itself and solidifies. This results in protrusions on the surface of the casting.
- Mold shift: this defect occurs due to misalignment or sliding of the cope and drag.
- Core shift: this is caused due to shifting of the core due to the buoyancy of molten metal.

- Mold crack: this occurs when the mold is not strong enough and thereby developing a crack through which the molten metal enters the sand and results in a "fin" in the casting.

An efficient way to improve casting quality is the numerical modeling of the filling process and solidification. As will be shown in the following section, defect formation can be predicted via means of computational fluid dynamics (CFD) simulations and casting quality can thus be enhanced.

7.2 Numerical modeling of casting processes

As discussed in the previous section, metal casting consists of two main stages: (1) filling a mold and (2) solidification. Thus the numerical modeling of casting processes can be split into two subtasks, that is, modeling of (1) the fluid flow and natural convection and (2) the heat transfer during solidification. In this section, the governing equations describing fluid flow and heat transfer in such problems will be presented and the numerical methods employed for solving the aforementioned equations will be discussed.

7.2.1 Fundamental equations

7.2.1.1 Fluid dynamics

The continuity equation states that mass can neither be created nor destroyed but is conserved within a control volume. This can be expressed in a differential form as follows (Winterscheidt and Huang, 2001):

$$\frac{\partial \rho}{\partial t} + \nabla \cdot (\rho \mathbf{u}) = 0, \tag{7.1}$$

where ρ is the fluid density and $\mathbf{u}$ the velocity vector.

The momentum equation relates the changes of the fluid momentum to the total force being exerted on the control volume:

$$\rho \left[\frac{\partial \mathbf{u}}{\partial t} + (\mathbf{u} \cdot \nabla)\mathbf{u} \right] = \nabla \cdot \boldsymbol{P} + \rho \boldsymbol{f}, \tag{7.2}$$

where $\boldsymbol{P}$ the stress tensor and $\boldsymbol{f}$ the external body forces. The elements of the stress tensor for a viscous fluid are given by

$$P_{ij} = -p\delta_{ij} + \tau_{ij}. \tag{7.3}$$

Substituting Eq. (7.3) into Eq. (7.2) yields the Navier–Stokes equations in conservation form:

$$\frac{\partial(\rho u_x)}{\partial t}+\frac{\partial(\rho u_x u_x)}{\partial x}+\frac{\partial(\rho u_y u_x)}{\partial y}+\frac{\partial(\rho u_z u_x)}{\partial z}=-\frac{\partial p}{\partial x}+\frac{\partial \tau_{xx}}{\partial x}+\frac{\partial \tau_{yx}}{\partial y}+\frac{\partial \tau_{zx}}{\partial z}+\rho f_x$$
$$\frac{\partial(\rho u_y)}{\partial t}+\frac{\partial(\rho u_x u_y)}{\partial x}+\frac{\partial(\rho u_y u_y)}{\partial y}+\frac{\partial(\rho u_z u_y)}{\partial z}=-\frac{\partial p}{\partial y}+\frac{\partial \tau_{xy}}{\partial x}+\frac{\partial \tau_{yy}}{\partial y}+\frac{\partial \tau_{zy}}{\partial z}+\rho f_y$$
$$\frac{\partial(\rho u_z)}{\partial t}+\frac{\partial(\rho u_x u_z)}{\partial x}+\frac{\partial(\rho u_y u_z)}{\partial y}+\frac{\partial(\rho u_z u_z)}{\partial z}=-\frac{\partial p}{\partial z}+\frac{\partial \tau_{xz}}{\partial x}+\frac{\partial \tau_{yz}}{\partial y}+\frac{\partial \tau_{zz}}{\partial z}+\rho f_z. \tag{7.4}$$

The energy equation can be derived from the first law of thermodynamics and states that the energy change rate inside a fluid element is equal to the net heat flux into the element and the work due to internal surface forces. The conservative form of the energy equation can be expressed as follows:

$$\frac{\partial(\rho E)}{\partial t}+\frac{\partial(\rho u_x E)}{\partial x}+\frac{\partial(\rho u_y E)}{\partial y}+\frac{\partial(\rho u_z E)}{\partial z}=\frac{\partial}{\partial x}\left(k\frac{\partial T}{\partial x}\right)+\frac{\partial}{\partial y}\left(k\frac{\partial T}{\partial y}\right)+\frac{\partial}{\partial z}\left(k\frac{\partial T}{\partial z}\right)-\frac{\partial(\rho u_x p)}{\partial x}-\frac{\partial(\rho u_y p)}{\partial y}-\frac{\partial(\rho u_z p)}{\partial z}+\Phi, \tag{7.5}$$

where E is the fluid energy, k the thermal conductivity, and Φ the dissipation function defined as follows:

$$\Phi=\frac{\partial(u_x\tau_{xx})}{\partial x}+\frac{\partial(u_x\tau_{yx})}{\partial y}+\frac{\partial(u_x\tau_{zx})}{\partial z}+\frac{\partial(u_y\tau_{xy})}{\partial x}+\frac{\partial(u_y\tau_{yy})}{\partial y}+\frac{\partial(u_y\tau_{zy})}{\partial z}+\frac{\partial(u_z\tau_{xz})}{\partial x}+\frac{\partial(u_z\tau_{yz})}{\partial y}+\frac{\partial(u_z\tau_{zz})}{\partial z}. \tag{7.6}$$

7.2.1.2 Free surface modeling

In order to model the filling process, it is very significant to track the free surface of the metal. The most widely employed method for this purpose is the volume of fluid (VOF) method proposed by Hirt and Nichols

(1981). This method assumes that the free surface $F = F(x,y,z,t)$ is a function of space and time and its value ranges between 0 (air) and 1 (liquid). All of the simulation cells with $0 < F < 1$ belong to the free surface. The VOF equation is given below:

$$\frac{\partial(\rho F)}{\partial t} + \nabla(\rho \nabla F) = 0. \tag{7.7}$$

7.2.1.3 Heat transfer

Heat transfer modeling lies in the core of the casting and solidification simulations. It is well known that there are three basic modes of heat transfer, namely, (1) conduction, (2) convection, and (3) radiation. Conduction is the main mechanism of heat transfer between the mold and cast material (liquid or solid). Convection refers to the heat transfer between a solid surface and a liquid flowing past it. Forced convection takes place when the fluid is set in motion by some external means, such as a piston in high pressure die casting, while natural convection occurs when the fluid moves because of gravity forces, temperature/density gradients, and so on. Radiation heat transfer is significantly different from the two aforementioned modes, as it does not rely on the contact between the source/sink and the heated object. This heat transfer mode is dominant in investment casting processes (Winterscheidt and Huang, 2001). In the following paragraphs, the aforementioned heat transfer modes as well as the corresponding heat transfer relationships are briefly presented.

Conduction

Conduction occurs when heat is transferred between two mediums, which are in contact due to molecular motion. The heat transfer direction is always from the hotter to the colder medium. The heat transfer rate in Cartesian coordinates is given by the Fourier's law:

$$q_i = -kA\frac{\partial T}{\partial x_i}, \tag{7.8}$$

where q is the rate of heat transfer, k the thermal conductivity, and A the cross-sectional surface area.

The temperature distribution over the simulation domain can be obtained using the heat conduction equation which states that the thermal energy stored in a control volume is equal to net transfer of thermal energy by conduction into the control volume plus the volumetric rate of heat generation:

$$\rho c_P \frac{\partial T}{\partial t} - \frac{\partial}{\partial x}\left(k\frac{\partial T}{\partial x}\right) - \frac{\partial}{\partial y}\left(k\frac{\partial T}{\partial y}\right) - \frac{\partial}{\partial z}\left(k\frac{\partial T}{\partial z}\right) = Q', \tag{7.9}$$

where c_P is the specific heat and Q' the volumetric rate of heat generation. As will be discussed in the following subsection, the solution of Eq. (7.9) requires prior knowledge of the initial and boundary conditions.

Convection

As discussed above, convection is a heat transfer mode that occurs when a fluid is flowing past a surface due to either external forces (forced convection) or force fields (gravity, temperature, or density gradients). The heat flux due to convection is given by Eq. (7.10) often referred to as the Newton's law of cooling:

$$q = hA(T_s - T_\infty), \tag{7.10}$$

where h is the heat transfer coefficient, A the surface area, T_s the temperature of the surface, and T_∞ the temperature of the bulk fluid, away from the surface. The estimation of the heat transfer coefficient is not trivial; in fact, its value depends on numerous factors (Incropera, Lavine, Bergman, & DeWitt, 2007), such as:

1. The fluid properties (density, viscosity, thermal conductivity, etc.)
2. Type of flow (laminar/turbulent)
3. Shape of duct/surface (rectangular/circular/flat surface)
4. Surface roughness
5. Surface orientation

For the estimation of the heat transfer coefficient, empirical relations have been established after performing a large number of experiments. These empirical relations are expressed in terms of three dimensionless numbers:

1. The *Nusselt number (Nu)* which is indicative of the enhancement of the heat transfer due to convection compared to conduction alone and can be defined as

$$Nu = \frac{q_{\text{conv}}}{q_{\text{cond}}} = \frac{h\Delta T}{k\frac{\Delta T}{L}} = \frac{hL}{k}, \tag{7.11}$$

where L is a characteristic length.

2. The *Reynolds number (Re)* which is a measure of the inertial forces over the viscous ones while higher Reynolds numbers are indicative of turbulent flows. The Reynolds number is defined as follows:

$$\mathrm{Re} = \frac{F_{\text{intertial}}}{F_{\text{viscous}}} = \frac{\rho u L}{\mu} = \frac{uL}{\nu}, \tag{7.12}$$

where u is the free stream velocity, μ the dynamic viscosity, and $\nu = \mu/\rho$ the kinematic viscosity.
3. The *Prandtl number (Pr)* which is indicative of the relative thickness of the velocity boundary layer over the thermal one and is defined as follows:

$$\mathrm{Pr} = \frac{\text{Momentum diffusivity}}{\text{Thermal diffusivity}} = \frac{\mu c_P}{k}, \tag{7.13}$$

where c_P is the specific heat.

Radiation

In the aforementioned heat transfer modes (conduction and convection), there is always a medium through which heat transfer occurs; this is not the case for radiation. According to the Maxwell theory, in radiation, heat is transferred via electromagnetic waves (Winterscheidt and Huang, 2001). The total emissive power or heat flux is given by the Stefan–Boltzmann Law:

$$q = \varepsilon \sigma T_s^4, \tag{7.14}$$

where $0 < \varepsilon < 1$ is the emissivity of the surface (dimensionless), $\sigma = 5.67 \times 10^{-8}$ W m^{-2} K^{-4} the Stefan–Boltzmann constant, and T_s the surface temperature. The net heat exchange between a surface and the surrounding environment can be estimated as follows:

$$q = F \varepsilon \sigma \left(T_s^4 - T_\infty^4\right), \tag{7.15}$$

where F the fraction of the radiative energy absorbed by the surrounding environment.

7.2.1.4 Solidification

One of the phenomena taking place during solidification is the release of the latent heat during phase change. The most common method to model the latent heat released during solidification is the enthalpy method.

Similarly, to Eq. (7.9), in this method the heat conduction equation is expressed in terms of enthalpy (H):

$$\rho \frac{\partial H}{\partial t} = \frac{\partial}{\partial x}\left(k\frac{\partial T}{\partial x}\right) + \frac{\partial}{\partial y}\left(k\frac{\partial T}{\partial y}\right) + \frac{\partial}{\partial z}\left(k\frac{\partial T}{\partial z}\right). \tag{7.16}$$

In this method, the latent heat (L) is accounted by the enthalpy term (H) which is considered to be dependent on temperature:

$$H = \int_{T_0}^{T} c_p dT + \Delta H, \tag{7.17}$$

where $\Delta H = (1 - f_s) \cdot L$ and f_s the solid fraction.

7.2.2 Initial and boundary conditions

The differential equations describing a casting simulation for an incompressible fluid ($\partial\rho/\partial t = 0$) are summarized below:

$$\begin{aligned} &\nabla \cdot \mathbf{u} = 0 && \text{Continuity equation} \\ &p\left[\frac{\partial \mathbf{u}}{\partial t} + (\mathbf{u}\cdot\nabla)\mathbf{u}\right] = \nabla \cdot \boldsymbol{P} + \rho f && \text{Momentum equation} \\ &p\left[\frac{\partial H}{\partial t} + (\mathbf{u}\cdot\nabla)H\right] = \nabla \cdot (k\nabla T) && \text{Heat conduction equation.} \end{aligned} \tag{7.18}$$

The most common examples of boundary conditions applied for the solution of this set of nonlinear differential equations are listed below (7.4,7.7):

1. No-slip/partial slip boundary condition ($\mathbf{u} = 0$): this type of boundary condition can be implemented to define the fluid velocity at the mold/molten metal interface.
2. Heat flux boundary condition (at the walls or free surface): the heat flux at the boundaries is already known and given by

$$\boldsymbol{q_b} = h\left(T_f - T_b\right) \cdot \boldsymbol{n}, \tag{7.19}$$

   where T_f and T_b are the fluid interface and boundary temperatures, respectively, and $\boldsymbol{n}$ the unit vector normal to the boundary or the fluid free surface. In the case of an adiabatic wall the flux is equal to 0.
3. Temperature boundary condition: the temperature at the boundary is already known.

4. Inflow boundary conditions: the fluid velocity/pressure/flow rate at the ingate/mass source is already known.
5. Molten metal free surface (dynamic boundary condition): the stress across the molten metal free surface has to be continuous, that is, the fluid pressure at the interface (p_{fs}) should be equal to the gas pressure (p_g) plus the surface tension and the viscous normal stress:

$$p_{fs} = p_g + \sigma K + 2\mu \frac{\partial u_n}{\partial n}, \tag{7.20}$$

where σ is the surface tension coefficient, K is the curvature of the free surface, μ the dynamic viscosity, $\boldsymbol{n}$ the normal to the free surface unit vector, and $u_n = \boldsymbol{u} \cdot \boldsymbol{n}$.

Besides boundary conditions, it is often required to prescribe initial conditions in order to solve the differential equations of Eq. (7.18). The initial conditions specify the velocity, temperature, or pressure field at the initial timestep and have the following form:

$$f(\mathbf{X}, 0) = f_0(\mathbf{X}). \tag{7.21}$$

7.2.3 Numerical models

There are three main methods employed to solve the set of partial differential equations of Eq. (7.18): (1) finite difference method (FDM), (2) finite volume method (FVM), and (3) finite element method (FEM) (Winterscheidt and Huang, 2001). These methods will be briefly discussed in the following paragraphs while a detailed review can be found in Garg (1998).

7.2.3.1 Finite difference method

This method is perhaps the oldest one employed to solve differential equations and its main characteristic is that it converts the partial derivatives contained in the governing equations to a set of algebraic equations that are solved for the discrete grid points of the simulation domain. FDM employs a structured rectangular grid while the derivative of a function $f(x)$ at the point is calculated using the central difference formula:

$$\frac{\partial f}{\partial x} = \frac{f(x + \Delta x) - f(x)}{\Delta x} + O(\Delta x), \tag{7.22}$$

where $O(\Delta x)$ the truncation error, which is defined as the difference between the value of the partial derivative and its central difference and is proportional to the mesh spacing Δx. Higher-order derivatives can be approximated using the Taylor expansion:

$$\frac{\partial^2 f}{\partial x^2} = \frac{f(x+\Delta x) - 2f(x) + f(x-\Delta x)}{\Delta x^2} + O(\Delta x^2). \tag{7.23}$$

The FDM method is suitable for castings with simple geometry where a rectangular grid can be used. However, due to its current limitations, complex geometries, which require an unstructured grid, cannot be efficiently solved using FDM. This issue can be solved using the FVM as it will be discussed in the following paragraph.

7.2.3.2 Finite volume method

This method can be applied to both structured and unstructured grids. In this method the conservation laws (in their integral form) are discretized in the flow domain. The conservation law for a conserved variable U (e.g., density, momentum, etc.) can be written as follows:

$$\frac{\partial}{\partial t}\int_{\Omega} U dT + \int_{S} \boldsymbol{F} d\boldsymbol{S} = \int_{\Omega} Q d\Omega, \tag{7.24}$$

where $\boldsymbol{F} = \boldsymbol{F}(U)$ is the flux of the conserved variable, Q a source term, and S the surface of the control volume. The main feature of this method is that the solution satisfies the conservation of states for all of the control volumes (cells) as well as for the entire computational domain. FVMs are suitable for flows around complex shapes and for computing discontinuities.

7.2.3.3 Finite element method

Although this method emerged from the field of structural/stress analysis in the 1950s, it has also been applied for studying fluid flows. In contrast to FDM which regards the solution as a set of grid points, the FEM solution consists of a group of interconnected subregions that form a grid. The weighted residual formulation of this method assumes a control volume Ω obeying the following differential equation:

$$L(\varphi) - f = 0, \tag{7.25}$$

where L is the differential operator, f is a known function of the independent variables. Functional behavior of φ can be obtained as follows:

$$\varphi \cong \overline{\varphi} = \sum_{i=1}^{n} c_i N_i, \tag{7.26}$$

where n is the number of the trial functions, N_i and c_i the unknown coefficients. Since $\overline{\varphi}$ is an approximation of φ, $L(\varphi)$ and $L(\overline{\varphi})$ are not equal while the result of their subtraction is equal to a residual R:

$$L(\overline{\varphi}) - L(\varphi) = R \overset{(25)}{\longleftrightarrow} L(\overline{\varphi}) - f = R. \tag{7.27}$$

The purpose of the weighted residual formulation method is to determine the values of coefficients c_i minimizing the value of R over the entire control volume (or solution domain) using a weighted average as follows:

$$\int_{\Omega} [L(\overline{\varphi}) - L(\varphi)] W_i d\Omega = \int_{\Omega} R W_i d\Omega = 0. \tag{7.28}$$

The selection of the most suitable numerical method (FDM/FVM/FEM) is not a trivial issue, as most of the times it is problem dependent and opinions may vary. As already reported, FDM is a straightforward technique and produces a system of algebraic equations with respect to the dependent variables on the grid points. However, the solution of this system is not always accurate or stable. Moreover, the application of the FDM is always limited to structured grids which are not suitable for simulating flows in or around complex geometries. This issue can be resolved with the implementation of the FEM which can be applied to unstructured grids as well. The weighted residual formulation of the FEM is widely applied to fluid dynamics problems. This approach allows for the natural integration of differential type boundary conditions, a significant advantage over the FDM. However, FEM is considered to be slower than FDM and does not always guarantee accurate results. Finally, the FVM method attempts to combine the advantages of both FDM and FEM, that is, it can be applied to unstructured grids and at the same time ensures that quantities such as mass, energy, and momentum are conserved at the discrete level.

7.2.4 Modeling casting defects

7.2.4.1 Introduction

Metal casting consists of two main stages: (1) pouring or forcing liquid metal into a mold cavity with the desired shape and (2) allowing for some additional time for solidification. Although this process sounds simple, producing high-quality castings is a challenging task. This is because of defects that may arise during filling or solidification. There are many reported types of defects in castings (Campbell, 2015). However, most of the defect types share some common characteristics and they will be here categorized according to their mechanism of formation. The discussion will be limited to entrainment, shrinkage defects, and gas porosity.

The two most common types of entrainment defects are oxide bifilms and bubbles while the main driving force for their formation is surface turbulence. Oxide bifilms are formed as the metal front surface folds or two melt fronts impinge (Mirak, Divandari, Boutorabi, & Campbell, 2007). These bifilms might float on the top surface of the melt or are often submerged into the bulk. In the latter case, they may trigger the initiation of cracks and shrinkage cavities during solidification. Gas bubbles are formed as air gets entrapped within the interstices of bifilms. Bubble entrapment might also deteriorate the quality of castings. Bubbles with a diameter smaller than 5 mm are considered to be responsible for microporosity formation. However, larger bubbles are occasionally buoyant enough to levitate themselves up to the top surface of the melt (sometimes even by passing through the dendrite structure). This counter gravity motion is accompanied by traces looking similar to long oxide bifilms and are addressed as "bubble trails." When the number of bubbles entrapped and consequently the density of trails is large, the end result is a complex network of bifilm trails and bubbles, which is called "bubble damage." Bubble damage has been identified as the most common defect in castings (Campbell, 2015).

Shrinkage defects can be categorized into two main types, namely, (1) open shrinkage defects and (2) closed shrinkage defects as illustrated in Fig. 7.5. Open shrinkage defects are located at the casting surface being in

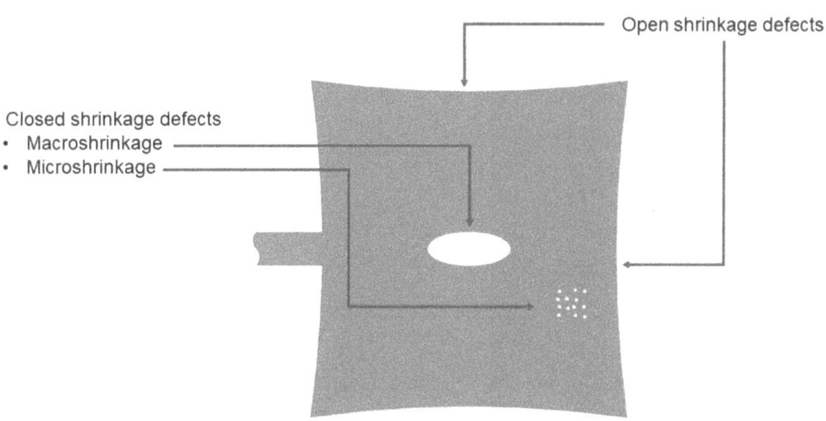

Figure 7.5 Types of shrinkage defects.

contact with the atmospheric air and occur during liquid metal contraction or solidification. Open shrinkage defects have the form of caved surfaces and pipes. On the contrary, closed shrinkage defects occur in the bulk of the casting and can be categorized into macro- and microshrinkage. Macroshrinkage refers to pores generated due to insufficient feeding during solidification and is visible during unaided radiographic inspection. The mechanism of microshrinkage formation is similar to the one of macroshrinkage. Micropores are formed in the bulk of the casting due to the blockage of liquid metal flow owing to bubble inclusions or solid metal in the interdendritic regions. Micropores are not visible without magnification during inspection.

Gas porosity is different from shrinkage defects and is formed when floating gas inclusions (air or hydrogen) are entrapped in the dendritic network. These pores are more common in aluminum castings due to the lower solubility of hydrogen in aluminum. During solidification, hydrogen dissolved in aluminum is expelled and gas bubbles are formed. Unlike microshrinkage pores which have a random shape, gas porosity is characterized by its spherical shape (Kenney et al., 1988).

7.2.4.2 Modeling entrainment defects

As discussed in the previous subsection, the majority of entrainment defects arise during the filling process. The modeling techniques for entrainment defects can be categorized into two main groups: (1) discrete and (2) indiscrete models (Reilly, Green, & Jolly, 2013a). Discrete models provide a rough indication of the final location of entrainment defects in castings, but no information on the entrainment mechanism or the level of entrainment. On the contrary discrete models are capable of identifying the entrainment mechanisms and predict the final location of defects in castings. However, the advantages offered by discrete models are offset by their high computational cost. In this subsection, some of the most common entrainment modeling techniques will be described.

Indiscrete methods

Cumulative entrained free surface area

This method makes use of the excess free surface area (EFSA) which is defined as the difference between the real free surface area of the liquid metal and the theoretical surface area that the melt should have if the mold was filled in a perfectly quiescent way. Higher EFSA values indicate increased amount of entrained oxide bifilms (Lai, Griffiths, & Campbell, 2003). The main drawback of this method is that the instantaneous free surface area for quiescent filling cannot be accurately computed.

Cumulative surface scalar method

Cumulative scalar models have been integrated into commercial CFD software such as Flow-3D and MAGMASOFT. In this technique, surface defect concentration is represented by a scalar value which is initially set to zero across the simulation domain. The physical interpretation of this variable is the mass of contaminated oxides per unit volume. The value of this scalar variable increases with a constant rate as the simulation progresses while it is allowed to advect with the melt. The distribution of oxide films across the casting at each timestep can be obtained by measuring the final value of the aforementioned variable in each cell of the simulation domain (Barkhudarov and Hirt, 1998).

Flow-3D air entrainment model

According to this model, air entrainment occurs if the turbulent kinetic energy per unit volume (P_t) exceeds the gravity and surface tension energy (P_d) which are given below:

$$\begin{aligned} P_t &= \rho Q, \\ P_d &= \rho g_n L_t + \frac{\sigma}{L_t} \end{aligned} \tag{7.29}$$

where ρ is the fluid density, Q the turbulent kinetic energy, g_n the gravitational acceleration normal to the free surface, and σ the surface tension coefficient. The characteristic size of turbulent eddies (L_t) can be calculated as follows:

$$L_t = cnu \frac{\sqrt{\frac{3}{2}}Q^{2/3}}{D}. \tag{7.30}$$

In Eq. (7.30), *cnu* is a constant and *D* is the dissipation function. The volume of air entrainment (δV) per unit time can be estimated according to the following equation:

$$\delta V = \begin{cases} C_{\text{air}} A_s \sqrt{\dfrac{2(P_t - P_d)}{\rho}} & \text{if } P_t > P_d \\ 0 & \text{if } P_t > P_d \end{cases}. \tag{7.31}$$

In Eq. (7.31) C_{air} is a coefficient of proportionality with an approximate value of 0.5 and A_s is the surface area.

Dimensionless number criteria

As mentioned in the previous section, the amount of air entrainment is strongly linked to the balance between the destabilizing effects of surface turbulence and the stabilizing forces of gravity and surface tension. This led to the use of dimensionless numbers such as the Weber number (*We*), which is indicative of the ratio of inertia forces to surface tension and the Froude number (*Fr*) which is equal to the ratio of inertial to gravitational forces in order to numerically quantify air and oxide entrainment during casting (Campbell, 2003; Hernández-Ortega, Zamora, Palacios, López, & Faura, 2010; Reilly, Green, Jolly, & Gebelin, 2009). The aforementioned dimensionless numbers can contribute toward assessing the extent of surface turbulence and consequently air entrainment in castings. However, they suffer from two significant drawbacks: (1) they are unable to provide information about the entrainment mechanism and (2) they cannot be used for tracking the locations of entrainment defects during casting.

Discrete methods

As reported by Reilly et al. (2013a), this family of models is characterized by higher computational cost compared to indiscrete methods. This is because they utilize Lagrangian particles in order to represent defects and track their movement during the filling process. A characteristic example is the oxide film entrainment model developed by Reilly, Green, Jolly, and Gebelin (2013b) which is a user-developed numerical routine integrated into the commercial software Flow-3D. This model was proven to be capable of tracking oxide films which were represented by particles introduced at the corresponding mesh cells whenever an entraining event occurred; particle introduction was based on Boolean logic criteria. A similar model was also developed by Melendez, Carlson, and Beckermann (2010).

The successful implementation of discrete methods with respect to the accurate estimation of the amount of entrained oxide films as well as their final location in the casting is dependent on the realistic representation of the behavior of oxides after their entrainment. More specifically, the major challenge associated with these methods is making accurate assumptions about (1) the behavior of defects in the liquid and solid states, (2) the interaction of defects with the mold and between themselves, (3) the evolution of entrainment during filling and solidification, and (4) the critical conditions under which entrainment occurs. Therefore future research should be directed toward the establishment of numerical models being capable of realistically representing the behavior of oxides in the melt and the validation against existing experimental results (Griffiths, Beshay, Parker, & Fan, 2008).

7.2.4.3 Modeling shrinkage porosity

The prediction of shrinkage porosity is not a trivial task. A wide range of numerical tools have been developed to predict the areas where macroshrinkage should be expected as well as the volume loss. The common characteristic between all of these methods is the need to have an estimate of the thermal history of the casting during its cooling phase. Macroshrinkage modeling methods can be grouped into three main categories, namely, (1) thermal models, (2) thermal/volume models, and (3) criteria functions. A brief overview of the aforementioned categories will be provided in the following subsections.

Thermal models

Thermal models have been widely used for modeling macroshrinkage. These models are based on the fact that macroshrinkage will most of the time occur in the regions which solidify last. These regions can be identified by mapping isotherms with a temperature equal to a value ranging between the liquidus and the solidus (Sheikh, Khan, Iqbal, & Al-Shaer, 2018). The most significant disadvantage of these methods is that the volume of shrinkage cavities cannot be easily quantified.

Thermal/volume models

This family of models has been grounded on the pioneering work of Imafuku and Chijiiwa (Griffiths et al., 2008). These models attempt to identify shrinkage locations by solving the heat transfer and solidification equations. The volume change in the simulation domain is calculated based on mass conservation. The main assumptions made by thermal/volume methods are listed below:

- Molten metal moves only and instantly downward under the effect of gravitational forces.
- The free surface remains always flat.
- The total volume of the shrinkage cavities is equal to the contraction volume.

- The metal fluidity becomes negligible when the solid fraction exceeds a critical value (f_{Scr}). This value is often set equal to 0.74, which is the atomic packing factor of a face centered cubic lattice.

The net volume change (ΔV) of a control volume V is given by

$$\Delta V = \beta \times V \times \Delta f_l, \tag{7.32}$$

where $\beta = (\rho_s - \rho_l)/\rho_l$ is the volume fraction change and Δf_l is the change of the liquid fraction over a specific time interval. Depending on the location of the control volume in the simulation domain, the volume change is either compensated with liquid metal or a void is introduced.

Criteria functions

The main advantage of criteria functions is their simplicity of implementation. They can be easily integrated into a computer code; however, they do not model the interaction of porous regions with the melt. Criteria functions attempt to predict shrinkage porosity by evaluating the feeding capability of the casting.

One of the most well-known criteria is the Pellini criterion (Pellini, 1953) which states that porosity should be expected in regions with a lower thermal gradient than a critical value ($G_T \leq G_T^{cr}$) close to the end of solidification. Pellini's criterion was further developed by Niyama, Uchida, and Morikawa (1982) who proposed that porosity should be expected when the pressure drop in the mushy region is high. They expressed this criterion as a function of the thermal gradient G and the cooling rate R ($N = G/R^{1/2}$). Similar criteria have been established by Lee, Chang, and Chieu (1990) and Suri, Paul, El-Kaddah, and Berry (1994).

7.2.4.4 Modeling gas porosity

The formation mechanism of gas porosity is different from shrinkage porosity and is related to either (1) dissolved gas (e.g., Hydrogen) rejected during solidification, or (2) gas bubbles entrained during turbulent filling which during solidification evolve into gas pores in the final casting. The formation of gas pores during solidification is a heterogeneous phenomenon while according to the literature inclusions (Mohanty, Samuel, & Gruzleski, 1995), bifilms (Rooy and Fischer, 1968) or microscopic size defects (Stefanescu and Catalina, 2011) might act as nucleation sites. Whether a pore will be formed or gas will remain dissolved in the melt is a matter concerning classical nucleation theory (CNT). According to

CNT, a gas pore with a radius r can exist if the internal gas pressure p_g is equal to the sum of the surface tension and the melt pressure p. This can be expressed in mathematical terms as follows:

$$p_g = p + \frac{2\sigma}{r}, \tag{7.33}$$

where σ is the gas–liquid surface tension. The pressure field p can be calculated using the mass balance equation in a control volume proposed by Kubo and Pehlke (1985)

$$\left(1 - \frac{\rho_s}{\rho_l}\right)\frac{\partial f_l}{\partial t} - \frac{\partial f_V}{\partial t} + \nabla \times (f_l \times \boldsymbol{v}) = 0, \tag{7.34}$$

where ρ_s and ρ_l are the densities of the solid and liquid phase, respectively, f_l the liquid fraction and f_V the fraction of the porosity volume. The velocity field $\boldsymbol{v}$ in the interdendritic regions can be obtained by the Darcy's law:

$$\boldsymbol{v} = -\frac{1}{K\mu f_l}(\nabla p - \rho \boldsymbol{g}), \tag{7.35}$$

where μ is the fluid viscosity and K is the permeability coefficient given by

$$K = \frac{f_l^3 d^2}{180(1-f_l)^2}. \tag{7.36}$$

In Eq. (7.36) d is the primary dendrite arm spacing which can be calculated as follows:

$$d = \alpha G^{-b} V_s^{-c}, \tag{7.37}$$

where a, c are constants, G is the local temperature gradient, and V_s is the local solidification velocity. The gas pressure p_g in Eq. (7.33) can be estimated using the conservation equation for gas content:

$$\begin{aligned} [Q_0] &= (1 - f_l)[Q_s] + f_l[Q_l] + \alpha_Q \frac{p_g f_V}{T}, \\ [Q_s] &= K_{sQ} p_g^{1/2} \\ [Q_l] &= K_{lQ} p_g^{1/2} \end{aligned} \tag{7.38}$$

where $[Q_0]$ is the initial gas content, $[Q_s]/[Q_l]$ is the gas content in the solid and liquid phase, respectively, a_Q is a constant related to gas solubility, and

K_{sQ}/K_{lQ} is the equilibrium constants for gas solubility in the solid and liquid phase, respectively. The presented methodology can be effectively used for the prediction of microporosity and the pore radii can be calculated according to Eq. (7.33) over the simulation domain. However, by observing Eq. (7.35) it can be concluded that the flow field is approximated by gravitational and frictional forces while momentum and energy equations are neglected.

7.3 Optimization of casting processes

7.3.1 Introduction

Decision making in manufacturing processes is based on four decision-making attributes, namely, (1) cost, (2) time, (3) quality, and (4) flexibility (Chryssolouris, 2013). This decision-making framework was later augmented by the addition of the sustainability component (Salonitis and Stavropoulos, 2013). With regard to casting, which is one of the most energy-intensive manufacturing processes, it is of utmost significance to produce defect-free castings, with a low cycle time, efficient energy usage, and reduced CO_2 emissions. Until the early 1990s the production of high-quality castings mostly relied on the experience and intuition of foundry engineers. Later on, Professor John Campbell summarized the main guidelines for producing high-quality castings with 10 rules (Campbell, 2015). The design of running and gating systems in castings was based on these rules which targeted the elimination of entrainment and shrinkage defects as well as other types of damage (Lyu, Papanikolaou, & Jolly, 2019; Papanikolaou, Pagone, Jolly, & Salonitis, 2020). Empirical mold and running system design considerations and recommendations for casting processes can also be found in Bralla (1999).

Process optimization refers to the implementation of a set of strategies in order to optimize a process outcome, such as energy efficiency or cost, by alternating one or more process parameters and considering some given constraints. During the last two decades the evolution of the computing power as well as the modeling techniques has allowed for the accurate numerical modeling of the filling and solidification stages in casting as well as the wide range of diverse physical phenomena involved, as discussed in the previous section. Moreover, during the last two decades numerical modeling has been effectively implemented for the optimization of casting processes. In the simplest case scenario, casting processes have been optimized by evaluating various feeding and gating systems designs with respect to their potential to eliminate or reduce defects in the final cast product (Iqbal, Sheikh, Al-Yousef, & Younas, 2012; Jezierski, Dojka, & Janerka, 2018; Kwon and Kwon, 2019; Prabhakar, Papanikolaou, Salonitis, & Jolly, 2019); this is a characteristic example of the traditional method of trial-and-error optimization. Alternative methods such as design of experiments (DoE) (Prabhakar, Papanikolaou, Salonitis, & Jolly, 2020; Verran, Mendes, & Valentina, 2008) have also been implemented for the optimization of casting processes.

However, traditional trial-and-error optimization or DoE include the danger of miscalculating the optimal solution. The need to obtain more accurate solutions in the casting process optimization led to the coupling of numerical optimization schemes with numerical models of the process. Such integrated models have been implemented to identify the optimum geometrical features of the casting (mold/runner/feeder) or process parameters in order to optimize one or more of the decision-making criteria in casting.

7.3.2 Numerical optimization of casting processes

Risers are one of the most significant components of casting methods while their design is one of the most challenging tasks. Risers are often placed on top of the mold in order to compensate for volume contraction during solidification and avoid the formation of hotspots in the casting. Moreover, their significance lies in the fact that their volume affects the cooling rate and consequently the mechanical properties of the cast product. Their design should be based on two main principles: (1) they should be large enough to compensate for shrinkage and simultaneously, (2) compact enough to ensure energy efficiency and minimize the amount of scrap (Jolly, 2003).

The design of risers has been one of the first applications of coupled numerical models of solidification and numerical optimization. In a pioneering study, Morthland, Byrne, Tortorelli, and Dantzig (1995) developed a framework consisting of (1) a finite element model of solidification, (2) explicit design-sensitivity analysis, and (3) the method of feasible directions for constrained minimization (Vanderplaats, 1984). The objective of this investigation was to minimize the volume of the risers and porosity in an L-shaped steel plate and a steel hammer. In this study, the Niyama's criterion was used for the prediction of porosity. The developed framework was able to converge to a solution (set of design parameters) leading to defect-free castings. A similar approach was also implemented by Lewis, Manzari, Ransing, and Gethin (2000). However, in this investigation the authors optimized the casting quality using the die interface heat transfer coefficients in addition to the dimensions of the riser as optimization variables. Moreover, they used the Broyden–Fletcher–Goldfarb–Shanno (BFGS) algorithm for minimizing their objective function.

Papanikolaou, Pagone, and Georgarakis (2018) developed an optimization framework consisted of a commercial CFD software and an optimization algorithm for the estimation of the optimum feeder dimensions tailored to the novel Constrained Rapid Induction Melting Single Shot Up-Casting (Salonitis, Jolly, Zeng, & Mehrabi, 2016). The objective of this investigation was the simultaneous minimization of microporosity in the casting and the maximization of yield. This multiobjective optimization problem was converted to a single objective one via a scalarization technique. The Constrained Optimization by Linear Approximation algorithm was implemented to identify an optimal set of riser dimensions leading to the minimization of the objective function and satisfy some dimensional constraints. As expected, larger riser dimensions led to decreased microporosity, but they also deteriorated the process yield. In order to illustrate the trade-off between the two conflicting objectives, a Pareto frontier was obtained using the Nondominated Sorting Genetic Algorithm-II.

Tavakoli and Davami (2008, 2009) dealt with the problem of automating the risers' design in aiming to minimize their volume and prevent the formation of hotspots in the casting. Regarding the numerical modeling of solidification, the authors modeled the unsteady nonlinear heat transfer equation via a Poisson-like equation. Although this approach neglects the initial temperature distribution at the end of the filling process as well as natural convection occurring during solidification, it has the significant advantage of lower computational cost; this is crucial when numerical modeling is coupled with optimization algorithms as many iterations might be required. Their simulation domain consisted of a unit box representing the mold and a riser located on top of the mold. Besides its location, no other geometric features were assigned to the riser whose final form was obtained via means of topology optimization using the simple isotropic material with penalization method. Topology optimization problems consider a characteristic function χ assigned to the cells of the simulation domain while its values vary depending on the location of the cell: 0 corresponds to void while 1 to material. The optimal riser design corresponds to the values of χ leading to the minimization of a predefined objective function provided that a set of constraints is satisfied. The authors managed to obtain the optimal riser design leading to the elimination of hotspots in the casting and minimum riser volume for two test cases examined.

The application of optimization schemes in conjunction with numerical modeling has not been limited to the design of risers. Singh, Mallinson, and Panton (2002) used the commercial solver ProCAST linked to an optimization package in order to accurately reproduce the experimental conditions of a low-pressure die casting process of an aluminum alloy wheel. For this reason, they installed 15 thermocouples in the casting in order to constantly monitor the temperature distribution during the filling process. The velocity component at the ingate was used as a design variable in order to realistically reproduce the filling process in terms of velocity profiles; thus the corresponding objective function was expressed as follows:

$$f(\boldsymbol{x}) = \sum_{i=1}^{N} \left(t_i^{\text{model}} - t_i^{\text{experimental}} \right), \tag{7.39}$$

where t_i^{model} and $t_i^{\text{experimental}}$ are the points in time when the ith thermocouple ($N = 15$) and the corresponding node of the FEM model came in contact with the molten metal. The BFGS algorithm was used for the minimization of the objective function. A similar process was followed for tuning the experimental and numerical interfacial heat transfer coefficient between the die and the melt as a function of the melt temperature. After calibrating the boundary conditions of the numerical model, the cycle time and casting quality were optimized under the constraint that unidirectional solidification should occur by adjusting the die material properties. The optimized design led to an 80% increase in the capacity of the manufacturing plant.

One of the most risk-inducing components of filling systems is the sprue or down-runner. This is because of the severe turbulence which might occur during the downfall of metal in gravity casting. A set of guidelines on the proper sprue design have been provided by Professor John Campbell (Stefanescu, 2015). However, as advocated by the author the onset of turbulence at the lowest level of the sprue is almost inevitable. Sama, Badamo, Lynch, and Manogharan (2019) tried to address this issue by introducing some novel designs such as the parabolic sprue casting

(PSC) and the conical-helix sprue casting (CHSC). These complex sprue designs cannot be manufactured using conventional manufacturing techniques; thus the 3D Sand-Printing method was used. PSC and CHSC were designed with the help of an optimization algorithm. The optimization objective was the minimization of the length of the sprue curve for the two cases while maintaining the velocity at the bottom of the sprue below a critical value of 0.5 m/s as suggested by Campbell (1993). The MATLAB fmincon constrained optimization function was implemented to solve this multivariable and single-objective optimization problem. Numerical modeling of the filling process using the FLOW-3D commercial software indicated a significant reduction of the entrained air volume for the PSC and CHSC designs. The obtained CT scan results indicated that casting defects are reduced by 56% and 99.5% for the PSC and CHSC designs, respectively, when compared to the traditional straight sprued casting (SSC) one, as illustrated in Fig. 7.6.

The continuous casting process has also been investigated via means of numerical modeling and optimization. However, due to a large number of process and design parameters involved in the process,

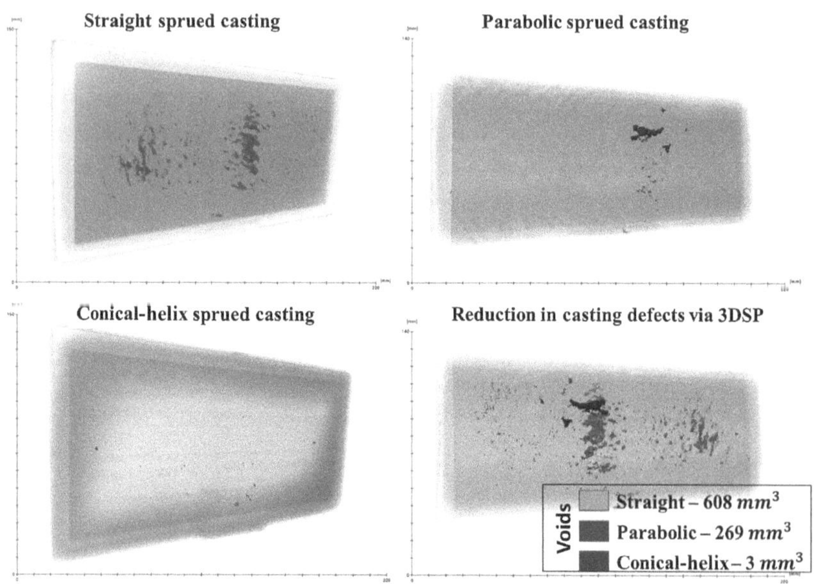

Figure 7.6 CT scan of (A) SSC, (B) PSC, (C) CHSC, (D) Comparison of void space for all three sprue designs (Sama et al., 2019). *PSC*, Parabolic sprue casting; *CHSC*, conical-helix sprue casting.

gradient-based optimization algorithms fail to achieve satisfactory convergence (Chakraborti and Mukherjee, 2000). Therefore evolutionary optimization algorithms have been coupled with numerical models for the optimization of this process. Santos, Spim, and Garcia (2003) coupled a mathematical solidification model with an evolutionary optimization algorithm and a knowledge base in order to optimize the quality as well as the production rate. The developed optimization scheme was capable of selecting the optimal set of cooling conditions and metallurgical criteria among a variety of available options and a set of given constraints. The quality of the casting was associated with more homogeneous strand surface profiles (lower thermal gradients). The simulated temperature profiles were compared against experimental ones and significant improvement was observed in terms of thermal gradients between the spray zones. A similar case study was also presented in Santos, Cheung, Garcia, and Spim (2005).

7.3.3 Optimization of casting processes using artificial neural networks

One of the major challenges in the casting industry is the development and integration of decision-making frameworks. This task is quite complicated due to the wide variety of process and design parameters involved in casting. During the last two decades, research has been focused on the development of computational frameworks integrating artificial neural networks (ANNs) in order to assist foundry engineers to optimize casting processes and select the optimum values for a wide range of process parameters. In this section, some of the main developments on computational frameworks integrating numerical models of casting processes and ANNs as a means of decision-making tools will be presented.

One of the first implementations of ANNs in conjunction with numerical modeling of solidification was by Lin (2003). The authors aimed at minimizing the deformation due to nonuniform temperature distribution during die casting of a free-form surface by modifying the geometric characteristics of a cooling channel. Solidification was modeled using a finite element software and an abductive neural network was trained using the results of 20 FEM simulations. The trained ANN was capable of predicting the maximum deformation value with an approximate error of 2%. The optimal cooling channel geometry was estimated using the simulated annealing technique.

In a study similar to Santos et al. (2003), which presented a framework consisted of a solidification heat transfer model with a knowledge base and an evolutionary optimization algorithm for optimizing the continuous casting process of steel, Santos et al. (2005) replaced the evolutionary optimization algorithm with a neural network–based algorithm. The employed neural network consisted of seven input nodes, three output nodes, and a hidden layer. The ANN was trained using a database produced using the results of the solidification numerical model with a gradient-descent method called delta rule (Zietsman, Kumar, Meech, Samarasekera, & Brimacombe, 1998). The trained ANN was capable of suggesting suitable values of the water flow rates across the strand, which resulted in more uniform strand surface temperature profiles with lower thermal gradients.

A strategic framework for the optimization of the pressure die casting process was developed by Krimpenis, Benardos, Vosniakos, and Koukouvitaki (2006). The authors trained two ANNs using data obtained from numerical simulations of the process using the commercial software ProCAST. The two aforementioned ANNs were trained for predicting defect formation and solidification time, respectively. The training datasets were generated with the help of Taguchi's orthogonal arrays in order to reduce the computational cost required by a full factorial experiment. The ANN used for defect prediction was trained using the learning vector quantization method; the trained ANN was 83.3% and 75% accurate with regard to the training and testing datasets, respectively. The output values of the ANNs were used during the optimization phase. More specifically, the optimization objective function was defined as follows:

$$f(\boldsymbol{x}) = 100[net1(\boldsymbol{x}) - 1] + net2(\boldsymbol{x}), \tag{7.40}$$

where $net1(\boldsymbol{x})$ and $net2(\boldsymbol{x})$ are the outputs of the defect and solidification time prediction ANNs, respectively, for a set of process parameters ($\boldsymbol{x}$). A higher weight was assigned to the defect prediction criterion as defects were considered more significant than the solidification time. The objective function was minimized via a genetic algorithm using the stochastic universal sampling technique with a 70% single point crossover rate and a 2.5% population mutation rate. The best chromosome (optimal set of process parameters) was identified after 55 iterations. The importance of training ANNs lies in the fact that the final computational cost is significantly reduced because there is no need to run additional simulations for each iteration of the optimizer; the objective function can be easily evaluated using the predictions of the trained neural networks.

7.4 Summary

The significant evolution of the computing power as well as the modeling techniques has enabled the numerical simulation of a variety of casting processes. Contemporary numerical models of solidification are capable of predicting the location as well as the extent of a variety of casting defects, including shrinkage porosity, gas porosity, and entrainment defects. Numerical models of solidification have been effectively coupled with optimization schemes and ANNs in order to optimize casting processes with respect to five decision-making manufacturing attributes, namely, cost, time, quality, flexibility, and sustainability. Successful optimization for quality improvement (elimination of defects), speeding up the process, reducing energy consumption, and tuning simulation parameters in order to accurately reproduce the real-time experiment are achieved. One of the future challenges is the integration of decision-making frameworks with the real-time production. Finally, the integration of techniques capable of providing atomistic resolution information (e.g., molecular dynamics simulations) on solidification and most importantly defect formation to the existing optimization frameworks would significantly enhance the accuracy and trustworthiness of the current models.

References

Barkhudarov, M.R., & Hirt, C.W. (1998) Tracking defects. In: *1st international aluminium casting technology symposium*.

Bralla, J. G. (1999). *Design for manufacturability handbook*. McGraw-Hill.

Campbell, J. (2015). *Complete casting handbook: Metal casting processes, techniques and design* (pp. 3–8). Butterworth-Heinemann.

Campbell, J. (1993). Invisible macrodefects in castings. *Journal de Physique IV Colloque, 3* (C7), C7-861–C7-872.

Campbell, J. (2003). *The melt. Castings* (pp. 1–16). Elsevier.

Chakraborti, N., & Mukherjee, A. (2000). Optimization of continuous casting mould parameters using genetic algorithms and other allied techniques. *Ironmak Steelmak, 27* (3), 243–247.

Chryssolouris, G. (2013). *Manufacturing systems: Theory and practice*. New York: Springer Science & Business Media.

Garg, V. K. (1998). *Applied computational fluid dynamics*. CRC Press.

Griffiths, W. D., Beshay, Y., Parker, D. J., & Fan, X. (2008). The determination of inclusion movement in steel castings by positron emission particle tracking (PEPT). *Journal of Materials Science, 43*(21), 6853–6856.

Groover, M. P. (2010). *Fundamentals of modern manufacturing: Materials, processes and systems* (4th ed.). John Wiley & Sons, Inc.

Hernández-Ortega, J. J., Zamora, R., Palacios, J., López, J., & Faura, F. (2010). An experimental and numerical study of flow patterns and air entrapment phenomena during the filling of a vertical die cavity. *Journal of Manufacturing Science and Engineering, Transactions of the ASME, 132*(5).

Hirt, C. W., & Nichols, B. D. (1981). Volume of fluid (VOF) method for the dynamics of free boundaries. *Journal of Computational Physics*, *39*(1), 201–225.

Incropera, F. P., Lavine, A. S., Bergman, T. L., & DeWitt, D. P. (2007). *Fundamentals of heat and mass transfer*. Wiley.

Iqbal, H., Sheikh, A. K., Al-Yousef, A., & Younas, M. (2012). Mold design optimization for sand casting of complex geometries using advance simulation tools. *Materials and Manufacturing Processes*, *27*(7), 775–785.

Jezierski, J., Dojka, R., & Janerka, K. (2018). Optimizing the gating system for steel castings. *Metals (Basel)*, *8*(4), 266.

Jolly, M. (2003). 1.18—Castings. In I. Milne, R. O. Ritchie, & B. Karihaloo (Eds.), *Comprehensive structural integrity* (pp. 377–466). Oxford: Pergamon. Available from:. Available from http://www.sciencedirect.com/science/article/pii/B0080437494010703.

Kenney, M. P., Courtois, J. A., Evans, R. D., Farrior, G. M., Kyonka, C. P., Koch, A. A., et al. (1988). Semisolid metal casting and forging. *Metals Handbook*, *15*, 327–338.

Krimpenis, A., Benardos, P. G., Vosniakos, G.-C., & Koukouvitaki, A. (2006). Simulation-based selection of optimum pressure die-casting process parameters using neural nets and genetic algorithms. *The International Journal of Advanced Manufacturing Technology*, *27*(5–6), 509–517.

Kubo, K., & Pehlke, R. D. (1985). Mathematical modeling of porosity formation in solidification. *Metallurgical and Materials Transactions B*, *16*(2), 359–366.

Kwon, H. J., & Kwon, H. K. (2019). Computer aided engineering (CAE) simulation for the design optimization of gate system on high pressure die casting (HPDC) process. *Robotics and Computer-Integrated Manufacturing*, *55*, 147–153.

Lai, N. W., Griffiths, W. D., Campbell, J., et al. (2003). Modelling of the potential for oxide film entrainment in light metal alloy castings. In D. M. Stefanescu, D. M. Warren, M. R. Jolly, & J. Campbell (Eds.), *Modeling of casting, welding and advanced solidification processes-X* (pp. 415–422). Destin, Florida: TMS (The Minerals, Metals & Materials Society).

Lee, Y. W., Chang, E., & Chieu, C. F. (1990). Modeling of feeding behavior of solidifying Al-7Si-0.3 Mg alloy plate casting. *Metallurgical and Materials Transactions B.*, *21*(4), 715–722.

Lewis, R. W., Manzari, M. T., Ransing, R. S., & Gethin, D. T. (2000). Casting shape optimisation via process modelling. *Materials and Design*, *21*(4), 381–386.

Lin, J. C. (2003). The optimal design of a cooling system for a die-casting die with a free form surface. *The International Journal of Advanced Manufacturing Technology*, *21*(8), 612–619.

Lyu, C., Papanikolaou, M., & Jolly, M. (2019). *Numerical process modelling and simulation of campbell running systems designs. Shape casting* (pp. 53–64). Springer.

Melendez, A. J., Carlson, K. D., & Beckermann, C. (2010). Modelling of reoxidation inclusion formation in steel sand casting. *The International Journal of Cast Metals Research*, *23*(5), 278–288.

Mirak, A. R., Divandari, M., Boutorabi, S. M. A., & Campbell, J. (2007). Oxide film characteristics of AZ91 magnesium alloy in casting conditions. *The International Journal of Cast Metals Research*, *20*(4), 215–220.

Mohanty, P. S., Samuel, F. H., & Gruzleski, J. E. (1995). Experimental study on pore nucleation by inclusions in aluminum castings. *Transactions of the American Foundrymen's Society*, *103*, 555–564.

Morthland, T. E., Byrne, P. E., Tortorelli, D. A., & Dantzig, J. A. (1995). Optimal riser design for metal castings. *Metallurgical and Materials Transactions B*, *26*(4), 871–885.

Niyama, E., Uchida, T., Morikawa, M., & Saito, S. (1982). *American Foundrymen's Society International Cast Metals Journal*, 7(3), 52–63.

Papanikolaou, M., Pagone, E., & Georgarakis, K. (2018). Design optimisation of the feeding system of a novel counter-gravity casting process. *Metals (Basel)*, *8*(10), 817.

Papanikolaou, M., Pagone, E., Jolly, M., & Salonitis, K. (2020). Numerical simulation and evaluation of campbell running and gating systems. *Metals (Basel)*, *10*(1), 68.

Pellini, W. S. (1953). Factors which determine riser adequacy and feeding range. *AFS Transactions*, *61*(67), 61–80.

Prabhakar, A., Papanikolaou, M., Salonitis, K., & Jolly, M. (2020). Minimising defect formation in sand casting of sheet lead: A DoE approach. *Metals (Basel)*, *10*(2), 252.

Prabhakar, A., Papanikolaou, M., Salonitis, K., & Jolly, M. (2019). Sand casting of sheet lead: Numerical simulation of metal flow and solidification. *The International Journal of Advanced Manufacturing Technology*, 1–13.

Reilly, C., Green, N. R., & Jolly, M. R. (2013a). The present state of modeling entrainment defects in the shape casting process. *Applied Mathematical Modelling*, *37*(3), 611–628.

Reilly, C., Green, N. R., Jolly, M. R., & Gebelin, J. C. (2013b). The modelling of oxide film entrainment in casting systems using computational modelling. *Applied Mathematical Modelling*, *37*(18–19), 8451–8466.

Reilly, C., Green, N. R., Jolly, M. R., & Gebelin, J. C. (2009). Using the calculated Fr number for quality assessment of casting filling methods. *Modeling of Casting, Welding, and Advanced Solidification Processes—XII*, *12*, 419–426.

Rooy, E. L., & Fischer, E. F. (1968). Control of aluminum casting quality by vacuum solidification tests. *AFS Transactions*, *54*(1), 97–100.

Salonitis, K., Jolly, M. R., Zeng, B., & Mehrabi, H. (2016). Improvements in energy consumption and environmental impact by novel single shot melting process for casting. *Journal of Cleaner Production*, *137*, 1532–1542.

Salonitis, K., & Stavropoulos, P. (2013). On the integration of the cax systems towards sustainable production. In: *Procedia CIRP* (pp. 115–120). Elsevier B.V.

Sama, S. R., Badamo, T., Lynch, P., & Manogharan, G. (2019). Novel sprue designs in metal casting via 3D sand-printing. *Additive Manufacturing*, *25*, 563–578.

Santos, C. A., Cheung, N., Garcia, A., & Spim, J. A. (2005). Application of a solidification mathematical model and a genetic algorithm in the optimization of strand thermal profile along the continuous casting of steel. *Materials and Manufacturing Processes*, *20*(3), 421–434.

Santos, C. A., Spim, J. A., & Garcia, A. (2003). Mathematical modeling and optimization strategies (genetic algorithm and knowledge base) applied to the continuous casting of steel. *Engineering Applications of Artificial Intelligence*, *16*(5–6), 511–527.

Saxena, P., Papanikolaou, M., Pagone, E., Salonitis, K., & Jolly, M.R. (2020). Digital manufacturing for foundries 4.0 (pp. 1019–1025). Retrieved from http://link.springer.com/10.1007/978-3-030-36408-3_138.

Sheikh, A. K., Khan, M. A. A., Iqbal, H., & Al-Shaer, B. S. (2018). Casting of adjuster bracket—Process optimization and validation. *Materials and Manufacturing Processes*, *33*(16), 1845–1850.

Singh, D. P. K., Mallinson, G. D., & Panton, S. M. (2002). *Applications of optimization and inverse modeling to alloy wheel casting. Numerical Heat Transfer; Part A: Applications* (pp. 741–756). Informa UK Ltd.

Stefanescu, D. M. (2015). *Science and engineering of casting solidification*. Springer.

Stefanescu, D. M., & Catalina, A. V. (2011). Physics of microporosity formation in casting alloys—Sensitivity analysis for Al–Si alloys. *The International Journal of Cast Metals Research*, *24*(3–4), 144–150.

Suri, V. K., Paul, A. J., El-Kaddah, N., & Berry, J. T. (1994). Determination of correlation factors for prediction of shrinkage in castings—Part I: Prediction of microporosity in castings; A generalized criterion (AFS research) (94-138). *Transactions of the American Foundrymen's Society*, *102*, 861–868.

Tavakoli, R., & Davami, P. (2009). Optimal riser design in sand casting process with evolutionary topology optimization. *Structural and Multidisciplinary Optimization*, *38*(2), 205–214.

Tavakoli, R., & Davami, P. (2008). Optimal riser design in sand casting process by topology optimization with SIMP method I: Poisson approximation of nonlinear heat transfer equation. *Structural and Multidisciplinary Optimization*, *36*(2), 193–202.

Vanderplaats, G.N. (1984). CODES/ADS-A FORTRAN control program for engineering synthesis using the ADS optimization program. Monterey, California.

Verran, G. O., Mendes, R. P. K., & Valentina, L. V. O. D. (2008). DOE applied to optimization of aluminum alloy die castings. *The Journal of Materials Processing Technology*, *200*(1–3), 120–125.

Winterscheidt, D. L., & Huang, G. X. (2001). Fundamentals of casting process modeling. In K.-O. Yu (Ed.), *Modeling for casting and solidification processing* (pp. 17–54). CRC Press.

Zietsman, J. H., Kumar, S., Meech, J. A., Samarasekera, I. V., & Brimacombe, J. K. (1998). Taper design in continuous billet casting using artificial neural networks. *Ironmak Steelmak*, *25*(6), 476.

CHAPTER 8

Diffusion of sustainability practices through supplier performance assessment

Handson Pimenta[1], Peter Ball[2] and Konstantinos Salonitis[3]
[1]Federal Institute of Education, Science and Technology Rio Grande do Norte, Brazil
[2]University of York, York, United Kingdom
[3]Sustainable Manufacturing Systems Centre, School of Aerospace, Transport and Manufacturing, Cranfield University, Bedfordshire, United Kingdom

8.1 Introduction

Manufacturing firms have a pivotal role in improving suppliers' sustainability performance. Enhancing supplier performance or, at least assuring the acceptable performance standard is a crucial objective of supply chain management (Gold et al., 2010a).

Supplier sustainability performance assessment (SSPA) allows buying firms to understand better the environmental impact of their suppliers (Simpson & Power, 2005), to identify what actions are needed (Gimenez & Tachizawa, 2012) and to control the implementation of the sustainability programs through the supply chain (Green, Zelbst, Bhadauria, & Meacham, 2011).

Despite the significant body of literature on supplier sustainability performance assessment, there is absence of a theory to explain organizational factors that affect the diffusion of both environmental and social sustainability practices across the supplier manufacturing base (Soundararajan & Brammer, 2018). Moreover, Beske-Janssen, Johnson, and Schaltegger (2015) stated that the literature is not clear about the responsibility for coordinating the supply chain performance assessment. Likewise, a better understanding on organizational factors that enhance the diffusion of sustainability practices through supplier performance assessment is also needed.

This chapter addresses the gap in the knowledge of how environmental and social sustainability practices are diffused across the supplier base through the performance assessment and what organizational factors are influential in this purpose. In attempting to answer these questions a conceptual framework is proposed and then tested through case application.

Sustainable Manufacturing
DOI: https://doi.org/10.1016/B978-0-12-818115-7.00017-1

The factors were rigorously identified and provided an orientation to diffuse both environmental and social sustainability practices according to different conditions and purposes (e.g. risk management, improvements in products, strengthening the relationship, collaborative approaches).

8.2 Sustainability practices diffusion through supplier performance assessment

8.2.1 Focus of the assessment, mechanisms used and sustainability practices measured

Supplier sustainability performance assessment includes activities of gathering and processing supplier information (Vachon & Klassen, 2006). This mostly focuses on supplier's compliance, supplier's sustainability performance, product life cycle perspective and products improvements.

An understanding on how well suppliers comply with local regulations and requirements of selection and contracts (compliance/risk) is a starting point for the supplier performance assessment (Beske & Seuring, 2014). Buying firms seek for a way to guarantee that the suppliers act according to laws and a set of requirements adopted in the supplier selection stage. This is often related to specific production sites (e.g. updated environmental authorization), working conditions and human rights (Pimenta, 2016). These aspects can influence the reputation of buying firms if suppliers do not comply with them. Indeed, risks originated from the suppliers can have effects on the buyer (Schaltegger & Burritt, 2014). For example, Mattel, Inc. carried out a recall of 20 million children's toys in 2007 due to materials found to contain traces of lead used by some suppliers. Therefore the focus of measuring supplier sustainability performance can be associated with risk management.

Improvements in the supplier's sustainability performance are also a common SSPA proposal. Buying firms seek to understand beyond the supplier compliance in order to create opportunities for improvements, innovation and reduce risk. This involves considerations of environmental factors and social aspects of the supplier activities, processes and facilities, such as the use of natural resource (e.g. water, energy, raw materials), waste management, carbon management (e.g. Greenhouse gas emissions, air pollution) and working conditions.

The product life cycle perspective in order to assess the impacts across the supply chain is another common sustainability performance assessment propose. This helps buying firms to understand the environmental performance of their products across the entire supply chain. The results can support buying firms to select suppliers and practices that can be implemented to reduce the life cycle impacts of the products. Analytical tools such as LCA, carbon footprint, material flow analysis are often employed for this purpose (Hagellar, Vorst, & Van Der, 2002; Lee & Cheong, 2011).

Finally, improvements in products are often reported in the literature as one of the core focus of SSPA (e.g. Handfield, Sroufe, & Walton, 2005; Vachon & Klassen, 2008). Not only improvements in the final product can be taken into consideration, but also processing technologies products' components and packaging. This requires collaborative approaches between suppliers and buying firms.

Different performance mechanisms have been adopted to assess suppliers' sustainability performance, which arise from self-assessment (Klassen & Vereecke, 2012), surveys (Nawrocka, Brorson, & Lindhqvist, 2009) and audits (Morali & Searcy, 2013) (Table 8.1). Second party audits are the most widely used for monitoring compliance/risk and improving the overall sustainability

Table 8.1 Performance mechanism according to the purpose of supplier performance assessment.

Performance mechanisms employed	Purpose of performance assessment			
	Compliance/ Risk	Improvements in supplier' performance	Lifecycle perspective	Improvements in product
2nd party audit	High	High	Medium	Medium
Self-assessment & Survey	High	Medium	Low	Medium
Follow-up certification	High	Low	Medium	Medium
Score-cards	Medium	Low	Medium	-
3rd party audit	High	High	Low	Medium
NGO	Medium	Medium	Low	-
Platforms	Low	Medium	Low	Low
LCA	High	High	High	High
Carbon footprint	Low	Low	High	Medium
Water footprint	Low	Low	Low	Medium
Material flow analysis	-	Low	High	-

Criteria: High – above the average and median of papers that mentioned the initiative and purpose of the supplier development initiative, Medium – Below average and above and Low – below the average and median. This analysis was done based on a systematic literature review (SLR) conducted by Pimenta (2016) on diffusion of sustainability practices through supply chain management activities. The SLR covered 140 papers published during 1992 and 2016 in peer reviewed journals. *NGO*, Non-governmental organization.

performance. It is important to note that the own buying firm team conducts second party audits what promotes more interaction with suppliers.

Third party audits, self-assessment and following-up suppliers are also commonly adopted with the focus on assessing suppliers' compliance/risk (Beske-Janssen et al., 2015; Chardine-Baumann & Botta-Genoulaz, 2014; Marshall, McCarthy, Heavey, & McGrath, 2015). Interestingly, some online platforms include a scoring of suppliers indicating their compliance level (Supplier Ethical Data Exchange – SEDEX) and identify all materials used within the current production process and the original manufacturer (International Materials Data System – IMDS – automotive sector). This allows buying firms to quickly assess the compliance performance of their suppliers and address to suppliers some opportunities to improve their sustainability performance.

Most suppliers are typically assessed on compliance with the sustainability requirements used during the supplier selection through self-assessment or audits mechanisms (Gold et al., 2010a). On the other hand, strategic suppliers usually receive more attention on their sustainability performance improvements (Nawrocka et al., 2009) or products improvement through life cycle perspective approaches (e.g. LCA) (Ashby, Leat, & Hudson-Smith, 2012). When the performance purpose is improvements in the product and life cycle management, the use of LCA were broadly employed (Hutchins & Sutherland, 2008; Nakano & Hirao, 2011) as can be seen observed in Table 8.1.

Table 8.2 presents the environmental practices measured by the buying firms in supplier performance assessment. The most common environmental practices measured were energy management in terms of the consumption of electricity, fuel and gas, efficiency, existence of renewable source and product recovery. Carbon management is also a common practice measured. It regards direct and indirect emission, control, reduction plan, disclosure information, GHG, air pollution. Overall, the environmental practices measured were related to compliance with regulation and improvements in the suppliers' sustainability performance, especially in terms of processes inputs and output and facilities.

Table 8.3 points out the social practices. The most measured practices were human rights, working conditions and health and safety conditions. Interestingly, most of the practices measured were also considered as requirements adopted to select suppliers.

This section has examined research works related to the environmental and social sustainability practices measured. More emphasis has been given to environmental sustainability practice diffusion but there is a lack of clarity on the conditions required, which will be the focus of the next section. Moreover,

Table 8.2 Environmental practices measured in supplier performance assessment.

Environmental practices measured	Purpose	Some examples of papers
Energy management	C/P/LC	Beamon (1999), Gimenez and Tachizawa (2012)
Carbon management (direct and indirect emission, control, reduction plan, disclosure information, GHG, air pollution)	ALL	Ageron, Gunasekaran, and Spalanzani (2012), Jira and Toffel (2013)
Waste management	ALL	Meacham, Toms, Green, and Bhadauria (2013), Schaltegger and Burritt (2014)
Compliance with regulations	ALL	de Bakker and Nijhof (2002), Gold, Seuring, and Beske (2010a)
Material intensity	C/P/LC	Mcintyre, Smith, Henham, and Pretlove (1998), Salvá, Jones, Marshall, and Bishop (2013)
Hazardous substances use	ALL	Handfield et al. (2005), Shaw, Grant, and Mangan (2010)
Environmental management system/ISO 14001	C/P/LC	Ahi and Searcy (2015), Tachizawa and Wong (2014)
Wastewater	P/LC	Gimenez and Tachizawa (2012), Hervani, Helms, and Sarkis (2005)
Resource (transparency and inventory)	C/P/LC	Dües, Tan, and Lim (2013), Shaw et al. (2010)
Waste packaging	ALL	Nagel (2003), Okongwu, Morimoto, and Lauras (2013)
Recycling	ALL	Ahi and Searcy (2015), Handfield et al. (2005)
Environmental policy	C	Dou, Zhu, and Sarkis (2014b), Salvá et al. (2013)
Transportation - efficiency, fuel consumption, miles traveled	LC	Ahi and Searcy (2015), Shaw et al. (2010)
Compliance with requirements	ALL	Beske and Seuring (2014), Rosen, Bercovitz, and Beckman (2001)
Packaging (volume and materials)	C/P	Nagel (2003), Okongwu et al. (2013)

(*Continued*)

Table 8.2 (Continued)

Environmental practices measured	Purpose	Some examples of papers
Biodiversity conservation	C/P	Chardine-Baumann and Botta-Genoulaz (2014), Shaw et al. (2010)
Number of regulatory violations	C	Ahi and Searcy (2015), Hervani et al. (2005)
Number of environmental accidents	C	Ahi and Searcy (2015), Hervani et al. (2005)
Costs of environmental management (e.g. compliance and control and management)	P/LC	Chardine-Baumann and Botta-Genoulaz (2014), Hervani et al. (2005)
Internal protocol - control and management, e.g. identify environmental aspects & regulations	C/P	Dou et al. (2014b), Rock, Angel, and Lim (2006)
Recycled materials	C/LC	Beamon (1999), Chardine-Baumann and Botta-Genoulaz (2014)
Recycled water	C/P	Chardine-Baumann and Botta-Genoulaz (2014), Dou et al. (2014b)
Soil degradation - land pollution	C/P	Chardine-Baumann and Botta-Genoulaz (2014), Okongwu et al. (2013)
Pollution prevention	C/LC	Nagel (2003)
Eco-systemic services	C/P	Chardine-Baumann and Botta-Genoulaz (2014), Okongwu et al. (2013)
Functions with environmental responsibilities	P	Hervani et al. (2005)
Capability	P	Akman (2015)
Re-usable pallets	C/P	Okongwu et al. (2013)
Hazardous wastes	C/P	Rock et al. (2006)
Volatile organic materials emissions	C/P	Rock et al. (2006)
Noise pollution	C/P	Okongwu et al. (2013)
Development of urban and rural areas	P	Chardine-Baumann and Botta-Genoulaz (2014)

ALL – Compliance/Risk, Improvements in performance, Lifecycle perspective & improvements in products; **C** – Compliance/Risk; **C/LC** – Compliance/Risk & Lifecycle perspective; **C/P** – Compliance/Risk & Improvements in performance; **C/P/LC**– Compliance/Risk, Improvements in performance & Lifecycle perspective; **P** – Improvements in performance.

Table 8.3 Social practices measured in supplier performance assessment.

Social practices measured	Purpose	Some examples of papers
Human rights	ALL	Hutchins and Sutherland (2008), Okongwu et al. (2013)
Working conditions	ALL	Chardine-Baumann and Botta-Genoulaz (2014), Pagell and Wu (2009)
Health and safety	C/P/LC	Ahi and Searcy (2015), Porteous, Rammohan, and Lee (2015)
SA 8000	C/P	Beske-Janssen et al. (2015), Hollos, Blome, and Foerstl (2012)
Community (Projects, investments)	P	Hollos et al. (2012), Okongwu et al. (2013)
Compliance with regulation	C	(Lee & Kim, 2011), Marshall et al. (2015)
Local jobs - jobs opportunity	C/P	Ahi and Searcy (2015), Beske-Janssen et al. (2015)
Human resource development - Education - qualification, skills	C/P	Chardine-Baumann and Botta-Genoulaz (2014), Porteous et al. (2015)
OHSAS 18001	C/P	Hollos et al. (2012), Okongwu et al. (2013)
Wages and working hours (including overtime)	C	Hutchins and Sutherland (2008), Porteous et al. (2015)
Community complaints	C	Beske-Janssen et al. (2015), Hervani et al. (2005)
Corruption/bribery	C/P	Chardine-Baumann and Botta-Genoulaz (2014), Okongwu et al. (2013)
Formal joint health and safety committees	C	Wang and Sarkis (2013)
Adult literacy	P	Hutchins and Sutherland (2008)
Healthcare	P	Hutchins and Sutherland, 2008
Philanthropy	P	Hutchins and Sutherland (2008)
Return on equity	P	Brandenburg, Govindan, Sarkis, and Seuring (2014)
Product safety	P	Varsei, Soosay, Fahimnia, and Sarkis (2014)
Fight against corruption	P	Chardine-Baumann and Botta-Genoulaz (2014)
Fair trade	P	Chardine-Baumann and Botta-Genoulaz (2014)

ALL – Compliance/Risk, Improvements in performance, Lifecycle perspective & improvements in products; **C** – Compliance/Risk; **C/P** – Compliance/Risk & Improvements in performance; **C/P/LC** – Compliance/Risk, Improvements in performance & Lifecycle perspective; **P** – Improvements in performance.

evidence suggested that more attention has been paid to engage key first tier suppliers leaving a lack of understanding of the diffusion across different tiers.

8.3 Influential organizational factors for supplier performance assessment: towards a framework

This session presents a set of organizational factors (OF) that assists the design and implementation of the supplier sustainability performance assessment. These organizational factors include for instance effort, initiatives and capabilities of both buying firms and suppliers (Sancha, Longoni, & Giménez, 2015).

The findings drive the development of the conceptual framework for supplier sustainability performance assessment. It covers, either graphically or in narrative form, the main things to be studied, including the key factors, variables, or constructs (Miles, Huberman, & Saldana, 2014).

The proposed framework depicts patterns in the factors and points out the most critical variables influencing the diffusion of environmental and social sustainability practices through SSPA. Each of these can be subdivided according to whether they relate to the process of design or the process of implementation.

8.3.1 Design of supplier sustainability performance assessment

Managing performance assessment for a large supplier base is challenging. For effective SSPA buying, firms need to be aware of which purpose will be addressed and who will be assessed (Gallear, Ghobadian, & Chen, 2012; Handfield et al., 2005). Definition of performance assessment purpose (OF1) and definition of supplier to be assessed (OF2) is therefore important.

Furthermore, performance measures need to be consistent, comparable, reliable, valid, and applicable across different economic activities and countries (Shaw et al., 2010). Definition of consistent measures (OF3) needs to be aligned with the SSPA purpose.

The sustainability requirements adopted in the supplier selection (OF4) and the buying firms' sustainability policy and strategy (OF5) are influential in the design of performance assessment. In turn, the sustainability requirements are more effective if systematically monitored by buying firms (Hoejmose & Adrien-Kirby, 2012). The buying firm's sustainability policy and strategies express the firm's own sustainability goals and clearly articulate its expectations with supplier's performance (Lippmann, 1999).

Moreover, the use of standards is recommended by Okongwu et al (2013). The use of industry code/guideline/principles/initiatives (OF7) brings a set of sustainability practices that might be diffused and monitored across the supply base. This includes FLA (Fair Labours Association on ethical and social sustainability), ILO (International Labour Association), Responsible Care Initiative, Fair trade and United Nation Global Compact (e.g. Ashby et al., 2012; Polgreen, 2002).

In addition, the expertise gained by the implementation of measurement system (ISO 14031, existing balanced scored card, GRI – Global Report Initiative) and management systems (certification scheme) (e.g. SA8000, ISO 14001, ISO 50001) can facilitate the design of consistent assessment measures to be used, in turn, for process management (Beske-Janssen et al., 2015). Such measurement (OF7) and management systems (OF8) are examples of practices implemented by buying firms to improve their capabilities and knowledge of sustainability, issues and potential solutions as well as barriers (Schaltegger & Burritt, 2014).

The importance of involvement and support of senior managers responsible for key interfaces with suppliers is also important in the design of supplier performance assessment. Hervani et al. (2005) stated that top managers played a vital role by helping to address the definition of purpose, measures and resources. Moreover, the authors claimed that the involvement of internal functional areas (e.g. operations, accounting, health, safety, and environment) avoided a lack of understanding of the multi-organizational measures. Therefore, support of top and middle managers (OF9) and cross-functional integration (OF10) complemented those factors uncovered so far.

Finally, Shaw et al. (2010) highlighted the engagement of external stakeholders as an influential factor to design the supplier sustainability performance measures (OF11). Engagement of stakeholders was also uncovered in the design requirements. Here, engagement of stakeholders is considered as the actor, which is external to the relationship between the buying firm and suppliers, such as non-governmental organizations (NGOs), customers, government, third party firms and independent platforms (Kogg & Mont, 2012).

To sum up, this section covered influential factors associated with the design of supplier sustainability performance assessment, which might affect the diffusion of sustainability. Performance assessment needs to be fairly consistent in order to assist suppliers in their compliance progress, compare suppliers overall sustainability performance between different supply chains (Ahi & Searcy, 2015) and provide them directions to select and implement the best sustainability practices (Mcintyre et al., 1998).

8.3.2 Implementation of supplier sustainability performance assessment

The implementation of supplier sustainability performance assessment requires buying firms to adopt mechanisms (OF12) (Nawrocka et al., 2009). As before, some mechanisms are related to external factors, such as NGOs, audit firms and independent platforms. In this case, engagement of external stakeholders (OF11) is also an influential factor for implementing the sustainability performance.

Gallear et al. (2012) found that the effort of buying firms to increase ethical and environmental awareness among their employees has a direct effect on suppliers' perception in terms of a fair treatment given by buyers. As with the development of internal capabilities earlier, training for purchasing staff needs to be emphasized (OF13). Moreover, internal capabilities (OF14), such as personal, financial, and technological resources, are influential in the implementation of performance assessment (Lee & Klassen, 2008; Paulraj, 2011). For instance, the availability of financial resources is influential on the selection of the SSPA mechanism (Shaw et al., 2010). It should be noted that the factors OF8 and OF9 are also evident.

Ongoing information exchange is a key for all supply chain partners (Wittstruck & Teuteberg, 2012), mainly in terms of performance improvements and fostering greater environmental supplier development (Vachon & Klassen, 2006). Jira and Toffel (2013) uncovered that unclear costs and benefits, as well as uncertainties on how buyers would interpret and use the information collected, were potential barriers to suppliers sharing information (e.g. impact in future purchasing decision). A clear understanding of the benefits and risk of sharing information is necessary (Gimenez & Tachizawa, 2012) (OF15).

de Bakker and Nijhof (2002) found that enhanced communication is key for implementing the supplier performance assessment (OF16). A strengthened relationship (OF17) is also presented in the literature as an influential factor for the diffusion of sustainability through performance assessment implementation. This includes traits of mutual commitment (Simpson & Power, 2005), trust (Gold, Seuring, & Beske, 2010b; Hassini, Surti, & Searcy, 2012), long-term relationship (Beske, Land, & Seuring, 2014), and sharing responsibilities (Hervani et al., 2005).

A better understanding of suppliers' capabilities to engage can lead to synergy for the implementation of environmental practices (Shaw et al., 2010). In addition, Dou et al. (2014b) found that buyers needed to verify the support of suppliers' top managers for improvement in performance.

Therefore, understanding suppliers' capabilities (OF18) and the support of senior managers (OF19) are considered as influential factors related to the suppliers' characteristics.

Of note from Vachon and Klassen (2008) was the value of collaboration, enabling ongoing information exchange and understanding of responsibilities. Collaboration impacted the suppliers' monitoring activities and consequently their performance (Green et al., 2011). Likewise, collaborative approaches with suppliers (OF20) facilitate monitoring supplier performance and consequently diffusion of sustainability practices.

Supporting and assisting suppliers in the assessment (OF21) can ensure the proper process to collect and assess data and understand potential sustainability practices for improvements (Jira & Toffel, 2013; Marshall et al., 2015). After gathering and assessing supplier data, the way that buying firms engage with their supplier for the performance assessment feedback (OF22) is an important aspect for enhancing the sustainability diffusion (e.g. Agan, Kuzey, Fatih, & Aç, 2016; Dou, Zhu, & Sarkis, 2014a). Indeed, the feedback might impact their behavior and improve performance.

Dou et al. (2014b) found that buying firms frequently implement a formal requirement of improved performance expectations. Porteous et al. (2015) suggested the establishment of procedure to analyze suppliers' non-compliances, including the analysis of trade-off (e.g. severity of the non-compliance and zero tolerance reaction – termination of the contract or reduce the volume of business), non-compliance treatment and reduction of the amount of non-compliance. The establishment of goals/targets was also presented by Schaltegger and Burritt (2014). Finally, Shaw et al. (2010) argued that buying firms need to establish supplier benchmarking to continuously improve. The authors found that this enables the implementation of best practice and better cross-firm involvement.

In summary, this section provides the influential organizational factors for implementing supplier sustainability performance assessment identified in the literature and enhancing the diffusion of sustainability. In general, the internal commitment (e.g. support from managers, training, capabilities), supplier commitment (capabilities, information sharing) as well as mutual responsibilities and benefits are considered.

8.3.3 Framework for diffusion of sustainability practices through supplier performance assessment

The proposed framework is designed by clustering organizational factors identified in the literature. The framework depicted in Fig. 8.1 captures

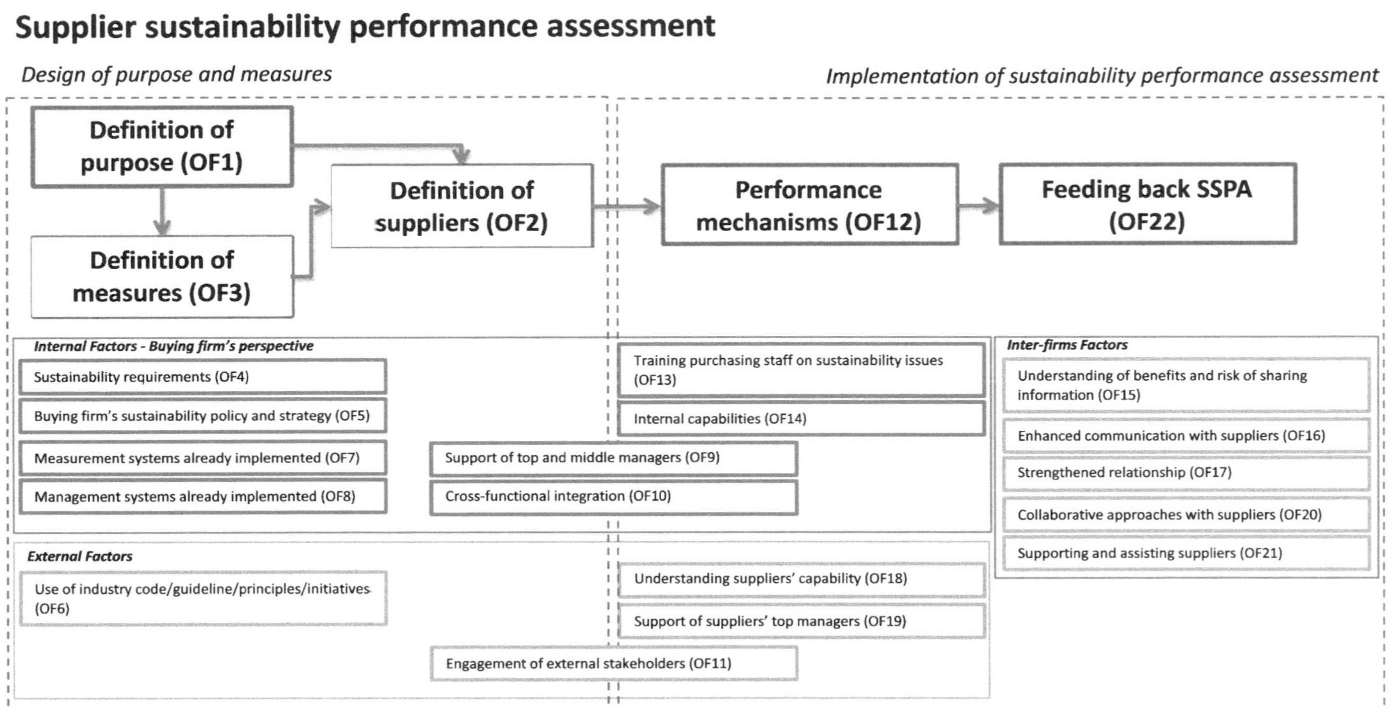

Figure 8.1 Framework for supplier sustainability performance assessment.

the two clusters of design and implementation of supplier sustainability performance assessment that align to the supply chain management process and emphasize organizational factors.

The design and implementation of the supplier sustainability performance assessment requires support of top and middle managers (OF9) and cross-functional integration (OF10). The engagement of stakeholders (OF11) can also be influential for both design and implementation of the performance assessment.

Particularly for the design of the supplier performance assessment, the purpose and measures can be influenced by the sustainability requirements (OF4), sustainability policy and strategy (OF5), the use of industry code (OF6) and the expertise gained by implementing internally measurement (OF7) and management (OF8) systems.

Supporting and assisting suppliers (OF21) during the implementation is also imperative. Furthermore, feedback (OF22) might impact supplier behavior, capability and performance over time. This might help both buying firms and suppliers to move forward with more commitment to sustainability and long-term relationship.

Even the proposed framework focuses on the role of buying manufacturing firms, understanding the influential factors related to the suppliers is also captured, such as understanding suppliers' capability (OF18) and support of suppliers' top managers (OF19). There are also common factors for both buyers and suppliers (inter-firm factors), namely, understanding of benefits and risk of sharing information (OF15), enhanced communication with suppliers (OF16), strengthened relationship (OF17) collaborative approaches with suppliers (OF20).

This framework extends therefore previous researches, which focus on one of these categories (Ashby et al., 2012; Gold et al., 2010b; Wong, Wong, & Boon-itt, 2015). Its contribution is therefore to depict patterns of the factors and points out the most critical variables influencing the supplier sustainability performance assessment.

8.4 Framework verification

Following the development of the framework, it was applied to four sustainability leading manufacturing firms from the beverage, cosmetic, and textile sector. The criteria for selecting firms for the case studies set out elements related to accessibility of sources of evidence (various level of employees, different organizational functions, and availability of documents, e.g. sustainability reports, supplier's sustainability audit reports, contracts, bill of materials, etc).

The case empirical design addressed the step-by-step approach proposed by Eisenhardt (1989) and Yin (2014), who are one of the best well known authors on case study methods (Easterby-Smith, Thorpe, & Jackson, 2012). The data analysis was carried out based on a triangulation of different sources of data (e.g. 30 interviews with directors or managers and an extensive array of documents).

This section therefore introduces the cases, presents the findings of the cases (sustainability practices assessed and the design and implementation of performance assessment) and then evaluates the results.

8.4.1 Firms studied description

8.4.1.1 Case 1 – "Beverage A"

The firm, called "Beverage A" here, is a bottler of soft drinks with more than ten manufacturing operations facilities in Europe. All sites are certified by ISO 14001 and the energy management standard (ISO 50001). The main materials supplied to "Beverage A" are ingredients (e.g. sugar beet, sugar cane, coffee and fruits), packaging (e.g. cans, glasses, cardboard) and gas (e.g. carbon dioxide). The firm has more than 12,000 suppliers, most of them located in Europe. Packaging is carbon-intensive for "Beverage A", representing 50% of carbon emission in the supply chain. Ingredients suppliers contribute for 18% of the carbon emission. Sugar is particularly critical for the business in terms of volume purchased and the water footprint (around 80% in the supply chain).

The supplier sustainability performance assessment is part of the "Beverage A"'s supplier relationship program. Two mechanisms are used in the performance assessment, an independent platform held by Ecovadis[1] scorecards and a supplier carbon program (questionnaire), called "carbon challenge". These two mechanisms annually assess the performance of key suppliers, which are covered based on the spending and the carbon risk. 140 key suppliers are currently engaged in performance assessment, representing more than 80% of the spending and 99% of the total carbon emissions in the supply chain. These suppliers are located in the first tier, including gas suppliers, packs suppliers, service providers, and ingredients processors.

[1] Ecovadis scorecard is an independent firm that assesses suppliers against 21 criteria divided into four areas: environment, social, ethics and sustainable procurement. These are designed based on international standards (e.g. ISO 26000, ISO 14001) and leading international sustainability standards (e.g. GRI, global compact principles e CERES principles). Source: http://www.ecovadis.com.

8.4.1.2 Case 2 – "Beverage B"

The second case study was carried out within a firm based in Italy specialising in making coffee products and espresso coffee machines. "Beverage B" was rewarded by the Ethisphere[2] as the world's most ethical company in the coffee sector. "Beverage B" adheres to the United Nations' global compact and ILO principles.

"Beverage B" uses more than 300,000 sacks of green coffee per year. The green coffee is exported from America (Brazil, Colombia, Costa Rica, El Salvador and Nicaragua), Africa (Ethiopia and Tanzania) and Asia (India and China). The industrial suppliers are responsible for providing primary and secondary packaging, inert gas, chemicals (e.g. inks and solvents), equipment and machineries, and coffee machines and accessories. The vast majority of industrial suppliers in terms of purchase spending consist of primary and secondary packaging manufacturers (most of them located in Italy, 100% in Europe), followed by coffee machine manufacturers (most of them in China).

"Beverage B" runs two supplier performance assessment programs for the green coffee and industrial suppliers. The green coffee assessment takes into consideration the sustainability practices established in the contract (requirements). The purpose of the assessment is to ensure compliance and improve the sustainability performance. In terms of the assessment of industrial suppliers, "Beverage B" has focused on assessing how industrial suppliers manage their environmental aspects.

8.4.1.3 Case 3 – "Cosmetics A"

"Cosmetics A" is a Brazilian firm specialized in producing cosmetics, perfumes, and personal care products, which are characterized by the use of ingredients from the Brazilian biodiversity, for example from the Amazon Forest. It is listed in the Down Jones Sustainability Index[3] and was rewarded by the Ethisphere as the world's most ethical company in the cosmetics sector. "Cosmetics A" is also a member of the Union for Ethical Bio Trade (UEBT)[4].

[2] Global initiative for defining and measuring corporate ethical standards and companies and promoting best practices in corporate ethics.

[3] Down Jones Sustainability Index (DJSI) releases annually the ranking of leading sustainability firms taking into consideration financial, environmental and social performance of the firms included in the Down Jones Global Index.

[4] UEBT is a non-profit association which promotes standards for sourcing biodiversity ingredients based on local development, biodiversity conservation, and business growth.

"Cosmetics A"'s supplier base is split into industrial suppliers (Pachaging, outsourced suppliers) and service suppliers (e.g. transport, marketing service, IT) and biodiversity ingredients suppliers. The industrial suppliers consist of biodiversity extract processors, chemicals, packaging, final product and organizational souvenirs (outsourced processes). Around 83% of the materials come from plants of the Brazilian biodiversity; the rest includes chemical and petrochemical oils. The firm uses 26 different biodiversity ingredients/raw-materials provided by cooperatives/associations of growers and agro-extractive families.

"Cosmetics A"'s supplier sustainability performance assessment involves three programs. A self-sustainability evaluation questionnaire to monitor the environmental and social performance of industrial suppliers. A total of 138 industrial suppliers (70% of the spending) are assessed through the self-sustainability evaluation program twice a year. The second program focuses on compliance with the sustainability requirements by using follow-up audits conducted by the quality department. All industrial suppliers are assessed. Finally, the third program undertakes biodiversity ingredients' cooperatives/associations and is run by the biodiversity supplier relationship and performance.

8.4.1.4 Case 4 – "Textile A"

"Textile A" is one of the world's leading sports performance and lifestyle brands with major business in footwear and apparel. It operates 54 businesses in over 30 countries in America, Europe and Africa, Asia and Oceania. The brand has employed regional and global sourcing strategies. For instance, garment manufacturers are sourced by regional offices, whereas fabrics are sourced by the global office. The scope of this case study is the European apparel office, responsible for designing and developing the products sold in Europe and Africa, and sourcing the garment suppliers.

The firm is also listed in the DJSI and a member of industry affiliations that promote sustainability across the textile sector, for instance the Sustainable Apparel Coalition (SAC)[5] and the Apparel and Footwear International Restricted Substances List Management Group (AFIRM Group)[6].

The regional apparel offices have a specific CSR team. The European apparel office has a seasonal cycle of production that includes the design process,

[5] The SAC is an industry-wide group working to reduce the environmental and social impacts of apparel and footwear products around the world.

[6] AFIRM group addresses the reduction of the use and impact of harmful substances in the apparel and footwear supply chain.

selection of fabric and sourcing garment suppliers. More than 300 different apparel products are produced every season. The supplier base consists of the garment manufacturer (outsourced process in the first tier) (around 25 suppliers, located especially in China and Indonesia) and the material suppliers (2nd tier).

"Textile A"'s supplier sustainability performance assessment focuses on garment suppliers (first tier) and more recently on fabric suppliers (second tier). For garment suppliers, two programs were identified. The first one aims at compliance with the policy of engagement and chemical standards. The second program is coordinated by the regional sourcing team and emphases efficient fabric usage in order to reduce fabric waste. The global CSR team will annually assess all fabric suppliers.

8.4.2 Sustainability practices diffused through the performance assessment

Compliance with environmental, human rights and health and safety regulations, working conditions, chemical management and investments in society were the sustainability practices commonly monitored and assessed by the firms studied across first and second tiers (Table 8.4). With the exception of the investment in society, those common practices monitored are compulsory and can be associated to the risk management.

Specific performance programs based on individual measures were also evidenced in "Beverage B" and "Textile A". In the first case, the program focused on environmental management through the existence of a function and procedures to control environmental aspects. "Textile A" introduced a marker efficiency program to garment suppliers in order to assess the efficiency of fabric usage.

Although not listed in the sustainability requirements used to select suppliers and listed in the contracts, some sustainability practices were monitored, such as the existence of environmental and social management system certifications (ISO 14001, SA8000) ("Beverage B", "Cosmetics A" and "Textile A"), investment in society and formal education of employee ("Beverage A" and "Cosmetics A"), use of recycled materials ("Cosmetics A"), carbon management ("Beverage A") and environmental management practices ("Beverage B"). It is important to note that these practices are not compulsory according to the environmental regulation. Most of these practices were considered in the sustainability policy and strategies of the firms.

By contrast, a few sustainability practices listed in the requirements (three of 38 sustainability practices) were not monitored, such as waste management and energy management ("Textile A") and the engagement with suppliers to

Table 8.4 Sustainability practices measured in the supplier performance assessment.

	Sustainability practices	Beverage A		Beverage B		Cosmetics A		Textile A	
		1st tier	2nd tier	1st tier	2nd tier	1st tier	2nd tier	1st tier	2nd tier
Environmental management	Compliance with regulation	–	–	yes	yes	yes	yes	yes	yes
	Environmental authorization	–	–	NE	NE	yes	yes	yes	yes
	Compliance with requirements	–	–	yes	yes	yes	yes	yes	yes
	EMS/Certification	–	–	NE	NE	yes	NE	NE	yes
	Identification of significant environmental aspects	–	–	yes	NE	yes	NE	NE	NE
	Procedures for environmental control	–	–	yes	NE	yes	NE	NE	NE
	Internal function for environmental management	–	–	yes	NE	NE	NE	NE	NE
	Water management[a]	–	–	NE	yes	yes	NE	NE	NE
	Efficiency of material usage	–	–	NE	NE	yes	yes	yes	NE
	Energy management[a]	yes	NE	NE	NE	yes	NE	NE	NE
	Use of recycled materials	–	–	NE	NE	yes	NE	NE	NE
	Wastewater treatment[b]	–	–	NE	yes	yes	yes	NE	NE
	Waste management[c]	–	–	NE	NE	yes	yes	NE	NE
	Carbon Management [d]	yes	NE	NE	NE	yes	yes	NE	NE
	Chemical management	–	–	yes	yes	yes	yes	yes	yes
	Fire safety[e]	–	–	NE	NE	yes	yes	yes	yes
	Good manufacturing practices	–	–	NE	yes	yes	yes	NE	NE
	Sustainable agriculture practices	–	–	NE	yes	NE	yes	NE	NE
	Soil management	–	–	NE	yes	NE	yes	NE	NE
	Average temperature rainfall	–	–	NE	yes	NE	NE	NE	NE
	Integrated agriculture	–	–	NE	yes	NE	NE	NE	NE

	Fertilization	–	–	NE	yes	NE	NE	NE	NE
	Use of pesticides and chemicals	–	–	NE	yes	NE	NE	NE	NE
	Natural reserves and protection of endemic species	–	–	NE	yes	NE	yes	NE	NE
	Biodiversity conservation	–	–	NE	yes	NE	yes	NE	NE
Human rights	Compliance with requirements	–	–	NE	NE	yes	yes	yes	yes
	Compliance with the labor regulation	–	–	yes	yes	yes	yes	yes	yes
	Certifications	–	–	yes	yes	yes	NE	yes	yes
	Human rights	–	–	NE	yes	yes	yes	yes	yes
	Investment in society (health, education, environment)	–	–	yes	yes	yes	yes	yes	yes
	Code of conduct	–	–	NE	NE	yes	NE	yes	NE
Working conditions	Compliance with regulation	–	–	yes	yes	yes	yes	yes	yes
	Working and labor conditions	–	–	yes	yes	yes	yes	yes	yes
	Wages	–	–	NE	yes	NE	yes	yes	NE
	Cost Analysis	–	–	NE	yes	yes	yes	yes	NE
	Workers competencies	–	–	NE	yes	NE	NE	NE	NE
	Investments in formal education and training for employees	–	–	NE	NE	yes	NE	NE	NE
	Inclusion of disable employees in the workplace	–	–	NE	NE	yes	NE	NE	NE
	Apprenticeship	–	–	NE	NE	yes	NE	NE	NE
	Labour relationship – families & cooperatives/association	–	–	NE	NE	NE	yes	NE	NE
	Public health facility	–	–	NE	yes	NE	NE	NE	NE
	Health and safety conditions	–	–	NE	yes	yes	yes	yes	yes

(Continued)

Table 8.4 (Continued)

	Sustainability practices	Beverage A		Beverage B		Cosmetics A		Textile A	
		1st tier	2nd tier	1st tier	2nd tier	1st tier	2nd tier	1st tier	2nd tier
	Rate of incidents (H&S)	–	–	NE	yes	yes	yes	yes	yes
	Risk of accidents or injures	–	–	NE	yes	NE	NE	NE	NE
	Machinery safety	–	–	NE	yes	NE	NE	yes	yes
	PPE	–	–	NE	NE	NE	NE	yes	NE
	Sanitary facilities, comfort and hygiene	–	–	NE	NE	NE	NE	yes	NE

NE – not employed.
[-] Not evidenced.
[a]source, volume and quality.
[b]technology employed, volume.
[c]volume, treatment & final disposal.
[d]carbon emission measurement and plans for reduction.
[e]equipment, evacuation routes, training, firefighting procedures.

comply with the regulation related to child labor and forced labor ("Cosmetics A"). This finding is related to the lack of significant risk to the buyers. For example, waste management is addressed in the contract with the supplier and is a compulsory practice, i.e. the garment suppliers need to demonstrate compliance with the regulation to the local environmental agency. Regarding the engagement of suppliers on human rights issues, which could induce the diffusion of human rights across different tiers, were any problem to occur in the 2nd tier or at the top of the supplier base, the first tier suppliers would receive most of the pressure and the buyer could defend itself based on the fact that the requirements and guidelines were communicated and the contract addressed this issues.

8.4.3 Design of supplier performance assessment

It was apparent that the organizational factors – support of top and middle managers (OF9), cross-functional integration (OF10) and use of industry code/guideline (OF6) impacted the inclusion of both environmental and social sustainability practices to be measured in the performance assessment. Managers from quality, procurement and sustainability departments were mostly involved.

In addition, the factor engagement of stakeholders (OF11) was supported by empirical evidence as being influential for designing both environmental and social measures. In the "Cosmetic A"'s case, the support of a consultancy firm was also critical in the design of the performance program (OF11).

Additionally, collaborative approaches with key suppliers representatives to design carbon performance program was evidenced in "Beverage A"'s case. The program aims at reducing the supplier carbon footprint. Integration between the internal environmental specialists, the procurement team and supplier representatives was carried out to understand better the carbon footprint challenges and, especially to develop the measures. Therefore, collaborative approaches with suppliers (OF20) to jointly discuss and design the carbon measures were critical in defining effective and consistent measures.

The supplier sustainability performance assessment in the firms studied mostly focused on compliance with the requirements and regulation, as well as, improvements in individual sustainability measures or set of sustainability measures, which included both environmental and social practices. The sustainability practices listed in the requirements (OF4) were commonly considered in the design of measures. Furthermore, the use of industry code/guidelines/principles/initiatives (OF6) was evidenced, such as WRI/WBCSD GHG protocol[7], ILO, UEBT, World Federation of the Sporting Goods Industry (WFSGI) and ILO. The sustainability policy and strategy (OF5) was also considered in the design of performance measures in "Beverage A", "Beverage B" and "Cosmetics A" (Table 8.5).

The SSPA adopted by the firms studied focused on suppliers in the first tier (e.g. packaging, outsourced suppliers and ingredients processors) and critical ingredients and material suppliers located in the second tier (e.g. coffee, biodiversity ingredients and fabrics). While "Beverage A and B"'s key industrial suppliers were assessed in terms of compliance through third party audits, "Cosmetics A" and "Textile A" employed second party audits for all industrial suppliers.

"Textile A" also addressed a specific program for the efficiency of fabric usage to all garment suppliers (first tier). The efficiency of material usage was also monitored by "Cosmetics A" through a self-evaluation program, which covered more measures. General sustainability performance programs were found for the critical ingredients and material suppliers in the second tier. Coffee and biodiversity ingredients' suppliers were monitored through second party audits by dedicated teams in "Beverage A" and "Cosmetics A". Not only compulsory measures in terms of compliance with regulation or requirements were adopted, but also measures associated with risk and feasibility of the suppliers' activities, such as biodiversity conservation, sustainable agriculture practices, wages calculation and production cost.

8.4.4 Implementation of supplier performance assessment

All organizational factors were confirmed in the implementation of supplier sustainability performance assessment. The performance assessment with focus on compliance were compulsory and the payment of the audit cost varied. For instance, "Beverage B" and "Textile A" paid costs, whereas suppliers were responsible for paying the "Beverage A" and "Cosmetics A"'s cases. The second party audits conducted by "Beverage B", "Cosmetic A" and "Textile A" across the second tier suppliers (i.e. coffee growers and biodiversity ingredient suppliers, respectively) were also compulsory, but the firms incurred the costs of assessment.

Particularly in "Beverage A"'s case, an independent database (Ecovadis) was adopted to annually monitor the key suppliers' compliance

[7] WRI/WBCSD greenhouse gas protocol was designed by the World Resource Institute in collaboration with World Business Council for Sustainable Development and provides guidelines for firms to measure, prepare and disclosure GHG emissions inventory.

Table 8.5 Organizational factors for designing supplier sustainability performance assessment.

Design and implementation of performance assessment	Beverage A	Beverage B	Cosmetics A	Textile A
Purpose (OF1)	Compliance with requirements & improvement in carbon programme	Compliance with requirements & improvement in sustainability performance assessment (SPA)	Compliance with requirements & improvement in SPA	Compliance with requirements & improvement in marker efficiency
Suppliers to be assessed (OF2) & Performance mechanisms (OF12)[a]	*Key IS suppliers—1st tier* (based on spending and carbon risk) - Compliance: Ecovadis & 3rd party audits - Carbon programme—lead by procurement team—self-evaluation	*All IS—1st tier* - Environmental programme—self-evaluation lead by quality team. *Key IS* (based on spending) - Compliance: 3rd party audit *All CG—2nd tier* Sustainability programme conducted by coffee procurement team	*All IS—1st tier* - Compliance—2nd party audits carried out by the quality team - SPA—Self-evaluation lead by procurement team *All BIS—2nd tier* - Compliance & SPP—2nd party audits carried out by BIS relationship team	****All Garment suppliers - 1st tier* - Compliance – 2nd party audits carried out by CSR team - Marker efficiency: self-evaluation lead by Sourcing team. *All Fabric suppliers - 2nd tier* - Compliance: 2nd party audits carried out by Global CSR team (in early stage)
Sustainability Requirements (OF4)	Yes	Yes	Yes	Yes

(Continued)

Table 8.5 (Continued)

Design and implementation of performance assessment	Beverage A	Beverage B	Cosmetics A	Textile A
Policy / strategies (OF5)	2020 Sustainability plan	Responsible supply chain certification	2050 Sustainability vision	Not evidenced
Use of industry code (OF6)	WRI/WBCSD GHG protocol (carbon challenge)	ILO	UEBT	ILO standards & WFSGI sustainability standards
Measurement systems (OF7)	ISO 14046 (Water footprint)	Not Evidenced	Not Evidenced	GRI
Management systems (OF8)	ISO 14001; ISO 50001	EMAS; ISO 14001	ISO 14001; B Corp accountability	Not Evidenced

IS – Industrial suppliers; CG – Coffee growers; BIS – Biodiversity ingredients suppliers; SSA – sustainability performance assessment.
[a]Organisational factor related to the implementation of performance assessment.

(i.e. requirements and regulation) (OF11). Essentially, "Beverage A" monitors the supplier progress through the platform, checks suppliers' resources and top manager support (OF18&19) and uses the database as benchmark tool. Low score suppliers in the Ecovadis platform could lose business if they do not improve their score in the final performance assessment. In this situation, they are responsible for defining and implementing a corrective action plan. For that, the procurement team, responsible for managing the assessment, was previously trained by the sustainability team to ensure they had the knowledge and skills needed to run assessment, monitoring and feeding back suppliers (OF13). Depending on the membership to which buyers and suppliers adhered, different buyers can access the full supplier assessment report, which might reduce the audit fatigue to suppliers.

Indeed, it was found in the "Textile A"'s case, based on data collected at a garment and fabric supplier level, a massive effort from the supplier to meet different requests and programs and audits preparation. Even where common subjects are covered, different measures are adopted. At fabric level (second tier), the manager reported a continuous engagement from the leading brand (core customers) to the firm to implement some apparel industry affiliations standards for sustainability performance assessment, such as SAC, Sedex, Global Social Compliance Program (GSCP), Outdoor Industry Association (OIA), and Better Cotton Initiative (BCI). This suggests a need of an integration between leading textile brands to design unified sustainability assessment tools.

The engagement of third party audit firms to assess suppliers was commonly employed to monitor the key industrial suppliers' compliance with the requirements in "Beverage A"'s case. In "Cosmetics A" and "Textile A" this mechanism was adopted when the department responsible for monitoring the industrial suppliers (second party audits) had high demand of audits. In terms of the second tier suppliers, it was apparent that there were only second party audits for green coffee, biodiversity ingredients and fabrics, covering a set of measures that go beyond compulsory sustainability practices. This slightly contrasts the research carried out by Grimm, Hofstetter, and Sarkis (2016), who found that second and third party audits had a similar effect on sub-suppliers' compliance with the requirements in the IT and retail sector. Perhaps, the amount of sub-suppliers justified the use and impact of both mechanisms.

All biodiversity ingredients cooperatives/associations in the "Cosmetics A"'s case are assessed annually through second party audits conducted by the biodiversity ingredients relationship and performance team. Non-compliances identified are treated by the development of action plans designed and implemented through collaborative approaches between the suppliers and

"Cosmetics A" (OF20). "Cosmetics A" has also developed a geographic information system (GIS) for real-time management of environmental and social data about the associations/cooperatives and to ensure traceability.

In the "Beverage B"'s case, the coffee supplier performance assessment is also periodically reviewed in order to cope with the specificities of different regions where the green coffee is supplied, such regulation and common practices adopted. This helps to ensure the adoption of realistic measures which are progressively more stringent. This is uncovered as a new organizational factor affecting the diffusion of both environmental and social sustainability practices across the green coffee suppliers (Review of supplier sustainability performance assessment – OF23).

Based on the findings from rich empirical data collected in the case studies, Fig. 8.2 presents an overview of the organizational factors influencing the diffusion of sustainability practices through the supplier performance assessment. It should be noted that in the design of purpose (OF1), suppliers to be assessed (OF2) and the definition of consistent measures (OF3) in all

Framework development*				Framework verification							
				Case studies							
Influential organisational factors for supplier sustainability performance assessment		Influential for diffusion of practices		Beverage A		Beverage B		Cosmetics A		Textile A	
		Env	Soc	Env	Soc	Env	Soc	Env	Soc	Env	Soc
Design	Definition of performance assessment purpose (OF1)	yes	yes	X	X	X	X	X	X	X	X
	Definition of suppliers to be assessed (OF2)	yes	yes	X	X	X	X	X	X	X	X
	Definition of consistent performance measures (OF3)	yes	yes	X	X	X	X	X	X	X	X
	Sustainability requirements (OF4)	yes	yes	X	X	X	X	X	X	X	X
	Buying firm's sustainability policy and strategy (OF5)	yes	yes	X	X	X	X	X	X	X	X
	Use of industry code/guideline/principles/initiatives (OF6)	yes	yes	X	X	X	X	X	X	X	X
	Measurement systems implemented (OF7)	yes	yes	X	NE	NE	NE	NE	NE	X	X
	Management systems implemented (OF8)	yes	yes	X	NE	X	NE	X	X	NE	NE
	Support of top and middle managers (OF9)	yes	NI	X	NE	X	X	X	X	X	X
	Cross-functional integration (OF10)	yes	NI	X	NE	X	NE	X	X	X	X
	Engagement of external stakeholders (OF11)	yes	NI	NE	NE	NE	NE	X	X	NE	NE
	Collaborative approaches with suppliers (OF20)	NI	NI	PI	NE	NE	NE	NE	NE	NE	NE
Implementation	Mechanisms to gather and assess suppliers performance (OF12)	yes	yes	X	X	X	X	X	X	X	X
	Engagement of external stakeholders (OF11)	yes	yes	X	X	X	X	X	X	X	X
	Training purchasing staff on sustainability issues (OF13)	yes	yes	X	X	X	X	X	X	NE	NE
	Internal capabilities (OF14)	yes	yes	X	X	X	X	X	X	X	X
	Support of top and middle managers (OF9)	yes	yes	X	X	X	X	X	X	X	X
	Cross-functional integration (OF10)	yes	yes	NE	NI	X	X	X	X	X	X
	Understanding of benefits and risk of sharing information (OF15)	yes	yes	X	X	X	X	X	X	X	X
	Enhanced communication with suppliers (OF16)	yes	yes	X	X	X	X	X	X	X	X
	Strengthened relationship (OF17)	yes	yes	X	X	X	X	X	X	X	X
	Understanding suppliers' capability (OF18)	yes	NI	X	X	X	X	X	X	X	X
	Support of suppliers' top managers (OF19)	yes	NI	X	X	NE	NE	X	X	X	X
	Collaborative approaches with suppliers (OF20)	yes	yes	X	X	X	X	X	X	X	X
	Supporting and assisting suppliers (OF21)	yes	yes	X	X	X	X	X	X	X	X
	Feeding back SSPA (OF22)	yes	yes	X	X	X	X	X	X	X	X
	Review of supplier sustanability performance assessment (OF23)	NI	NI	NE	NE	NFI	NFI	NE	NE	NE	NE

Env-Environmental practices, Soc-Social practices, NI-Not identified in the literature, X - Evidenced in the case, NE - Not evidenced in the case, PI-Previously identified and NFI-New factor identified.
*Based on Pimenta (2016)

Figure 8.2 Organizational factors for sustainability practices diffusion through supplier sustainability performance assessment.

cases were confirmed as being critical in the design of performance assessment, adding empirical evidences to the literature. These factors were influenced by OF4, OF5, OF6, OF7, OF8, OF9, OF10, OF11 and OF20. In general, the performance assessment focused on compliance with the requirements and improvements in individual environmental measures.

8.5 Discussion and conclusion

The results presented in Table 8.4 extends previous researches by presenting both environmental and social sustainability practices measured in multi-tiers (e.g. Ahi & Searcy, 2015; Chardine-Baumann & Botta-Genoulaz, 2014; Hassini et al., 2012; Schaltegger & Burritt, 2014; Varsei et al., 2014). Moreover, this research adds some social performance measures which have not been systematically covered in the current literature, such as the inclusion of disabled employees in the workplace. The majority of measures adopted by the firms studied to assess their supplier performance took into consideration environmental and social practices listed in the sustainability requirements. In general, the performance assessments are based on individual measures (e.g. carbon management and marker efficiency) and a set of measures covering different sustainability themes.

Most of the existing literature on diffusion of sustainability practices through performance assessment simply analyses the practices measured in general, without distinguishing where the practices were measured (i.e. tiers) and there is little empirical evidence. Moreover, even the supplier sustainability performance assessment enforces the implementation of sustainability practices listed in the requirements, and few papers empirically presented links between the sustainability requirements or policies and the measures employed (Hoejmose & Adrien-Kirby, 2012; Testa & Iraldo, 2010).

Several studies argued that supplier performance assessment is influential for buying firms to build long-term relationship with suppliers, especially due to the improvements in their performance (e.g. Beske & Seuring, 2014; Gold et al., 2010a). Likewise, this chapter adds rich empirical data to the literature on diffusion of sustainability practices through supplier performance assessment. For instance, the findings suggest that the supplier performance programs were helpful for transferring knowledge and improving not only certain deficient measures or compliance with the regulation and requirements, but also individual and general performance measures. This is important for suppliers to form an attitude toward the sustainability practices measured based on an understanding of the advantage (benefits to measure), complexity (challenges to measure), compatibility (measures in line with suppliers' capability) and observability (common measures in the supplier base) (Rogers, 2003).

Moreover, the feeding back process and collaborative approaches with suppliers to build improvement plans helped suppliers to improve not only their performance but also to build up capabilities to manage sustainability performance. Therefore, this chapter is in line with Agan et al. (2016), who stated that supplier performance assessment could be influential for suppliers to become more efficient and qualified by implementing sustainability practices.

In essence, this chapter establishes the conditions for enhancing the diffusion of environmental and social sustainability practices across the supplier base through sustainability performance assessment.

References

Agan, Y., Kuzey, C., Fatih, M., & Aç, A. (2016). The relationships between corporate social responsibility, environmental supplier development, and firm performance. *Journal of Cleaner Production, 112*, 1872–1881.

Ageron, B., Gunasekaran, A., & Spalanzani, A. (2012). Sustainable supply management: an empirical study. *International Journal of Production Economics, 140*, 168–182.

Ahi, P., & Searcy, C. (2015). An analysis of metrics used to measure performance in green and sustainable supply chains. *Journal of Cleaner Production, 86*, 360–377.

Akman, G. (2015). Evaluating suppliers to include green supplier development programs via fuzzy c-means and VIKOR methods. *Computers & Industrial Engineering, 86*, 69–82.

Ashby, A., Leat, M., & Hudson-Smith, M. (2012). Making connections: a review of supply chain management and sustainability literature. *Supply Chain Management An International Journal, 17*, 497–516.

Beamon, B. M. (1999). Designing the green supply chain. *Logistics Information Management, 12*, 332–342.

Beske, P., Land, A., & Seuring, S. (2014). Sustainable supply chain management practices and dynamic capabilities in the food industry: a critical analysis of the literature. *International Journal of Production Economics, 152*, 131–143.

Beske, P., & Seuring, S. (2014). Putting sustainability into supply chain management. *Supply Chain Management An International Journal, 19*, 322–331.

Beske-Janssen, P., Johnson, M. P., & Schaltegger, S. (2015). 20 years of performance measurement in sustainable supply chain management – what has been achieved? *Supply Chain Management An International Journal, 20*, 664–680.

Brandenburg, M., Govindan, K., Sarkis, J., & Seuring, S. (2014). Quantitative models for sustainable supply chain management: developments and directions. *European Journal of Operational Research, 233*, 299–312.

Chardine-Baumann, E., & Botta-Genoulaz, V. (2014). A framework for sustainable performance assessment of supply chain management practices. *Computers & Industrial Engineering, 76*, 138–147.

de Bakker, F., & Nijhof, A. (2002). Responsible chain management: a capability assessment framework. *Business Strategy and the Environment, 11*, 63–75.

Dou, Y., Zhu, Q., & Sarkis, J. (2014a). Evaluating green supplier development programs with a grey-analytical network process-based methodology. *European Journal of Operational Research, 233*, 420–431.

Dou, Y., Zhu, Q., & Sarkis, J. (2014b). Integrating strategic carbon management into formal evaluation of environmental supplier development programs. *Business Strategy and the Environment, 24*, 873–891.

Dües, C. M., Tan, K. H., & Lim, M. (2013). Green as the new lean: how to use lean practices as a catalyst to greening your supply chain. *Journal of Cleaner Production, 40*, 93–100.

Easterby-Smith, M., Thorpe, R., & Jackson, P. (2012). *Management research* (4th ed). London: SAGE.

Eisenhardt, K. M. (1989). Building theories from case study research. *Academy of Management Review, 14*, 532–550.

Gallear, D., Ghobadian, A., & Chen, W. (2012). Corporate responsibility, supply chain partnership and performance: an empirical examination. *International Journal of Production Economics, 140*, 83–91.

Gimenez, C., & Tachizawa, E. M. (2012). Extending sustainability to suppliers: a systematic literature review. *Supply Chain Management: An International Journal, 17/5*, 531–543. Available from https://doi.org/10.1108/13598541211258591.

Gold, S., Seuring, S., & Beske, P. (2010a). The constructs of sustainable supply chain management – a content analysis based on published case studies. *Progress in Industrial Ecology – An International Journal*, 7, 114–137.

Gold, S., Seuring, S., & Beske, P. (2010b). Sustainable Supply Chain Management and Inter-Organizational Resources: A Literature Review. *Corporate Social Responsibility and Environmental Management, 17*, 230–245.

Green, K. W., Jr, Zelbst, P. J., Bhadauria, V. S., & Meacham, J. (2011). Do environmental collaboration and monitoring enhance organizational performance? *Industrial Management & Data Systems, 112*, 186–205.

Grimm, J. H., Hofstetter, J. S., & Sarkis, J. (2016). Exploring sub-suppliers' compliance with corporate sustainability standards. *Journal of Cleaner Production, 112*, 1971–1984.

Hagellar, G. J. L. F., Vorst, J. G. A. J., & Van Der. (2002). Environmental supply chain management: using life cycle assessment to structure supply chains. *International Food and Agribusiness Management Review, 4*, 399–412.

Handfield, R., Sroufe, R., & Walton, S. (2005). Integrating environmental management and supply chain strategies. *Business Strategy and the Environment, 14*, 1–19.

Hassini, E., Surti, C., & Searcy, C. (2012). A literature review and a case study of sustainable supply chains with a focus on metrics. *International Journal of Production Economics, 140*, 69–82.

Hervani, A. A., Helms, M. M., & Sarkis, J. (2005). Performance measurement for green supply chain management. *Benchmarking: An International Journal, 12*, 330–353.

Hoejmose, S. U., & Adrien-Kirby, A. J. (2012). Socially and environmentally responsible procurement: a literature review and future research agenda of a managerial issue in the 21st century. *Journal of Purchasing and Supply Management, 18*, 232–242.

Hollos, D., Blome, C., & Foerstl, K. (2012). Does sustainable supplier co-operation affect performance? Examining implications for the triple bottom line. *International Journal of Production Research, 50*, 2968–2986.

Hutchins, M. J., & Sutherland, J. W. (2008). An exploration of measures of social sustainability and their application to supply chain decisions. *Journal of Cleaner Production, 16*, 1688–1698.

Jira, C. F., & Toffel, M. W. (2013). Engaging Supply Chains in Climate Change. *Manufacturing & Service Operations Management, 15*, 559–577.

Klassen, R. D., & Vereecke, A. (2012). Social issues in supply chains: capabilities link responsibility, risk (opportunity), and performance. *International Journal of Production Economics, 140*, 103–115.

Kogg, B., & Mont, O. (2012). Environmental and social responsibility in supply chains: the practise of choice and inter-organisational management. *Ecological Economics: the Journal of the International Society for Ecological Economics, 83*, 154–163.

Lee, K., & Kim, J. (2011). Integrating Suppliers into Green Product Innovation Development: an Empirical Case Study in the Semiconductor Industry. *Business Strategy and the Environment, 20*, 527–538. Available from https://doi.org/10.1002/bse.714.

Lee, K.-H., & Cheong, I.-M. (2011). Measuring a carbon footprint and environmental practice: the case of Hyundai Motors Co. (HMC). *Industrial Management & Data Systems, 111*, 961–978.

Lee, S.-Y., & Klassen, R. D. (2008). Drivers and enablers that foster environmental management capabilities in small- and medium-sized suppliers in supply chains. *Production and Operations Management, 17*, 573–586.

Lippmann, S. (1999). Supply chain environmental management: elements for success ath. *Corporate Environmental Strategy, 6*, 175–182.

Marshall, D., McCarthy, L., Heavey, C., & McGrath, P. (2015). Environmental and social supply chain management sustainability practices: construct development and measurement. *Production Planning & Control, 26*, 673–690.

Mcintyre, K., Smith, H., Henham, A., & Pretlove, J. (1998). Case study environmental performance indicators for integrated supply chains: the case of Xerox Ltd. *Supply Chain Management, 3*, 149–156.

Meacham, J., Toms, L., Green, K. W., Jr, & Bhadauria, V. S. (2013). Impact of information sharing and green information systems. *Management Research Review, 36*, 478–497.

Miles, M. B., Huberman, A. M., & Saldana, J. (2014). *Qualitative data analysis: a methods sourcebook, qualitative data analysis*. London: SAGE Publications.

Morali, O., & Searcy, C. (2013). A review of sustainable supply chain management practices in Canada. *Journal of Business Ethics: JBE, 117*, 635–658.

Nagel, M. H. (2003). Managing the environmental performance of production facilities in the electronics industry: more than application of the concept of cleaner production. *Journal of Cleaner Production, 11*, 11–26.

Nakano, K., & Hirao, M. (2011). Collaborative activity with business partners for improvement of product environmental performance using LCA. *Journal of Cleaner Production, 19*, 1189–1197.

Nawrocka, D., Brorson, T., & Lindhqvist, T. (2009). ISO 14001 in environmental supply chain practices. *Journal of Cleaner Production, 17*, 1435–1443.

Okongwu, U., Morimoto, R., & Lauras, M. (2013). The maturity of supply chain sustainability disclosure from a continuous improvement perspective. *International Journal of Productivity and Performance Management, 62*, 827–855.

Pagell, M., & Wu, Z. (2009). Building a more complete theory of sustainable supply chain management using case studies of 10 exemplars. *Journal of Supply Chain Management, 45*, 37–56.

Paulraj, A. (2011). Understanding the relationships between internal resources and capabilities, sustainable supply management and organizational sustainability. *Journal of Supply Chain Management, 47*, 19–37.

Pimenta, H. C. D. (2016). *Diffusion of environmental and social sustainability practices across the supplier base*. Crafiled University, PhD Thesis.

Polgreen, K. (2002). Social and environmental supply-chain management: an overview. *Small Enterprise Development, 13*, 25–33.

Porteous, A. H., Rammohan, S. V., & Lee, H. L. (2015). Carrots or sticks? Improving social and environmental compliance at suppliers through incentives and penalties. *Production and Operations Management, 24*, 1402–1413.

Rock, M. T., Angel, D. P., & Lim, P. L. (2006). Impact of firm-based environmental standards on subsidiaries and their suppliers – evidence from Motorola-Penang. *Journal of Industrial Ecology, 10*, 257–278.

Rogers, E. M. (2003). *Diffusion of innovations* (5th ed.). New York: The Free Press.

Rosen, C. M., Bercovitz, J., & Beckman, S. (2001). Environmental supply-chain management in the perspective. *Journal of Industrial Ecology*, *4*, 83–103.

Salvá, M., Jones, S., Marshall, R. J., & Bishop, C. F. H. (2013). An audit tool for environmental measurement in the UK food sector. *International Journal of Food Science & Technology*, *48*, 1509–1518.

Sancha, C., Longoni, A., & Giménez, C. (2015). Sustainable supplier development practices: Drivers and enablers in a global context. *Journal of Purchasing and Supply Management*, *21*, 95–102.

Schaltegger, S., & Burritt, R. (2014). Measuring and managing sustainability performance of supply chains. Review and sustainability supply chain management framework. *Supply Chain Management An International Journal*, *19*, 232–241.

Shaw, S., Grant, D. B., & Mangan, J. (2010). Developing environmental supply chain performance measures. *Benchmarking: An International Journal*, *17*, 320–339.

Simpson, D. F., & Power, D. J. (2005). Use the supply relationship to develop lean and green suppliers. *Supply Chain Management An International Journal*, *10*, 60–68.

Soundararajan, V., & Brammer, S. (2018). Developing country sub-tier responses to social sustainability requirements of intermediaries: exploring the influence of framing on fairness perception and reciprocity. *Journal of Operations Management*, *59-59*, 42–58.

Tachizawa, E. M., & Wong, C. Y. (2014). Towards a theory of multi-tier sustainable supply chains: a systematic literature review. *Supply Chain Management An International Journal*, *19*, 643–663.

Testa, F., & Iraldo, F. (2010). Shadows and lights of GSCM (Green Supply Chain Management): determinants and effects of these practices based on a multi-national study. *Journal of Cleaner Production*, *18*, 953–962.

Vachon, S., & Klassen, R. D. (2006). Extending green practices across the supply chain: The impact of upstream and downstream integration. *International Journal of Operations & Production Management*, *26*, 795–821.

Vachon, S., & Klassen, R. D. (2008). Environmental management and manufacturing performance: The role of collaboration in the supply chain. *International Journal of Production Economics*, *111*, 299–315.

Varsei, M., Soosay, C., Fahimnia, B., & Sarkis, J. (2014). Framing sustainability performance of supply chains with multidimensional indicators. *Supply Chain Management An International Journal*, *19*, 242–257.

Wang, Z., & Sarkis, J. (2013). Investigating the relationship of sustainable supply chain management with corporate financial performance. *International Journal of Productivity and Performance Management*, *62*, 871–888.

Wittstruck, D., & Teuteberg, F. (2012). Understanding the Success Factors of Sustainable Supply Chain Management: Empirical Evidence from the Electrics and Electronics Industry. *Corporate Social Responsibility and Environmental Management*, *19*, 141–158.

Wong, C. Y., Wong, C. W. Y., & Boon-itt, S. (2015). Integrating environmental management into supply chains. *International Journal of Physical Distribution & Logistics Management*, *45*, 43–68.

Yin, R. K. (2014). *Case study research: design and methods*. London: SAGE.

CHAPTER 9

Dry cutting: a sustainable machining technology

Senol Bayraktar
Department of Mechanical Engineering, Faculty of Engineering and Architecture, Recep Tayyip Erdogan University, Rize, Turkey

9.1 Introduction

Human history is generally regarded as the development of the means utilized by sorts to survive. The capability to make superior weapons using animal bones gave them superiority among sorts in the nutrient chain. The demand for weapons, and so on, led human to the innovation of metals. Some tools and equipment were needed for forming metals to be usable. All these equipment were required for urgent technological advances in cutting tools. The development of new cutting tools and production techniques began at this point.

Traditional metal cutting operations are those in which the material is removed by a cutting tool whose cutting edge remains in contact with the workpiece material. It requires higher hardness of the tool material compared to the workpiece material. The essential force to move the tool on the surface of the workpiece/part can be decreased by appropriately designing the tool geometry. Cutting force (CF), relative speed between tool and part, tool angles, specific fluids for lubrication, and cooling are the basic factors affecting the dimensional stability and surface integrity of the machined part (Ganesh, Sidpara, & Sankha, 2017). Due to rapidly increasing environmental concerns, the reduction in usage of cutting fluids is being recommended. Moreover, the recycling of chip during cutting becomes difficult with the use of coolant. For the recycling of chip, the chemical residue on the chip must be removed. This leads to additional time consumption and increased cost.

The prominent alternative techniques are dry and near-dry machining. Dry cutting or machining is done without the use of coolant and found to be environmentally friendly. Since there is no waste on sawdust in dry cutting, chips are easier and cheaper to recycle (Dixit, Sarma, & Davim,

Sustainable Manufacturing
DOI: https://doi.org/10.1016/B978-0-12-818115-7.00004-3

2012). The advantages of dry cutting are the absence of any danger that may affect the operator's health in terms of not damaging the skin and creating an allergic condition and the lack of expenses such as storage, maintenance, cost, and labor required for the coolant (Landgraf, 2004). The costs associated with the usage, storage, and supply of cutting fluids range from 7% to 17% of the total manufacturing cost (Dixit et al., 2012; Kissler, 2000; Weinert, Inasaki, Sutherland, & Wakabayashi, 2004). This cost factor can be significantly decreased by dry cutting and near-dry machining type technologies. In addition to increasing performance in the manufacturing, such a technology change contributes to the protection of manpower and the atmosphere (Gunter & Sutherland, 1999; Hands, Sheehan, Wong, & Lick, 1996; Rossmoore, 1995; Weinert et al., 2004). In addition, reducing the exposure of operators to coolants in enterprises increases work efficiency. Enterprises can utilize economically appropriate manufacturing processes for advertisement, which contributes to increased competition in the market (Fig. 9.1) (Aronson, 1995; Sutherland, Kulur, & King, 2000; Weinert et al., 2004).

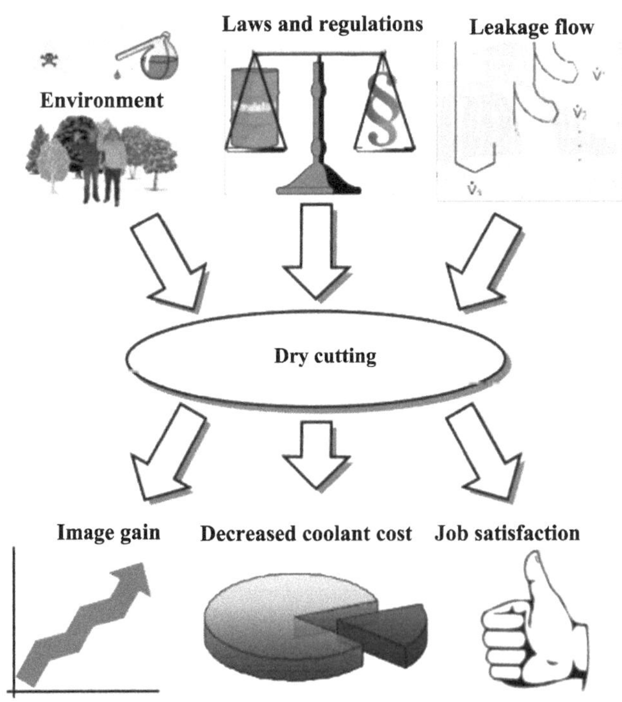

Figure 9.1 Advantages of dry cutting (Weinert et al., 2004). *With kind permission of Elsevier.*

9.1.1 Mechanics of machining

The orthogonal cutting method, which is a two-dimensional cutting method, makes the calculations easier to explain the cutting theory in a simple way rather than three-dimensional cutting (Fig. 9.2A and B).

The tool cutting edge is vertical to the cutting direction in Fig. 9.2. The volume of the rectangular area is specified by the uncut chip thickness (t_1) and the cutting width (b) removed from the workpiece. Chips are consisted on the cutting edge in the normal plane during cutting. Thus no change in chip formation along the cutting edge is observed.

The cutting model is put forward by the shear plane angle in orthogonal cutting. The shear plane angle is specified by t_1, α, and t_2,

$$\tan = \frac{(t_1/t_2)\cos a}{1-(t_1/t_2)\sin a} = \frac{r_c\cos a}{1-r_c\sin a}. \tag{9.1}$$

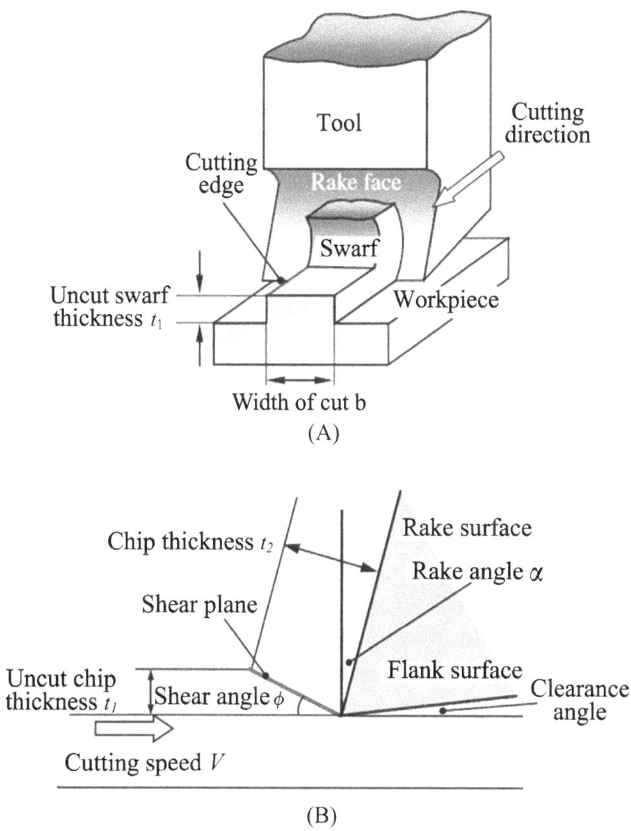

Figure 9.2 Illustration of orthogonal cutting (Yoshimi & Matsumura, 2017). *With kind permission of Springer.*

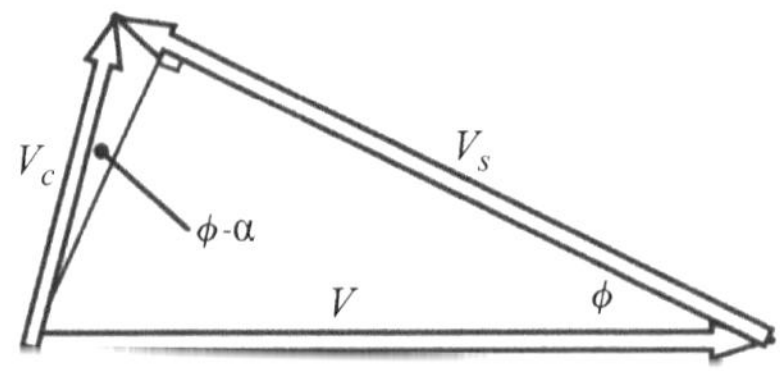

Figure 9.3 Cutting, chip and shear velocity in orthogonal cutting model.

In Eq. (9.1), r_c is the cutting rate and is calculated as in Eq. (9.2).

$$r_c = \frac{t_1}{t_2}. \tag{9.2}$$

The shear stress γ can be calculated as in Eq. (9.3).

$$\gamma = \frac{\cos\alpha}{\sin\phi\cos(\phi - \alpha)}. \tag{9.3}$$

According to Eq. (9.3), it can be said that large shear stresses occur with small shear plane angle and small rake or negative rake angle, the shear and chip speed along with the vectors of the CS should form a closed diagram attribute on the basis of plasticity as shown in Fig. 9.3.

The amount of cutting and chip speed vectors is equal to the shear speed. Accordingly, chip speed V_c and shear speed V_s values are calculated according to Eqs. (9.4) and (9.5), respectively (Yoshimi & Matsumura, 2017).

$$V_c = \frac{\sin\phi}{\cos(\phi - \alpha)} V = r_c V \tag{9.4}$$

$$V_s = V_c \sin(\phi - \alpha) + V\cos\phi = \frac{\cos\alpha}{\cos(\phi - \alpha)} V = \gamma \sin\phi V. \tag{9.5}$$

9.1.2 Machining parameters

Surface quality is an important machinability indicator for the machined parts. Surface roughness (SR) is one of the important surface quality parameters and of major concern for the machined parts used in different areas such as sealing, hydrodynamic, electrical and heat conduction as well as wear, friction, and lubrication (Pathak & Jain, 2017). Parameters such as cutting speed—CS (V_c), feed rate—FR (f), and depth of cut—DOC (a_p) directly affect the surface integrity. If these parameters are not well determined, the workpiece surface quality is reduced and the notch effect causes crack formation, fatigue, and corrosion resistance. Therefore determination of ideal or optimum parameters is important for surface quality. It is found that during cutting, the CFs are reduced by raising the CS. The increase in CS increases the temperature in the deformation zones and reduces the shear forces by decreasing the yield strength of the material due to thermal softening in the workpiece. However, CFs increase due to the increase in stress and strain rate at higher CS (Gopal & Prakash, 2018; Subramanian, Sakthivel, Sooryaprakash, & Sudhakaran, 2013). As the FR rises, more power is needed for the cutting process, which increases the CFs per unit time. Increasing the DOC raises the contact surface area between the cutting tool and the workpiece and increases the CFs, as more volume of chip is required per unit time (Gopal & Prakash, 2018).

9.1.3 Tool material

Nowadays, due to their superior properties, high carbon steel, ceramic, and diamonds are widely used for cutting tool manufacturing. Investigations have been conducted to know the effectiveness of various tool materials for machining of a wide range of engineering materials (Davim, 2008; Davis, 2005). In the early 20th century, Taylor demonstrated that high-speed steel (HSS) cutting tools can be used instead of carbon steel. In 1926 Krupp developed the sintered carbide material Widia, which is used in cutting and shaping. Thus CSs were increased by 3–6 times of what used to be with HSS (Youssef, 2016). Ceramics have been introduced as tool materials that can use CSs of 6–8 times higher than HSS tools after the 1950s. The most recent studies on cutting tool materials have been determined on coated carbide and cubic boron nitride (CBN). To achieve very high CS during machining, polycrystalline diamond tools are preferred for finishing operations in turning (Youssef, 2016; Youssef & El-Hofy, 2012). Under certain cutting conditions, any cutting tool material performs well, but change in conditions may affect the performance. For instance, ceramic tool material performs well at high speeds, while at low CS, it can break due to shock and overload. In general terms, it is expected that the higher the hardness of the tool material, the better the cutting performance is. However, tungsten carbide (WC) successfully used in hard metal carbides has a lower hardness value than titanium carbide (TiC). This shows that there may be some exceptions for

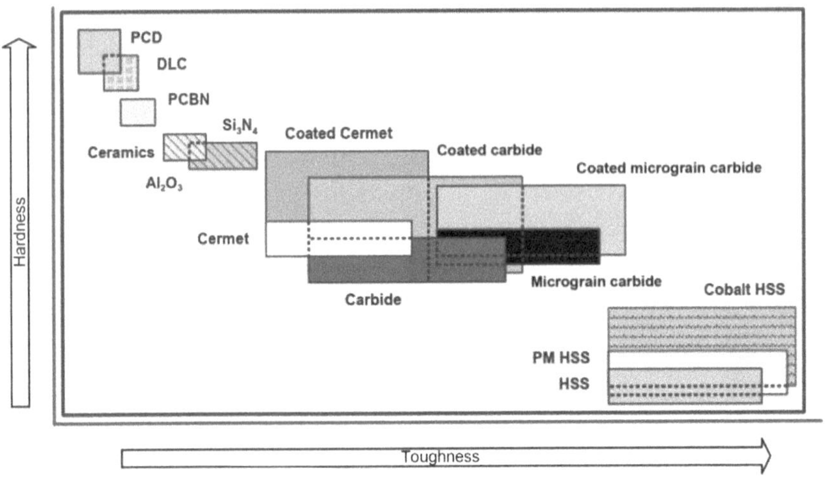

Figure 9.4 Hardness and toughness diagram for different tool materials (Astakhov & Davim, 2008). *With kind permission of Springer.*

the cutting tool materials. Cost of TiC is less than WC and more abundant in nature. However, WC is the main material of hard metal industry. Because the binder phase, in combination with cobalt, generally provides the best cutting properties. Different cutting tool materials are used to cut metals or alloys. As shown in Fig. 9.4, a good cutting tool needs to provide many features, both physically and metallurgically. These are high yield strength at cutting temperatures (CTs), high fracture toughness, and fatigue and wear resistance, high thermal capacity and conductivity, low solubility in workpiece, high thermal shock and oxidation resistance (Davim, 2008; Mills & Redford, 1983). The features of different tool materials with the hardness–toughness relationship are given in Fig. 9.4 and Table 9.1, respectively.

Carbon steels are low cost cutting tool materials used for low CSs. It contains 0.6%–1.5% carbon and a small amount (<0.5%) of Mn and Si. Metal elements such as Cr and V can also be added to modify the hardness (65 HRC) and grain size. They are not preferred at high CSs under dry cutting conditions as they lose their hardness properties at about 250° C. HSSs contain alloying elements such as molybdenum, chromium, and tungsten to improve hardness (67 HRC), toughness, and wear resistance. It is recommended to use coolant as they lose their hardness properties at approximately 650°C during dry cutting. In addition, worn cutting edges

Table 9.1 Properties of different tool materials.

Hard materials	Hardness (HV)	Energy formation (kcal/g atom)	Solubility in iron (%. 1250 r)	Thermal conductivity (W/m k)	Thermal[a] expansion ($\times 10^{-6}$/k)	Tool material
Diamond	>9000	–	Highly soluble	2100	3.1	Sintered diamond
CBN	>4500	–	–	1300	4.7	Sintered CBN
Si_3N_4	1600	–	–	100	3.4	Ceramics
Al_2O_3	2100	−100	0	29	7.8	Ceramics-cemented carbide
TiC	3200	−35	<0.5	21	7.4	Cermet-coated carbide
TiN	2500	−50	–	29	9.4	Cermet-coated carbide
TaC	1800	−40	0.5	21	6.3	Cemented carbide
WC	2100	−10	7	121	5.2	Cemented carbide

[a]1 W/m K = 2.39×10^{-3} cal/cm s r.

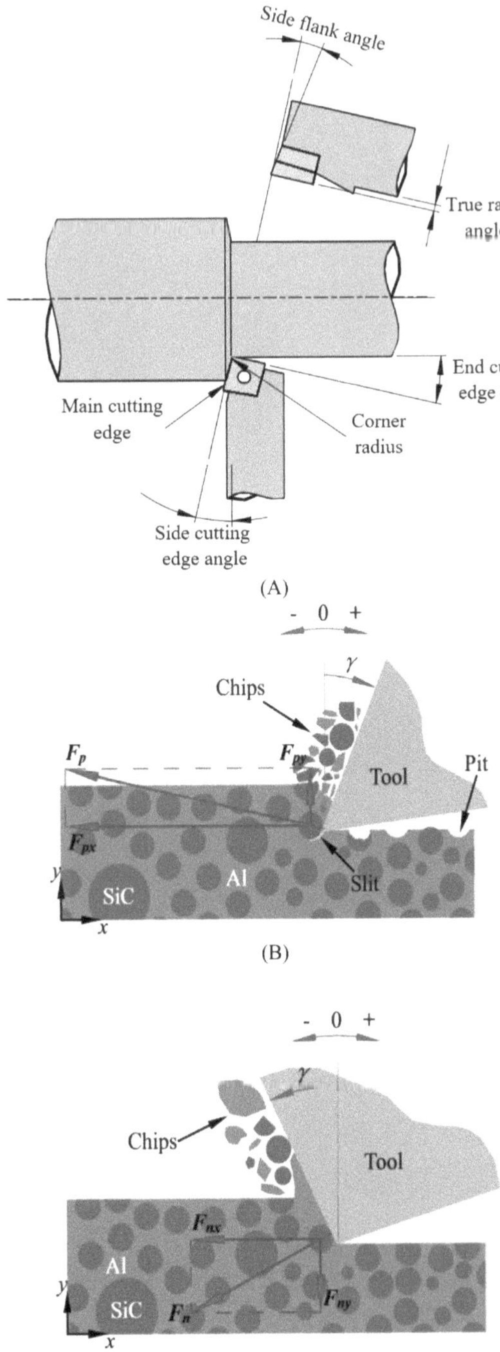

(C)

Figure 9.5 Tool geometries in cutting, (A) end cutting edge angle and (b) positive back rake (Liu & Zong, 2019), and (C) Negative back rake angle (Liu & Zong, 2019). *With kind permission of Springer Nature.*

can be reused by sharpening. Cast iron cobalt alloys have been developed to fill the gap between HSS and carbides. It is used at CSs up to 20% faster than HSS. Although HSS's are heat-treated to increase their hardness, cast cobalt alloys are hard and cannot be softened and hardened by heat treatment (Davis, 1989). Cemented carbide tools are produced by powder metallurgy technique and generally consist of W, Ta, and TiC using cobalt binder. Hardness values (90 HRC) are very high but can be used at high CSs. They also maintain their hardness properties up to 1000°C. Ceramic materials are generally Al_2O_3 and Si_3N_4. They are produced by sintering the compressed ceramic powder material at high temperatures. Ceramics exhibit properties such as chemically inert and corrosion resistance. In addition to being stable up to 1800°C during dry cutting, they maintain their hardness to 93 HRC and have low friction and thermal conductivity at the tool–chip interface. CBN tools are the second hardest (higher than 95 HRC) after diamond. In addition to CSs of 600–800 m/min, they are also used as an abrasive material in the grinding process due to their high abrasion resistance. Diamond is the hardest material known as cutting tool for a long time. Diamond tools have superior features such as high thermal conductivity, low coefficient of friction, and nonstick to many materials. However, single-crystal diamond applications are limited due to factors such as brittleness and high cost. Therefore they are generally preferred in finishing operations to improve surface quality in applications (Sheikh-Ahmad, 2009).

9.1.4 Tool geometry

Tool geometry has a significant impact on machinability. It is one of the most important variables that affect the machining performance of materials. With the increase of negative rake angle, the tool–chip contact surface area increases and causes the CFs to increase with the effect of high friction forces at the tool–chip interface (Nalbant, Altın, & Gökkaya, 2007).

The end cutting edge angle is the angle between the machined surface and the leading edge of the tool as shown in Figs. 9.5 and 9.6. The side clearance (flank) angle is the angle required to cut the workpiece (Fig. 9.5A). Without these angles, the cutting tool negatively affects the surface integrity by rubbing the machined surface. The back rake is the angle between the center of the workpiece and the horizontal line through the upper surface of the cutting edge. It can be positive, negative

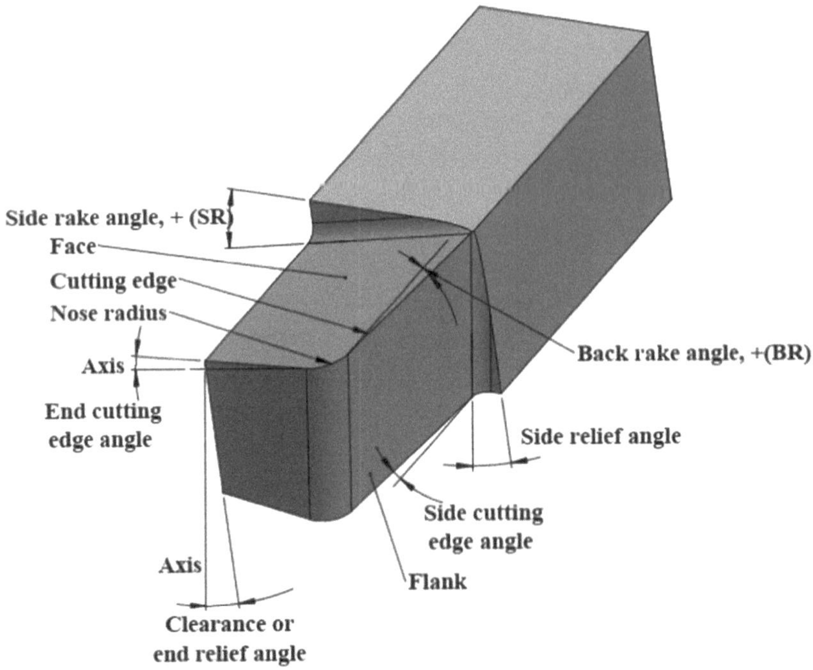

Figure 9.6 Side rake and other angles on the cutting tool.

(Fig. 9.5B and C), and neutral (0°) (Hoffman, Hopewell, Janes, & Sharp, 2012).

The side rake is the angle perpendicular to the top surface of the cutting tool. It can be positive, negative, or neutral. Positive chip angles make the chip thinner and require less force for cutting. The tool radius is the radius at the end of the tool where the leading edge meets the end cutting edge. Radius directly affects the integrity of the machined surface. Generally, the large radius value improves the surface quality, while the extremely large radius values cause vibration as it increases the contact surface area between the cutting tool and the workpiece. This situation reduces the surface quality. On the contrary, when the radius value is extremely small, the tool starts to wear down by weakening the tip (Hoffman et al., 2012). The most important elements of drill geometry are lips, chisel edge, web and helix, point, chisel edge, and clearance angles. The helix angle determines the chip angle at the cutting edge of the drill bit. The helix angle is specified by considering the strength of the cutting edge and that the chip can be efficiently removed along the channel. The point angle is the angle between the two side surfaces forming the cutting edge of the drill. The chisel edge angle is the angle between the chisel edge and the cutting lip (Selvam & Sujatha, 1995). It is stated that the cutting movement on the chisel edge is similar to the orthogonal cutting method when high rake angle and low CSs are used resulting in discontinuous chip formation (Audy, 2008). The clearance angle is the angle between the two surfaces tangential to the flank and revolution surface at any point on the cutting edge (Selvam & Sujatha, 1995). The two-edged drill is mainly composed of chisel and lips. The lips form the main cutting edge of the drill for cutting the material and deform the material and determine the cutting direction. Thus the axial component of the CF in drilling process depends on the geometry of the chisel. The greater the thrust force in drilling, the harder it is to obtain good hole surface quality. In order to improve the cutting efficiency by decreasing the thrust force, chisel thinning is applied (Fig. 9.7). Point thinning is suitable for general use. With the improvement of drilling technology and the development of materials that are difficult to cut (DTC), X and XR type methods have been used. X type is suitable for drilling of

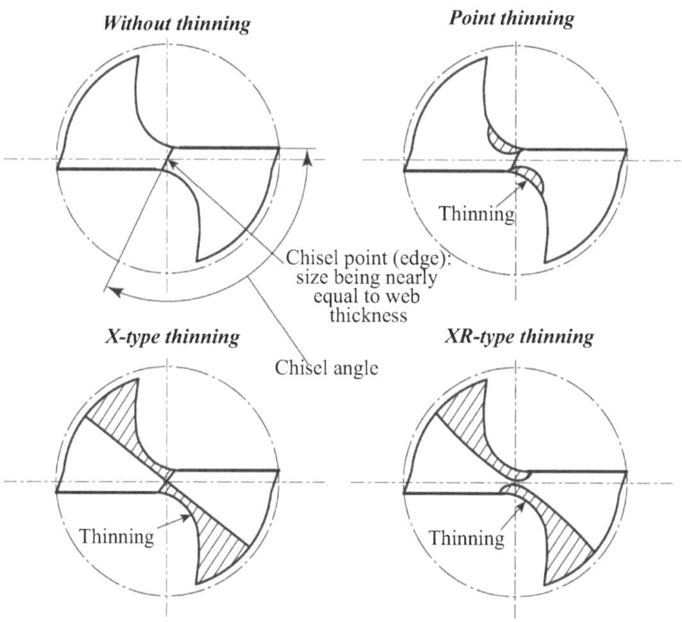

Figure 9.7 Drill thinning techniques (Yoshimi & Matsumura, 2017). *With kind permission of Springer.*

DTC materials when the chisel is thick and for gun drilling. It also makes it easier for chisel to sink into the piece. The XR type is more lip resistant than the X type but has less ability to sink the workpiece material (Yoshimi & Matsumura, 2017).

The drill point angle has an important effect on factors such as thrust force, surface integrity, and burr formation. The increase in point angle in drilling of steels results in a smaller cutting edge, reducing undeformed chip thickness, width, and thrust force. For the dimensional stability and surface quality of the hole, a combination of high point angle and CS is recommended (Demir, 2018). High point angle drills reduce burr height, especially with increasing FR (Çelik, 2014).

9.1.5 Tool coating

To achieve high productivity at high CSs during machining is based on automatics and better tool materials that can keep the machining operational until desired form/shape is made or required amount of material is removed. Tool modifications such as coatings also play a major role in the success of the aforementioned (Gupta & Laubscher, 2017). With time, there have been tremendous advancements in tool coatings. In particular, multilayer composite, nanostructured, gradient and multicomponent elemental compositions have been extensively used and improved day by day (Faga et al., 2007; Fox-Rabinovich et al., 2009; PalDey & Deevi, 2003; Vereschaka, Grigoriev, Sitnikov, & Batako, 2017; Yamamoto, Sato, Takahara, & Hanaguri, 2003). Hard ceramic coatings with a thickness of several micrometers are made by depositing on cemented carbides to develop crater wear resistance, especially due to the effect of thermal and abrasive loads. Coatings are made by two different methods. In the first type, the coating is formed in the range of approximately 900°C–1050°C by chemical vapor deposition (CVD). Single and multilayer TiN, TiCN, TiC, and Al_2O_3 coatings are made with this technique. The second coating method, physical vapor deposition (PVD), is a later improved method to benefit from low temperatures such as 400°C–450°C. The thinner microstructure is preferred for high toughness and less damage to the carbide in the lower layer at high temperatures. Residual stresses are affected by tensile stress in CVD method and compression in PVD method. Therefore PVD coatings have higher chipping resistance than CVD coatings. However, the adhesion and abrasion resistances of CVD coatings are superior. Therefore CVD coatings are generally used in different machining operations, while PVD coatings are used where high CFs and chipping resistance occur. Engineering materials and metal alloys, which have high strength and toughness, generally interact abrasively and chemically with tool materials. The difficulty in efficiently processing these materials has led to significant improvements in coated tools. The coatings have many advantageous properties. Some of them are low friction, resistance to abrasion and breakage, high hot hardness and impact resistance, providing a diffusion layer between tool and chip. TiN, TiC, TiCN, and Al_2O_3 coatings are commonly used. The thickness of these coatings varies between 2 and 15 μm. TiN has low coefficient of friction, high hardness and temperature resistance, and good adhesion to the substrate. Coatings enhance the tool service life and make them capable to machine at high speeds and feeds. TiCN coating on cemented carbide has high flank wear resistance in the processing of abrasive materials. Al_2O_3 ceramic coatings are preferred due to their chemical stability, low thermal conductivity, high temperature, free surface, and crater wear resistance. However, they are very stable and have poor adhesion to the cutting tool surface. Multiphase coatings are effective in machining cast iron and steel. For example, in these coatings, TiC is first deposited onto the substrate, followed by TiN and Al_2O_3 coating. The first layer should have good adhesion to the substrate, the outer layer should be abrasion resistant and have low thermal conductivity, while the intermediate layer should be well bonded and compatible with both layers. Diamond coatings are made on silicon nitride and carbide tools. They are effective in the processing of nonferrous metals, silicon-containing abrasive materials, fiber-reinforced and metal matrix composites (MMCs), and graphite. Diamond coatings increase tool life by 10 times compared to other coatings (Kalpakjian & Schmid, 2014).

9.2 Evaluation of machining performance in dry cutting

9.2.1 Cutting force

CF is one of the important machinability indicators (Bayraktar & Afyon, 2020; Colligan, 1994; Kim & Ramulu, 2005). Generally, low CFs are desired during machining. The CFs affect the quality of the machined surface as well as the tool life. Increasing the CS causes the generation of heat into tool−workpiece interface. The CFs generated during the machining of different materials such as alloy steels, aluminum, metal, or polymer matrix composites vary due to the different structural and mechanical properties of these materials as well as the cutting parameters. For example (Ezugwu, 1997; Yang & Liu, 1999), it is stated that the CFs in the drilling of titanium are higher than aluminum alloys. In the machining of titanium, due to the contact at the tool–chip interface, higher stresses occur around the cutting edge compared to other metal materials. The power consumption in the drilling of titanium is approximately the same or lower than that used in the drilling of low hardness steels (Kahles, Field, Eylon, & Froes, 1985). During the machining of the materials, the CF reduces with the effect of the temperature due to the yield strength of the workpiece in the primary deformation zone with the increase of the CS (Bayraktar, 2020). However, as the temperature increases, the CF increases due to wear formation in the cutting tool (Bayraktar & Turgut, 2020). As the FR increases, the CF increases with the increase in the chip volume that must be removed per unit time (Bayraktar, Çamkerten, & Salihoğlu, 2020). If the DOC increases, the contact surface length between the cutting tool and chip also increases. Thus depending upon the increased contact length, more power is required for cutting. Due to that CF is increased. CF depends upon the material load on the tool–chip surface. Due to the changes in cutting depth, variation in CF also occurs. The maximum DOC is achieved when the cutting tool reaches below the microdimple. When the cutting of the microdimple begins, the transient force is zero. When the cutting tool reaches below the microdimple, the CF reaches its maximum value. When the tool cuts upward, the transient force gradually reduces and reaches zero at the end of the cut (Fig. 9.8) (Kurniawan, Kiswanto, & Ko, 2016). Transient and principle CFs are calculated as per Eqs. (9.6)–(9.11). A_{cut} is the cut surface area. R and $\text{TOC}_{(t)}$ are tool nose radius and transient thickness of cut, respectively, in Eq. (9.6).

$$A_{\text{cut}}(t) = \left\{ R^2 \cos^{-1}\left(\frac{R - \text{TOC}_{(t)}}{R}\right) \right\} - \left\{ \left(R - \text{TOC}_{(t)}\right)\sqrt{2R\text{TOC}_{(t)} - \text{TOC}_{(t)}^{\ 2}} \right\}. \tag{9.6}$$

$F_S(t)$ is transient CF. τ shows the constant shear stress along the shear plane achieved using the Johnson–Cook theorem and $\sin\phi_\alpha$ The principal shear angle according to Merchant theorem in Eq. (9.7).

$$F_S(t) = \frac{\tau A_{\text{cut}}(t)}{\sin\phi_\alpha} \tag{9.7}$$

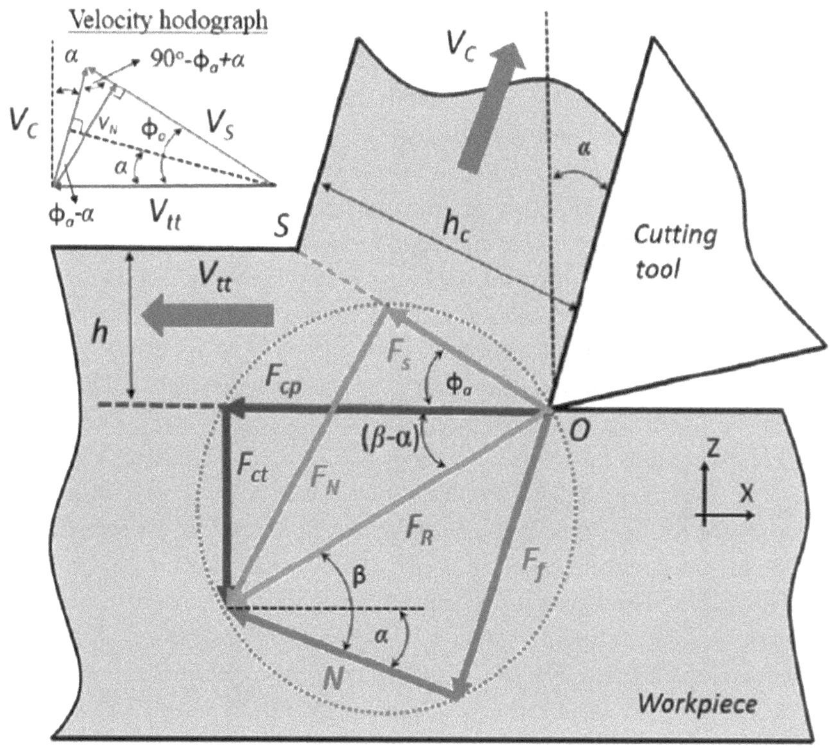

Figure 9.8 Merchant force diagram in orthogonal cutting.

$F_R(t)$ is the transient resultant force, β is the angle of friction, and α is the rake angle in Eq. (9.8). When Eqs. (9.7) and (9.8) are replaced, Eq. (9.9) is obtained.

$$F_R(t) = \frac{F_s(t)}{\cos\left(\phi_\alpha + \beta - \alpha\right)} \tag{9.8}$$

$$F_R(t) = \frac{\tau}{\sin\phi_\alpha \cos(90^\circ - \phi_\alpha)} A_{\text{cut}}(t). \tag{9.9}$$

Transient principle cutting force $F_{cp}(t)$, transient thrust force $F_{ct}(t)$, $F_R(t)$, β and α considering along x and z directions, considering the variables are calculated according to Eqs. (9.10)–(9.11).

$$F_{cp}(t) = F_R(t)\cos(\beta - \alpha) \tag{9.10}$$

$$F_{ct}(t) = F_R(t)\sin(\beta - \alpha). \tag{9.11}$$

9.2.2 Surface roughness

Surface quality of the machined materials heavily influences fatigue life, corrosion resistance, and wear strength. It is generally stated that the machined surface efficiency is related to the surface structure. Irregularities in the form of coves and protrusions on the machined surface trigger stress concentrations that cause plasticization and crack propagation in the workpiece. As a result, the smooth surface reduces the risk of crack. In general applications, engineers design according to the criteria appropriate to the property of each surface. Some surfaces are in mechanical interaction, others are in contact with factors such as air and oil. As a result, the properties of each surface differ depending upon their property. These are mechanical (load carrying capability), thermal (heat resistance or temperature conductivity), tribological (sealing, sliding), optical (light reflection behavior), and flow (effect on fluids flow) functions (Buzio, Boragno, Biscarini, De Mongeot, & Valbusa, 2003; Rech, Hamdi, & Valette, 2008). The better the surface quality of the workpiece, the more it can perform these functions. The SR that occurs in the machining of materials depends on factors such as cutting parameters, workpiece, and tool material. In normal course, the average SR (R_a) is generally used to determine the surface quality. R_a (μm) is calculated according to Eqs. (9.12) and (9.13). Here L represents the sampling length, the arithmetic mean of the y ordinate values, f (mm/rev) the feed and, R (mm) the cutting tool radius.

$$R_a = \frac{1}{L}\int_{x=0}^{x=L} |y|\,dx \tag{9.12}$$

$$R_a = \frac{f^2}{32R}. \tag{9.13}$$

In machining operations, the yield strength of the material is reduced and the cutting process is facilitated by the effect of thermal softening in the primary deformation area due to the increase of the CS. Therefore SR decreases. However, SR increases as a result of different wear mechanisms in the tool at high CSs (Khellaf et al., 2017). According to Eq. (9.13) SR increases with rising FR. This is due to the increased volume of chip that must be cut per unit time. On the other hand, it was stated that DOC had no effect on SR as much as CS and FR parameters and SR decreased with increasing tool radius (Asiltürk, Neşeli, & Ince, 2016; Fnides, Boutabba, Fnides, Aouici, & Yallese, 2013; Khellaf et al., 2017).

9.2.3 Tool wear

Tool wear (TW) has a significant impact on the economy of cutting operations. TW is difficult to predict due to the complexity of cutting operations. TW in cutting starts with the effect of contact and shear between the cutting tool and the workpiece. The CT can exceed 1000°C under extreme conditions in the cutting area between the tool and the workpiece and may lead to an increase in TW (Li, 2012). TW occurs at different levels when cutting, as shown in Fig. 9.9. Crater and flank wear are represented in the figure. Crater wear K_T depth and flank wear are represented with V_B. Crater wear originates from the abrasion of the microlevel abrasive particles broken from the tool and workpiece during chip formation in machining and friction by contact with the rake face (Sharma & Gupta, 2019), while flank wear occurs on the flank face due to the sliding effect between the cutting tool and the machined surface (Yoshimi & Matsumura, 2017). Notch wear occurs on the boundary of the cutting zone as a result of stress hardening and chemical interaction with the atmosphere.

Cutting TW is the criterion for determining tool life, especially the maximum flank wear. Since TW is largely dependent upon the CT, the

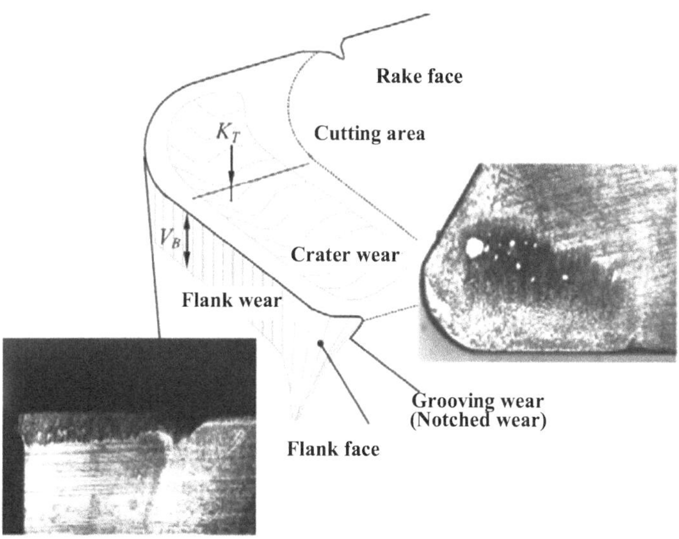

Figure 9.9 Crater, flank, and notched wears (Yoshimi & Matsumura, 2017). *With kind permission of Springer.*

TW rate rises with the CS. The tool life is calculated according to Taylor's theorem as in Eq. (9.14).

$$VT^n = C \tag{9.14}$$

where n and C are the parameters stated as an interaction of tool and workpiece in Eq. (9.14). In machining of hard materials, cutting tools with a large insert radius give better results in terms of surface quality. However, the small insert radius shows better resistance to flank face wear. In both cases, as the CS increases, SR and flank face wear increase with the effect of high temperature and pressure on the insert. This induces thermal softening of the cutting tool (Gekonde & Subramanian, 1995; Manivel & Gandhinathan, 2016). Wear rate at high CS is higher than the rate at low CS, because the friction between the tool and the chip is decreased due to the increased CS. TW is reduced due to the effect of decreased interaction time between tool and material with increasing FR. However, the combination of FR with high CS increases SR and TW (Manivel & Gandhinathan, 2016). Interaction stresses cutting tool–chip interface, chip compression ratio, and average interaction temperatures do not change if the DOC rises and the uncut thickness remains the same. Therefore if the optimum cutting regime is achieved, no change in TW rate is observed with increasing DOC (Astakhov, 2007; Astakhov & Shvets, 2004). Built up edge (BUE) in cutting tools is another type of wear. In BUE formation, a strongly deformed material is adhered firmly to the cutting tool–chip interface during dry cutting (Bayraktar & Hekimoğlu, 2020; Corrêa, Schroeter, & Machado, 2017). Not only dry cutting, in fact during other machining conditions as well it occurs. Due to the bond formed between the BUE and the cutting tool, the compressive stresses are high enough to prevent the cutting tool shearing on the workpiece (Wang & Liu, 2016). BUE emerges during continuous chip formation in the machining of ductile materials such as aluminum, steel, and titanium. It is known to affect SR and TW. In many studies, it has been stated that BUE reduces with increasing CS and weaker and brittle BUE formation is observed at low CSs (Thamizhmanii & Hasan, 2008). Even at high CSs, a small amount of BUE can appear. Another parameter that has an impact on BUE is FR. BUE formation increases when FR exceeds a certain value (Bayraktar & Demir, 2020; Cronjäger & Meister, 1992). If the parameters are not well selected, particle adhesion to the tool rises due to the rise in thermal stresses due to loads. The use of cooling fluids contributes to the minimization of BUE formation as it reduces both friction and temperature between the tool and chip.

9.3 Case studies on dry cutting

9.3.1 Case study 1

Kümmel et al. (2014) performed a study on the impact of BUE formation on microstructural properties during dry turning of SAE 1045 steel conducted using an uncoated carbide tool. It was found that BUE formation reduced with rising CS during dry cutting (Fig. 9.10) and hardness of the material adhered to the cutting edge at 100 m/min increased 2–3 times more than the hardness of the workpiece in the first case. Different wear types (flank, crater, notch, and radius wear) occurred in the same cutting parameters. Flank and radius wear mechanisms occurred with increasing CS. It is stated that surface quality rises with rising CS and BUE height decreases. In addition, residual stress increased at high CSs and wear rates on the machined surface (Kümmel et al., 2014).

9.3.2 Case study 2

Dosbaeva, El Hakim, Shalaby, Krzanowski, and Veldhuis (2015) investigated the impact of CT on polycrystalline cubic boron nitride (PCBN) and CVD-coated carbide cutters in dry turning of D2 tool steel. TiN (40%), PCBN with low content (60%), and TiN/Al_2O_3 multilayer-coated cutters with TiN on carbide tool were used as binders. The analysis of results showed that Ti-O and Cr-O tribo-film layers were formed on the surface of the worn CVD-coated carbide exposed to CTs up to 923°C. At a CS of 60 m/min, which corresponds to a temperature of 824°C, the tool life increase compared to PCBN in the coated carbide can reach up to 330%. The latter temperature is considered suitable for exhibiting

Figure 9.10 BUE formation at, (A) 50 m/min, (B) 75 m/min, and (C) 100 m/min.

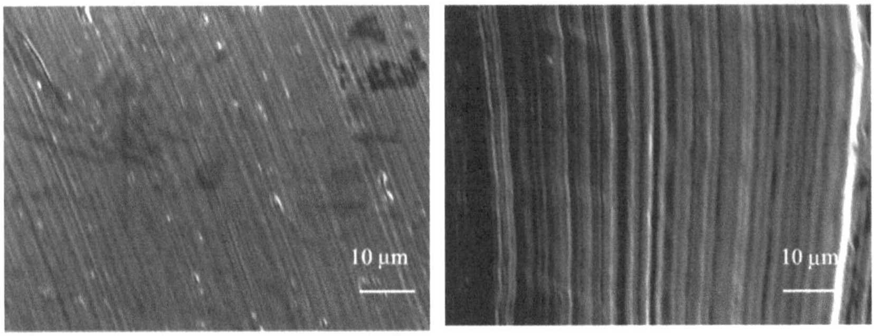

Figure 9.11 SEM images of chip underside (CS:175 m/min), (A) PCBN and (B) coated carbide.

lubricant properties for Cr-O tribo-films. Long-term oxidation of chromium can help to disintegrate this oxide in alumina as a middle layer in a CVD-coated carbide and has also been found to improve wear resistance. When the CS increases to 175 m/min and the corresponding CT above 1100°C, these tribo-films become inefficient and the tool life is longer due to the high hot hardness of the PCBN. This situation can be explained by the density of high adhesion marks formed by the coated tool on the chip underside at a CS of 175 m/min (Fig. 9.11). In the PCBN tool used, it was determined that the main wear mechanisms were adhesion and chemical wear (Dosbaeva et al., 2015).

9.3.3 Case study 3

In dry cutting, friction and adhesion are more severe due to the lack of coolant and lubricant (Xing, Deng, Zhao, Zhang, & Zhang, 2014). However, high CTs and critical wear require much better performance characteristics from cutting tools. In this respect, ceramic cutters are widely used for dry cutting for high hot hardness and abrasion resistance (Broniszewski et al., 2015; Waqar, Akhtar, Arif, & Hakeem, 2018). Yin et al. researched the cutting tool performance in dry turning of hardened 40Cr material with microwave sintered subcrystal Al_2O_3/SiC (AS) ceramic tool. Microwave Al_2O_3, SiCp (AS), sintered with hot-pressing Al_2O_3, SiCw (SW 500) and pressure-less WC, Co tools were used. AS tools were found to give better results in cutting performance and wear resistance at high and low CSs. At the low shear rate, a small amount of crater and groove wear was observed due to the effect of abrasion on the AS tool, while at the high shear rate, slight chipping was observed with

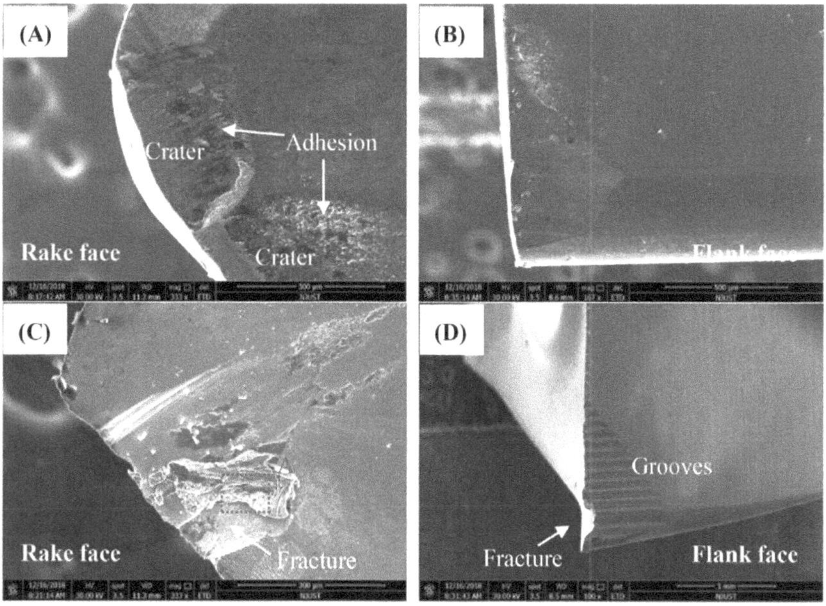

Figure 9.12 Wear formation of (A, B) SW500 and (C, D) YG8 ($v = 270$ m/min, $f = 0.1$ mm/r, $ap = 0.1$ mm) (Yin et al., 2019). *With kind permission of Elsevier.*

crater and channel wear due to abrasion and adhesion mechanisms. The SW 500 tool had severe crater and groove wear at a CS of $v \geq 270$ m/min (Fig. 9.12A and B), while the HV8 tool suffered breakage and groove wear due to poor temperature resistance and adhesion (Fig. 9.12C and D) (Yin, Yan, Ye, Zhu, & Yuan, 2019).

9.3.4 Case study 4

Salur et al. investigated the drillability of CuSn10 and GGG40 MMCs having different reinforcement rates (10%–20%–30%–40%) under dry cutting conditions with carbide drill. As the reinforcement ratio is increased, the SR is also increased, whereas the feed force is decreased. During cutting process, GGG 40 metallic chips act as fillers with different hardness values than the matrix. These chips did not show much resistance during drilling, because these chips formed weak zones on the composites. In other words, the formation of stress concentration or gap in the structure of GGG 40 facilitates the machining of MMC. The same experimental procedure was performed on cast CuSn10 to compare with the supplement. Generally, SR values in cast CuSn10 were found to be

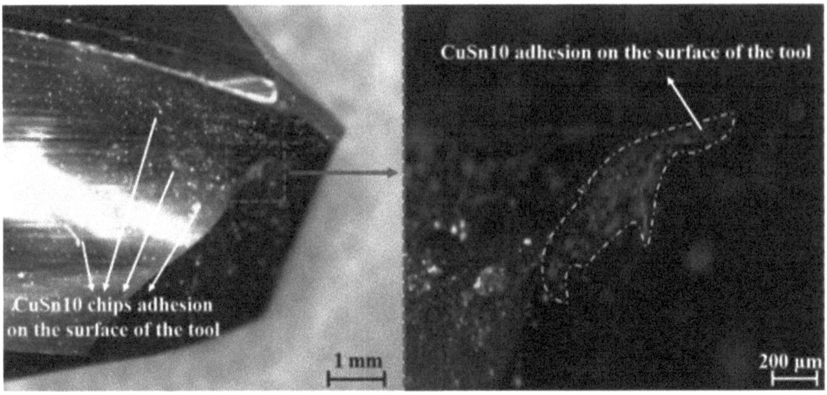

Figure 9.13 Adhesion formation in drilling of CuSn10 (Salur et al., 2019). *With kind permission of Elsevier.*

more uniform than those of reinforced. The high hardness values of the reinforced GGG 40 caused more feed force than cast CuSn10. CuSn10 material smeared on the drill surface facilitated the cutting process by reducing friction and SR (Salur, Aslan, Kuntoglu, Gunes, & Sahin, 2019). Fig. 9.13 presents the wear of drill tool observed in that study.

9.3.5 Case study 5

Jadhav et al. conducted a study on the effect of reinforcement ratio, drill point angle, and cutting parameters on surface quality in dry drilling of Al6061/Al_2O_3p MMC. They found that abrasive wear occurs on the cutting tool depending on the reinforcement ratio and this wear increases the SR. Al_2O_3p supplementation adversely affects the cylindricality of the hole. Smaller drill point angle means greater cutting edge surface area. Therefore the interaction zone between the tool and the workpiece increases. Consequently, the feed force and torque values are reduced. With increasing feed, the feed force increases due to TW and more power is required for cutting. A small point angle, low FR, and high CS are recommended for surface quality, while a small point angle, high FR, and CS are recommended for optimum hole cylindricality (Jadhav, Kakde, Patil, & Sankpal, 2018).

9.3.6 Case study 6

Zheng et al. investigated TW, SR, and CF in high-speed dry milling of AISI 4340 steel with CVD-Al_2O_3/TiCN-coated carbide cutter. Abrasive, diffusion, adhesion, and oxidation wear occurred in the cutting tool

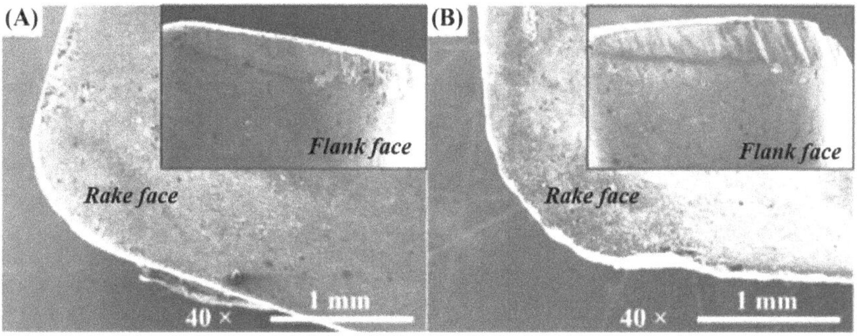

Figure 9.14 Wear formation in Al_2O_3/TiCN coated tool (Zheng, Cheng, Li, Xu, & Tian, 2019). *With kind permission of Springer.*

(Fig. 9.14). They reported that SR was mainly affected by FR, and CFs by DOC and further obtained optimum parameters.

9.3.7 Case study 7

Sahoo et al. performed dry turning of AISI 4340 steel with uncoated, MTCVD-TiN/TiCN/Al_2O_3/TiN and CVD-TiN/TiCN/Al_2O_3/ZrCN multilayer-coated cutters. They found that the CF was lower than that of uncoated cutters. Severe friction and abrasion between the workpiece and the surface caused TW in case of uncoated tools. It was determined that the CFs of ZrCN-coated cutter were lower than TiN-coated cutter during 10 s machining. The lubricant property of TiN coating decreases friction and interface temperature. Therefore the CFs increase as temperature formation is not sufficient for softening of the material. Both coated tools exhibited less wear than the uncoated tool. This is due to the reduction of friction between the tool–chip and the chip–workpiece, as well as coatings that act as thermal barriers. This barrier prevents the thermal conduction of heat to the tool and ensures that most of the heat is thrown out with chip. While the Al_2O_3 coating provides thermal barrier and oxidation resistance at the tool–chip interface, the multilayer TiN coating exhibits more tool life as it reduces the heat formation due to the lubrication effect of the TiCN coating (Sahoo & Sahoo, 2012).

9.4 Summary

This chapter presented dry cutting technology with fundamental understanding and case studies conducted by various researchers. Dry cutting method is becoming more attractive these days in machining sector to minimize the adverse effects of conventional lubricants/coolants and cooling techniques. The most important problem in dry cutting is the formation of temperature and excessive TW. To overcome this, many techniques have been developed and investigated. Selection of appropriate tool material and coating is the most prominent solution. Optimization of cutting parameters is also one of the strategies applied by various researchers. Dry cutting is still in developmental stage, and future research is required to be done in the areas of developing more robust machine tools, novel cutting tool materials, their treatment, coatings, hybrid machining technologies, and so on. Textured and cryogenic-treated tools, hybridization of dry cutting with heat-assisted or vibration-assisted machining can be some of the solutions if investigated in detail by the researchers.

References

Aronson, R. B. (1995). Why dry machining? *Manufacturing Engineering, 114*(1), 33–36.

Asiltürk, I., Neşeli, S., & Ince, M. A. (2016). Optimisation of parameters affecting surface roughness of $Co_{28}Cr_6Mo$ medical material during CNC lathe machining by using the Taguchi and RSM methods. *Measurement, 78*, 120–128.

Astakhov, V., & Davim, J. (2008). *Tools (geometry and material) and tool wear. Machining*. London: Springer.

Astakhov, V. P. (2007). Effects of the cutting feed, depth of cut, and workpiece (bore) diameter on the tool wear rate. *The International Journal of Advanced Manufacturing Technology, 34*(7-8), 631–640.

Astakhov, V. P., & Shvets, S. (2004). The assessment of plastic deformation in metal cutting. *Journal of Materials Processing Technology, 146*(2), 193–202.

Audy, J. (2008). A study of computer-assisted analysis of effects of drill geometry and surface coating on forces and power in drilling. *Journal of Materials Processing Technology, 204*(1-3), 130–138.

Bayraktar, Ş. (2020). *Cryogenic cooling-based sustainable machining. High speed machining*. London, United Kingdom: Elsevier Academic Press.

Bayraktar, Ş., & Afyon, F. (2020). Machinability properties of Al-7Si, Al-7Si-4Zn and Al-7Si-4Zn-3Cu alloys. *Journal of the Brazilian Society of Mechanical Sciences and Engineering, 42*(4), 1–12.

Bayraktar, Ş., Çamkerten, Ç., & Salihoğlu, N. (2020). Investigationof the effect on copper and silicon additives on machinability in turning of Al-25Zn Alloy with CVD-Al_2O_3 coated tools. *Gazi University Science Journal, Part C Design and Technology, 8*(1), 79–93.

Bayraktar, Ş., & Demir, O. (2020). Processing of T6 heat-treated Al-12Si-0.6 Mg alloy. *Materials and Manufacturing Processes, 35*(3), 354–362.

Bayraktar, Ş., & Hekimoğlu, A. P. (2020). Effect of zinc content and cutting tool coating on the machinability of the Al-(5–35) Zn alloys. *Metals and Materials International, 26*(4), 477–490.

Bayraktar, Ş., & Turgut, Y. (2020). Determination of delamination in drilling of carbon fiber reinforced carbon matrix composites/Al 6013-T651 stacks. *Measurement, 154*, 107493.

Broniszewski, K., Wozniak, J., Kostecki, M., Czechowski, K., Jaworska, L., & Olszyna, A. (2015). Al_2O_3-V cutting tools for machining hardened stainless steel. *Ceramics International*, *41*(10), 14190–14196.

Buzio, R., Boragno, C., Biscarini, F., De Mongeot, F. B., & Valbusa, U. (2003). The contact mechanics of fractal surfaces. *Nature Materials*, *2*(4), 233–236.

Çelik, H. Y. (2014). Investigating the effects of cutting parameters on the hole quality in drilling the Ti-6Al-4V alloy. *Materiali in Tehnologije*, *48*(5), 653–659.

Colligan, K. (1994). New tool drills both titanium and carbon composites. *American Machinist*, *138*(10), 56–58.

Corrêa, J. G., Schroeter, R. B., & Machado, Á. R. (2017). Tool life and wear mechanism analysis of carbide tools used in the machining of martensitic and supermartensitic stainless steels. *Tribology International*, *105*, 102–117.

Cronjäger, L., & Meister, D. (1992). Machining of fibre and particle-reinforced aluminium. *CIRP Annals Manufacturing Technology*, *41*(1), 63–66.

Davim, J. P. (2008). *Machining-fundamental and recent advances*. London: Springer Verlag.

Davis, J. R. (1989). (9th Ed.). *Metals handbook-machining*, (16). Metals Park, OH, USA: ASM International.

Davis, J. R. (2005). *Tool materials (ASM specialty handbook)*. Materials Park, OH, USA: ASM International.

Demir, Z. (2018). An experimental investigation of the effects of point angle on the high-speed steel drills performance in drilling. *Measurement and Control*, *51*(9-10), 417–430.

Dixit, U. S., Sarma, A. K., & Davim, J. P. (2012). *Environmentally friendly machining*. New York: Springer, Chapter 3.

Dosbaeva, G. K., El Hakim, M. A., Shalaby, M. A., Krzanowski, J. E., & Veldhuis, S. C. (2015). Cutting temperature effect on PCBN and CVD coated carbide tools in hard turning of D2 tool steel. *International Journal of Refractory Metals and Hard Materials*, *50*, 1–8.

Ezugwu, E. Q. (1997). Titanium alloys and their machinability—A review. *Journal of Materials Processing Technology*, *68*(3), 262–274.

Faga, M. G., Gautier, G., Calzavarini, R., Perucca, M., Aimo Boot, E., Cartasegna, F., & Settineri, L. (2007). AlSiTiN nanocomposite coatings developed via Arc Cathodic PVD: Evaluation of wear resistance via tribological analysis and high speed machiningoperations. *Wear*, *263*, 1306–1314.

Fnides, B., Boutabba, S., Fnides, M., Aouici, H., & Yallese, M. A. (2013). Cutting tools flank wear and productivity investigation in straight turning of X38CrMoV5-1 (50 HRC). *International Journal of Applied Engineering and Technology*, *3*(1), 1–10.

Fox-Rabinovich, G. S., Kovalev, A. I., Aguirre, M. H., Beake, B. D., Yamamoto, K., Veldhuis, S. C., … Rashkovskiy, A. Y. (2009). Design and perfor-mance of AlTiN and TiAlCrN PVD coatings for machining of hard to cut materials. *Surface and Coatings Technology*, *204*, 489–496.

Ganesh, M., Sidpara, A., & Sankha, D. (2017). *Fabrication of micro-cutting tools for mechanical micro-machining*. Springer. (Chapter 1).

Gekonde, H. O., & Subramanian, S. V. (1995). High speed machining of ductile iron. *AFS Transactions*, *120*, 309–317.

Gopal, P. M., & Prakash, K. S. (2018). Minimization of cutting force, temperature and surface roughness through GRA, TOPSIS and Taguchi techniques in end milling of Mg hybrid MMC. *Measurement*, *116*, 178–192.

Gunter, K. L., & Sutherland, J. W. (1999). An experimental investigation into the effect of process conditions on the mass concentration of cutting fluid mist in turning. *Journal of Cleaner Production*, 7(5), 341–350.

Gupta, K., & Laubscher, R. F. (2017). Sustainable machining of titanium alloys: A critical review. *Proceedings of the Institution of Mechanical Engineers, Part B: Journal of Engineering Manufacture*, *231*(14), 2543–2560.

Hands, D. M., Sheehan, J., Wong, B., & Lick, H. B. (1996). Comparison of metalworking fluid mist exposures from machining with different levels of machine enclosure. *American Industrial Hygiene Association Journal*, *57*(12), 1173–1178.

Hoffman, P. J., Hopewell, E. S., Janes, B., & Sharp, K. M. (2012). *Precision machining technology*. Delmar Cengage Learning.

Jadhav, S. S., Kakde, A. S., Patil, N. G., & Sankpal, J. B. (2018). Effect of cutting parameters, point angle and reinforcement percentage on surface finish, in drilling of Al6061/Al_2O_3p MMC. *Procedia Manufacturing*, *20*, 2–11.

Kahles, J. F., Field, M., Eylon, D., & Froes, F. H. (1985). Machining of titanium alloys. *Journal of Metals*, *37*(4), 27–35.

Kalpakjian, S., & Schmid, S. R. (2014). *Manufacturing engineering and technology* (7th ed.). Pearson Education.

Khellaf, A., Aouici, H., Smaiah, S., Boutabba, S., Yallese, M. A., & Elbah, M. (2017). Comparative assessment of two ceramic cutting tools on surface roughness in hard turning of AISI H11 steel: including 2D and 3D surface topography. *The International Journal of Advanced Manufacturing Technology*, *89*(1-4), 333–354.

Kim, D., & Ramulu, M. (2005). Cutting and drilling characteristics of hybrid titanium composite laminate(HTCL), In *Proceedings of Materials and Processing Technologies for Revolutionary Applications Fall Technical Conference* (pp. 1–8). Seattle, Washington.

Kissler, H. (2000). KSS-bedingte Kosten in derspanenden Metallbearbeitung als Anreiz für dieTrockenbearbeitung. In *12th International Colloquium Tribology-Plus, Conference Proceedings* (pp. 901-913), Esslingen, Germany.

Kümmel, J., Gibmeier, J., Müller, E., Schneider, R., Schulze, V., & Wanner, A. (2014). Detailed analysis of microstructure of intentionally formed built-up edges for improving wear behaviour in dry metal cutting process of steel. *Wear*, *311*(1-2), 21–30.

Kurniawan, R., Kiswanto, G., & Ko, T. J. (2016). Micro-dimple pattern process and orthogonal cutting force analysis of elliptical vibration texturing. *International Journal of Machine Tools and Manufacture*, *106*, 127–140.

Landgraf, G. (2004). Dry goods-factors to consider when dry or near-dry machining. *Cutting Tool Engineering*, *56*(1), 28–35.

Li, B. (2012). A review of tool wear estimation using theoretical analysis and numerical simulation Technologies. *International Journal of Refractory Metals and Hard Materials*, *35*, 143–151.

Liu, H. Z., & Zong, W. J. (2019). Prediction model of tool wear volume in precision turning of ceramic particle reinforced aluminum matrix composites. *The International Journal of Advanced Manufacturing Technology*, *100*(9-12), 2689–2700.

Manivel, D., & Gandhinathan, R. (2016). Optimization of surface roughness and tool wear in hard turning of austempered ductile iron (grade 3) using Taguchi method. *Measurement*, *93*, 108–116.

Mills, B., & Redford, A. H. (1983). *Machinability of engineering materials*. Springer Science & Business Media.

Nalbant, M., Altın, A., & Gökkaya, H. (2007). The effect of cutting speed and cutting tool geometry on machinability properties of nickel-base Inconel 718 super alloys. *Materials & Design*, *28*(4), 1334–1338.

PalDey, S., & Deevi, S. C. (2003). Single layer and multilayer wear resistant coatings of (Ti, Al)N: A review. *Materials Science Engineering A*, *342*, 58–79.

Pathak, S., & Jain, N. K. (2017). Modeling and experimental validation of volumetric material removal rate and surface roughness depth of straight bevel gears in pulsed-ECH process. *International Journal of Mechanical Sciences*, *124*, 132–144.

Rech, J., Hamdi, H., & Valette, S. (2008). In J. P. Davim (Ed.), *Workpiece surface integrity, machining-fundamental and recent advances*. London: Springer Verlag.

Rossmoore, H. W. (1995). Microbiology of metal-working fluids: Deterioration, disease, and disposal. *Lubrication Engineering*, *51*(2), 113–130.

Sahoo, A. K., & Sahoo, B. (2012). Experimental investigations on machinability aspects in finish hard turning of AISI 4340 steel using uncoated and multilayer coated carbide inserts. *Measurement*, *45*(8), 2153–2216.

Salur, E., Aslan, A., Kuntoglu, M., Gunes, A., & Sahin, O. S. (2019). Experimental study and analysis of machinability characteristics of metal matrix composites during drilling. *Composites Part B: Engineering*, *166*, 401–413.

Selvam, S. M., & Sujatha, C. (1995). Twist drill deformation and optimum drill geometry. *Computers & structures*, *57*(5), 903–914.

Sharma, N., & Gupta, K. (2019). Influence of coated and uncoated carbide tools on tool wear and surface quality during dry machining of stainless steel 304. *Materials Research Express*, *6*(8), 086585.

Sheikh-Ahmad, J. Y. (2009). *Machining of polymer composites*. New York: Springer.

Subramanian, M., Sakthivel, M., Sooryaprakash, K., & Sudhakaran, R. (2013). Optimization of cutting parameters for cutting force in shoulder milling of Al7075-T6 using response surface methodology and genetic algorithm. *Procedia Engineering*, *64*, 690–700.

Sutherland, J. W., Kulur, V. N., & King, N. C. (2000). An experimental investigation of air quality in wet and dry turning. *Annals of the CIRP*, *49*(1), 61–64.

Thamizhmanii, S., & Hasan, S. (2008). Investigating flank wear and cutting force on hard steels by CBN cutting tool by turning. In: *Proceedings of the world congress on engineering WCE 2008*, London, UK.

Vereschaka, A. A., Grigoriev, S. N., Sitnikov, N. N., & Batako, A. D. (2017). Delamination and longitudinal cracking in multi-layered composite nano-structured coatings and their influence on cutting tool life. *Wear*, *390*, 209–219.

Wang, B., & Liu, Z. (2016). Evaluation on fracture locus of serrated chip generation with stress triaxiality in high speed machining of Ti6Al4V. *Materials & Design*, *98*, 68–78.

Waqar, T., Akhtar, S. S., Arif, A. F. M., & Hakeem, A. S. (2018). Design and development of ceramic-based composites with tailored properties for cutting tool inserts. *Ceramics International*, *44*(18), 22421–22431.

Weinert, K., Inasaki, I., Sutherland, J. W., & Wakabayashi, T. (2004). Dry machining and minimum quantity lubrication. *CIRP Annals*, *53*(2), 511–537.

Xing, Y., Deng, J., Zhao, J., Zhang, G., & Zhang, K. (2014). Cutting performance and wear mechanism of nanoscale and microscale textured Al2O3/TiC ceramic tools in dry cutting of hardened steel. *International Journal of Refractory Metals and Hard Materials*, *43*, 46–58.

Yamamoto, K., Sato, T., Takahara, K., & Hanaguri, K. (2003). Properties of (Ti, Cr, Al) N coatingswith high Al content deposited by new plasma enhanced arc-cathode. *Surface and Coatings Technology*, *174-175*, 620–626.

Yang, X., & Liu, C. R. (1999). Machining titanium and its alloys. *Machining Science and Technology*, *3*(1), 107–139.

Yin, Z., Yan, S., Ye, J., Zhu, Z., & Yuan, J. (2019). Cutting performance of microwave-sintered sub-crystal Al_2O_3/SiC ceramic tool in dry cutting of hardened steel. *Ceramics International*, *45*(13), 16113–16120.

Yoshimi, I., & Matsumura, T. (2017). *Theory and practice in machining systems*. Springer International Publishing.

Youssef, H., & El-Hofy, H. (2012). *Principles of traditional and nontraditional machining*. Dar Elfath Press.

Youssef, H. A. (2016). *Machining of stainless steel and super alloys-traditional and nontraditional techniques*. John Wiley & Sons.

Zheng, G., Cheng, X., Li, L., Xu, R., & Tian, Y. (2019). Experimental investigation of cutting force, surface roughness and tool wear in high-speed dry milling of AISI 4340 steel. *Journal of Mechanical Science and Technology*, *33*(1), 341–349.

CHAPTER 10

Cryogenic cooling-based sustainable machining

Suleyman Cinar Cagan and Berat Baris Buldum
Engineering Faculty, Department of Mechanical Engineering, Mersin University, Mersin, Turkey

10.1 Introduction

Today, machining sector is busy investigating, developing, and adopting innovative and sustainable machining strategies due to design, processing, and manufacturing-related challenges; cost concerns; and strict environmental regulations (Gupta, 2020). Metal removing operations such as turning, milling, and drilling are the most widely used machining operations in the manufacturing industry (Ghosh & Rao, 2015; Jawahir & Balaji, 2000). Factors such as cutting fluid, cutting tools, and cutting process parameters largely affect performance characteristics and cost of the product development. Cutting fluid (coolant or lubricant) is one of the most important factors to improve machining processes (Vukelic et al., 2020). It accounts for about 15%–20% of the total machining cost. The basic function of the cutting fluid is to dissipate heat from the cutting zone during the machining process (Khanna & Agrawal, 2020). Traditionally, water- and oil-based cutting fluids are utilized throughout machining, but industries are forced to use less harmful cutting fluids due to many technological and environmental issues (Hsien, 2015). Another way to get the better of this problem is to utilize the near dry machining technology, in which, minimum quantity lubrication (MQL) came into existence first (Boswell, Islam, Davies, Ginting, & Ong, 2017). In addition to the advantage of sustainability, the MQL technique also produces better surface quality and longer tool life. Although this method is effective in providing lubrication, it is not very effective in removing heat from the cutting area. Therefore recent research has been focused towards innovations in flood cooling and cryogenic machining. Cryogenic machining is one of the most sustainable machining methods (Gupta, 2020). In order for a production to be sustainable, energy and resource consumption, production cost, environmental degradation, etc. should be at minimum

Sustainable Manufacturing
DOI: https://doi.org/10.1016/B978-0-12-818115-7.00008-0

levels (Hegab, Darras, & Kishawy, 2018). Increased product quality and operator safety will provide the social advantage of sustainability.

Due to its abundance availability and effectiveness, the liquid nitrogen (LN_2) is widely used in cryogenic applications, such as cryogenic cooling in machining (Dinesh, Senthilkumar, & Asokan, 2017; Jawahir et al., 2016; Kaynak & Gharibi, 2018; Morkavuk, Köklü, Bağcı, & Gemi, 2018; Suhaimi, Yang et al., 2018; Wang et al., 2020). According to the National Institute of Standards and Technology (NIST), temperature below −180°C has recognized to be cryogenic (Jawahir et al., 2016; Suhaimi, Yang et al., 2018). The first use of liquified gases as the coolant in machining operations goes back to the 1900s, where carbon dioxide (CO_2) gas was used as a coolant during machining (Isakson, 2018). During World War II, scientists discovered that frozen metals down to low temperatures demonstrated greater abrasion resistance (Balasubramanian, Gupta, & Singh, 2012). This discovery carries with the development of cryogenic science. The term cryogenic machining began to be spoken about for the first time in 1966 by CryoTech Company operating in the state of Detroit in the United States (Wolf, 2019). The first guide to cryogenic applications for engineers and scientists was launched in the 1970s. When the old studies are examined, it is seen that various researchers use the term cryogenic to associate temperatures below 0°C, mostly in CO_2 applications. The term "cryogenic machining" was first used by Uehara and Kumagai in 1968 (Bruschi & Ghiotti, 2018). In cryogenic machining, liquid gases such as N_2, CO_2, and He are used as alternative refrigerants to conventional oil- and water-based coolants/lubricants. Cryogenic machining also helps to change material properties of the cutting tool and workpiece and to control the heat generated in the cutting zone (Shokrani, Dhokia, Muñoz-Escalona & Newman, 2013). In this way, cryogenic machining has been shown to have the ability to produce-treated surfaces while reducing/eliminating the heat-induced damage from the machining process.

10.2 Cryogenic machining

The schematic representation of CO_2-based cryogenic machining is presented in Fig. 10.1 (Jerold & Kumar, 2012). Nitrogen can also be used in place of CO_2. As seen in the figure, this system consists of a cylindrical tank containing CO_2 gas, pressure regulators, flow meter, rubber pipe, and nozzle. Cylindrical tank carries out CO_2 storage, and pressure

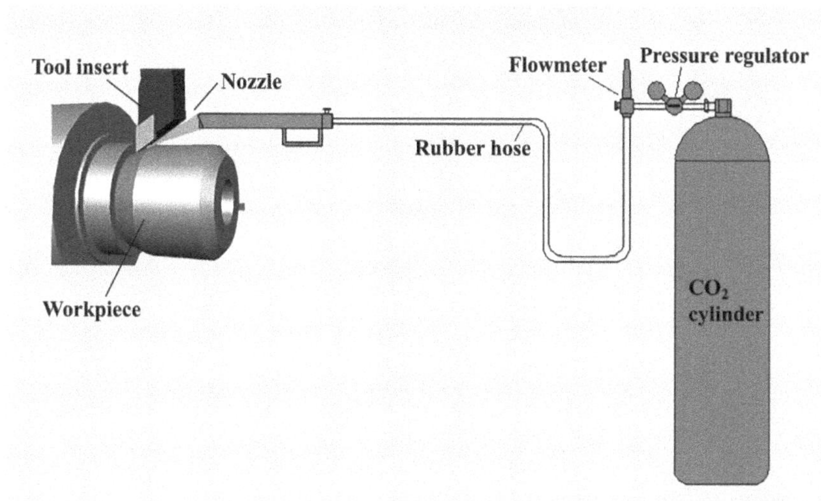

Figure 10.1 Schematic representation of the CO_2-based cryogenic machining system. *From Jerold, BD, & Kumar, MP. (2012). Experimental comparison of carbon-dioxide and liquid nitrogen cryogenic coolants in turning of AISI 1045 steel.* Cryogenics, 52*(10), 569–574, with kind permission of Elsevier.*

regulator is placed in the mouth of the tank. One of the two regulators used here consists of two quadrant gauges, one showing the pressure inside the cylinder and the other showing the feed pressure. Next to the regulator, a flow meter was added to measure the quantity of CO_2 coolant. One end of a rubber hose is attached to the outlet of the flow meter, and the other end is connected to a torch that can be replaced whenever the nozzle is required (Balaji, Ravi, & Chandran, 2018). It also comprises a valve to control flow or stop flow when machining on the torch is not performed. From the nozzle to the tip of the torch, CO_2 is directed to the cutting zone or tool to ensure its supply while machining. Since nitrogen (N_2) gas is an inert gas lighter than air, it disperses and does not harm the workers in the factory if used in place of CO_2. Conversely, CO_2 is considered as an air pollutant, heavier than air, and causes lack of O_2 problems in the factory (Shokrani, Dhokia, & Newman, 2012). For this reason, it is more common to use nitrogen gas instead of CO_2.

In Fig. 10.2, a schematic view of the liquid nitrogen supported cryogenic processing mechanism is presented (Jerold & Kumar, 2012). Liquid nitrogen tank, sized appropriately for user requirements, stores cryogen gas as per the user requirement. The compressor is used to develop the pressure of the gas to be sent from the liquid nitrogen tank (Lu, Kudaravalli, & Georgiou, 2018).

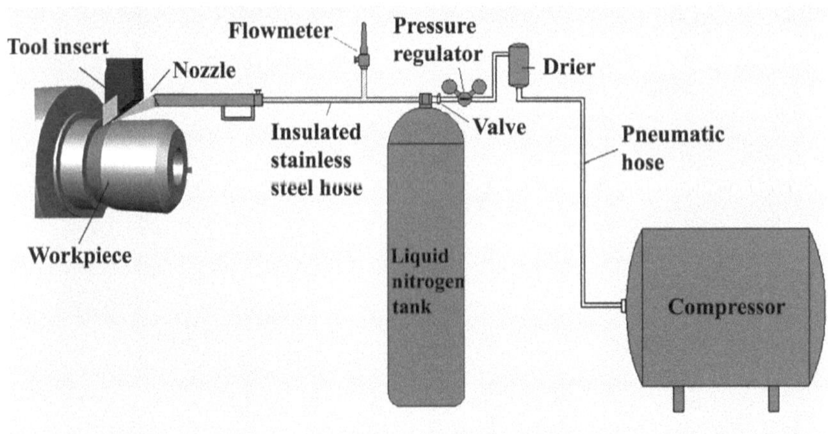

Figure 10.2 Schematic representation of the liquid nitrogen-based cryogenic machining system. *From Jerold, BD, & Kumar, MP. (2012). Experimental comparison of carbon-dioxide and liquid nitrogen cryogenic coolants in turning of AISI 1045 steel. Cryogenics, 52(10), 569–574, with kind permission of Elsevier.*

The device in which the tank is located is equipped to maintain pressure build-up and the source pressure requirements. In addition, a cryogen quantity monitor (weight or level) can be equipped with devices such as primary pressure relief valve and secondary pressure relief rupture disk and remote controlled main flow on/off valve and filling valve (Lu et al., 2018). It is in the state of cryogen-saturated liquid stored in the tank, and it is drawn from inside the tank and takes heat from the outside while going to the tool tip, and there is a drop in pressure. As a result of pressure drop, both along the heat input and flow path, the quality of the liquid reaching the tool changes. Liquid quality is defined as the percentage or fraction of the liquid mass in the fluid. While in transit to the tool tip, the cryogen is cooled under the apparatus cooler so the liquid conveyed to the instrument is of high capability, for example, high fluid stage content. Liquid channels are made of industry standard adaptable cryogenic hoses (Lu et al., 2018). The cryogenic fluids from the funnels are moved to the apparatus tip, permitting the instrument tip to be cooled. This cryogenic unit framework activity is checked by different sensors and constrained by an extraordinary programmable logic controller. There are two methods of activity, manual activity and computerized numerical control mode (Lu et al., 2018). The tool holders utilized in the cryogenic machining framework are good with standard cutting tool holders and device frameworks. They can be utilized as conventional holders without rolling out any improvements. The cryogenic cooling system can be used both internally and externally. In the internal system, cryogenic liquid is applied by cooling the tool tip from the channels extending to the cutting tool tip. The channels are designed taking into account the thermal resistance and strength between the leading edge and the liquid channel depending on the nitrogen flow condition. Since N_2 has a high liquid–gas expansion rate and an excessive low-boiling point (−196°C), it is rapidly transformed into vapor form when it is sprayed to the machining area and the pressure increases rapidly in closed containers due to the high-expansion rate (Fratila, 2013). Therefore appropriate safety valves and isolation are used for each component to prevent any damage to the transmission system. In the external system, cryogenic liquid is sprayed directly through the insulated stainless steel (SS) hose on the tool tip. In this external system, the nozzle can be placed in different positions to locate different purposes. For instance, it can steer away from the workpiece to prevent dimensional changes due to overcooling during cryogenic exhaust flow. The nozzles may also direct the cryogenic exhaust stream to spray on the treated surface to scale back local thermal expansion and improve the surface integrity of the part. Although conventional tooling experience is employed in tool development for cryogenic machining, differences in temperature distribution, cutting mechanics, tribological behavior, and material characteristics should be considered compared to traditional cooling scenarios (Lu et al., 2018).

10.2.1 Cryogenic treatment

In the cryogenic treatment, materials are cooled down to very low temperatures (−196°C) to accomplish the ideal metallurgical and microstructural properties (Gill, Singh, Singh, & Singh, 2009; Kalsi, Sehgal, & Sharma, 2010). The drop to these temperatures is possible by feeding computer-controlled liquid nitrogen and using the most suitable insulation materials (Pusavec, Courbon, Rech, & Kopac, 2014). The cryogenic treatment is divided into two as "deep" and "shallow" (Deshpande, Andhare, & Padole, 2018). In the shallow cryogenic treatment (SCT), the temperature of the materials is lowered to around −150°C. Despite the remarkable improvement in the mechanical properties of the materials, the desired properties are not fully achieved. In the deep cryogenic treatment (DCT), the materials are made cold to −196°C, and the effects of this process increase significantly compared to the SCT (Senthilkumar, Rajendran, Pellizzari, & Siiriainen, 2011). The cryogenic machining is often applied to a good range of materials like steels, cast irons, nonferrous metals, alloys, carbides, plastics, and ceramics. The process is cooled at speeds of 1–2°C/min and is performed between 36 and 72 h and varies according to the material type and weight. The cycles of the DCT process are depicted in Fig. 10.3 (Akincioğlu, Gökkaya, & Uygur, 2015).

These are the cycles in the cryogenic treatment process (Akincioğlu et al., 2015):

- Controlled cooling
- Soaking time
- Controlled heating
- Tempering

Since the system prevents computer-controlled and liquid nitrogen-part contact, any changes in part sizes are prevented and the risk of cracking in the part is eliminated. This special process is not a surface treatment, but the same effect is created at every point of the material. The improvement theories of cryogenic processing are briefly described as follows.

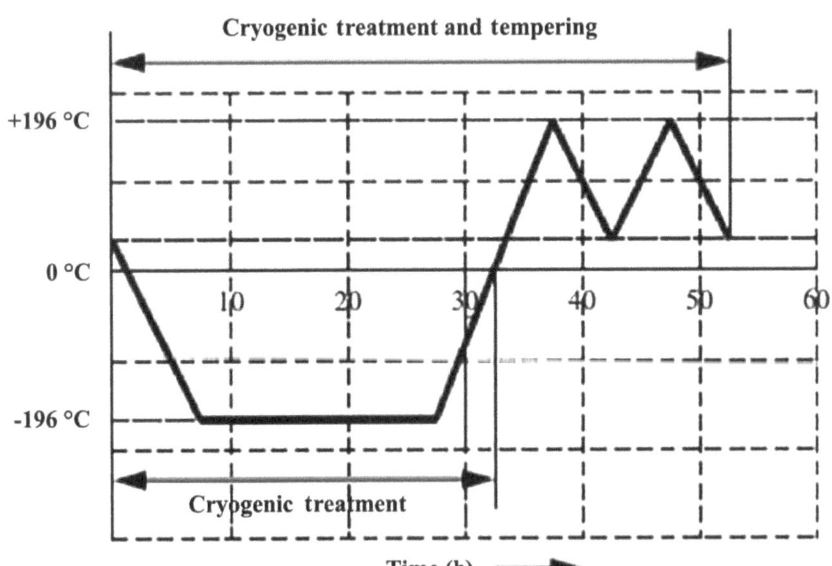

Figure 10.3 The deep cryogenic treatment and tempering process. *From Gill, SS, Singh, R., Singh, H., & Singh, J. (2009). Wear behaviour of cryogenically treated tungsten carbide inserts under dry and wet turning conditions.* International Journal of Machine Tools and Manufacture, *49(3–4), 256–260, with kind permission of Elsevier.*

10.2.1.1 Converting residual austenite phase to martensite

In conventional heat treatment applications, firstly, austenite phase is obtained to harden the steels, and this phase is rapidly cooled, and it is desired to obtain the martensite phase. However, not all austenite phase can turn into martensite as a result of traditional heat treatment. At this point, the cryogenic cooling takes over, converting approximately the entire residual austenite phase to martensite (Li, Xiao, Ye, Zhao, & Yang, 2018; Sert & Celik, 2019).

10.2.1.2 Change in carbide forms

In conventional hardening methods, carbides enter the crystal structure, causing stresses in the crystal structure. During cryogenic machining, carbide particles are freed from the crystal lattice and dispersed in the material. In this way, residual stresses are significantly reduced and high hardness carbide structures occur, leading to high increases in wear resistance (Özbek, ÇİçeK, Gülesİn, & Özbek, 2016).

10.2.1.3 Thinning in grain structure

The atoms that make up the alloy, try to remain in the state where they are most stable in the structure. These atoms capture the optimum sequence due to the cryogenic machining and cause thinning in the grain structure (Gill & Singh, 2013). In this way, molecular bonds are strengthened and an increase in the wear resistance of the material occurs (Baldissera & Delprete, 2008).

10.2.1.4 Elimination of internal stresses

Due to the rearrangement of atoms and therefore thinning in the grain structure, the atomic spaces (errors) in the crystal structure are removed and the atom density in the unit volume increases. In this way, the internal stresses in the material are significantly reduced (Kalia, 2010). As a result of microstructural changes, the strength, toughness, and hardness of the materials increase, and the abrasion resistance improves. Especially, the increase in abrasion resistance can reach up to 800%. As a result, material life is prolonged significantly (Gill et al., 2009).

10.2.2 Cryogenic cooling trends

The cryogenic cooling techniques being applied by the researchers during machining of various materials are: cooling the workpiece to a predetermined temperature, indirect cryogenic cooling, and cryogenic spraying and jet cooling, and different cryogenic cooling trends, as presented in (Fig. 10.4) (Pervaiz, Kannan, & Kishawy, 2018; Stachurski, Sawicki, Wójcik, & Nadolny, 2018; Tahir et al., 2013).

10.2.2.1 Cryogenic precooling the workpiece

The precooling method of the workpiece is carried out by immersing the workpiece in a cryogen by applying a cryogenic bath or spraying cryogen gases on the workpiece immediately before machining (Yildiz & Nalbant, 2008). The cryogenic bath precooling method increases the workability of high elasticity and adhesion materials such as viscoelastic polymer; it is stated that cryogen spraying before the workpiece is processed along, with the penetration of cryogen, brittle chips are formed during machining, thus facilitating chip formation in the machining of ductile materials (Schoop et al., 2016; Shukla & Bajpai, 2019).

10.2.2.2 Indirect cryogenic cooling

In the indirect cooling method, the aim is to cool the cutting tool and indirectly the cutting region or workpiece bypassing the cryogenic gas through the cutting tool (Suhaimi, Park et al., 2018). In this method,

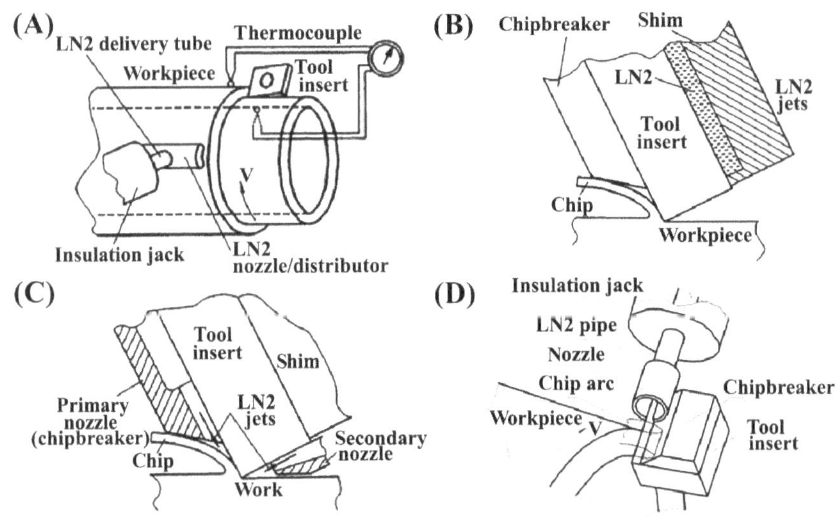

Figure 10.4 Cryogenic cooling trends in machining (A) workpiece precooling, (B) indirect cooling, (C) cryogenic spraying, and (D) cryogenic delivery at chip breaker. *From Yildiz, Y, & Nalbant M. (2008). A review of cryogenic cooling in machining processes.* International Journal of Machine Tools and Manufacture, *48(9), 947–964, with kind permission of Elsevier.*

cryogenic gas decreases the temperature generated in the cutting zone and evaporates. Since cryogenic gas is not in direct contact with the workpiece, it is aimed to prevent negative effects such as dimensional differences that may occur with the effect of cryogenic cooling (Kaushal, Vardhan, Tiwari, & Saluja, 2016). Notwithstanding, it is expressed that the adequacy of this cryogenic cooling technique is profoundly identified with the material properties of the cutting device. This strategy is likewise gets back to cryogenic device cooling and conductive remote cooling (Bayraktar, 2020). In this apparent cryogenic cooling trend, the point is to cool the cutting point by heat conduction from a LN_2 chamber located on the insert surface or tool holder. At the end, the LN_2 is not driven into the instrument or workpiece (Yildiz & Nalbant, 2008). The machining execution can be improved by the roundabout cryogenic cooling strategy since cooling is constrained uniquely by the addition, the LN_2 does not come into contact with the workpiece and does not cause a huge change in the qualities of the workpiece, additionally, the cooling impact is steady. Nevertheless, the impact of this methodology generally relies upon the warm conductivity of the cutting instrument material, the most noteworthy temperature point on the bleeding edge from the LN_2 source, and the addition thickness. It might be progressively successful if a bigger zone of the apparatus connection is in contact with the LN_2 (Yildiz & Nalbant, 2008).

10.2.2.3 Cryogenic gas spraying

This method is accomplished by cooling the cutting zone using two or more external nozzles, by spraying the cryogen into the cutting zone or by sending cryogen through the cutting tool (Senevirathne & Fernando, 2012). In this method, the main purpose is to reduce the temperature of the cutting tool's chemical affinity with the workpiece material by reducing the temperature in the cutting area, thus increasing the machining performance by providing less adhesion and diffusion and preventing the damage to the workpiece material due to high temperatures (Bermingham, Kirsch, Sun, Palanisamy, & Dargusch, 2011; Nalbant & Yildiz, 2011). Consumption of LN_2, and hence production, can be high due to generally overflowing or spraying cooling water into the general cutting area in a machining operation (Pu et al., 2012). Cryogenic jet cooling has obvious advantages (Josyula & Narala, 2018). Cooling power is not squandered in pointless zones, and thus the workpiece is at consistent temperature and is not exposed to dimensional mistake and geometric contortion; this limited cryogenic cooling brings down the instrument surface temperature, expands its hardness and hence decreases the pace of wear; this methodology additionally wrinkles the chips at a chilly temperature and curves the chips with the chip breaker (Liwen, Zhaocheng, Minjie, & Riguang, 2019). This cryogenic processing trend prevents the occurrence of built-up edge (BUE) or built-up layer, which is likely to occur because it reduces the temperature of the chip and provides high-pressure cryogenic flow. As a solution to this problem in materials with high probability of BUE formation during processing, it provides a better surface quality in materials. (Shokrani, Dhokia, Muñoz-Escalona & Newman, 2013; Yildiz & Nalbant 2008).

10.2.2.4 Comparison between cryogenic cooling and other cooling/lubrication systems

It is known that in cryogenic machining, which could be a sustainable choice to traditional metalworking fluids, cryogen, a liquefied gas, is employed in situ of traditional cutting fluids (Muhamad, Ghani, & Haron, 2018). It is delivered mechanically by fragmentary procedure of air and is normally communicated by the shortened form LN_2. Nitrogen liquids are converted from solid phase to liquid phase under medium pressure at −210°C in isolated tanks and begin to boil at −198.79°C (Yildiz & Nalbant, 2008). Nitrogen gas, which makes up four quarters of the air, is a harmless, nonflammable, neutral, and odorless gas. Since it is lighter than air, it scatters into the air after functionality. Thus, any upkeep, cleaning, and removal prerequisites in the wake of machining are dispensed with. Additionally, low temperatures cause expanded hardness and durability in the cutting tool material, prompting higher material removal rates (consequently lower energy consumption) and better efficiency. Recent studies report that cryogenic cooling significantly improves the functional performance of treated components in terms of wear and corrosion resistance and fatigue life (Revuru, Posinasetti, Venkata Ramana, & Amrita, 2017). Process performance indicators such as friction, temperature, cutting forces, surface quality, etc. greatly depend upon the cutting fluid application conditions. Conventional wet or flood cooling that utilizes hydrocarbon oil-based cutting fluids is not effective to attain the best values of the aforementioned indicators (Bordin, Sartori, Bruschi, & Ghiotti, 2017). The use of the cryogenic medium offers the advantage with the fast cooling impact. The most significant cryogenic cooling gases are CO_2 and LN_2 gases (Jawahir et al., 2016). The new hybrid cooling strategy, which uses a small number of oil particles in combination with cryogenic CO_2 or LN_2, has also been developed to ensure the lubrication of the cutting zone in recent years (Çetindağ, Çiçek, & Uçak, 2020). Table 10.1 shows effectiveness of various cooling techniques and sustainability concerns.

10.2.3 Economic aspect of cryogenic machining

If the economic aspect of the cryogenic assisted machining process is to be examined, then the most important parameters are the initial installation cost of the cryogenic machining system and the cost of cryogenic gas, the cost of chip management, and the cost and quality cost of worker health and safety (Anand, Haq & Raina, 2020; Khanna & Agrawal, 2020; Sivaiah & Chakradhar, 2019).

- Cryogenic gas cost: Although the price of cryogenic gas varies from country to country, it is seen that the price of a liter of LN_2 is approximately (\$ 0.75−1.25) if the cost is compared in liters, and these prices will be even lower if produced within companies. LN_2 is a gas that is available worldwide for an easy production. This causes commercial suppliers to be filled and supplied at a competitive price all over the world.
- Chip management cost: Oil-based cutting fluids must be used during the manufacture of hard-to-work materials where dry machining is not possible. With the use of these liquids, the chips exposed during machining will accumulate oil on them. With the heat released during machining, these oils will burn and become harmful to human and environmental health. According to the laws brought on by governments, these chips need to be cleaned and adapted to the medium before disposal. This causes additional expenses for organizations. In any case, on account of cryogenic machining, LN_2 rapidly assimilates heat from the cutting zone and vanishes into the air at typical temperature and weight, guaranteeing that the sawdust is totally dry and clean. It will be simpler to pulverize dry chips as opposed to sleek and overwhelming chips. Hence, expelling the chips, which is an inconvenient procedure for organizations, can be eliminated all the more effectively by utilizing a cryogenic machining strategy.
- Worker health and safety cost: A spotless, sound, and safe working medium expands the proficiency of the workforce and propels it. During machining, it is created in the high-temperature cutting zone.

Table 10.1 Viability and utilization of different cooling/lubricating strategies (Jawahir et al., 2016; Lu et al., 2018).

Effects of the cooling/lubricating strategies	Dry	Flood	MQL	Cryogenic	Hybrid (Cryogenic + MQL)
Cooling	Poor	Good	Average	Outstanding	Outstanding
Lubrication	Poor	Outstanding	Outstanding	Moderate	Outstanding
Metal removing	Poor	Good	Average	Good	Good
Cooling of machine	Poor	Good	Poor	Average	Average
Cooling of workpiece	Poor	Good	Poor	Good	Good
Product quality	Poor	Good	Average	Outstanding	Outstanding
Sustainability concerns	Poor surface integrity due to thermal damage	Water pollution, microbial infection, high cost	Harmful oil vapor	High initial cost	Harmful oil vapor, High initial cost

At this high temperature, the chemical affinity between tool, workpiece, and coolant increases and creates toxic fumes and flood medium. These poisonous vapor and flood mediums are the reasons for dermatitis, bronchitis, and lung issues for machine instrument operators. As a coolant, LN_2 offers a supportable alternative because of its idle nature and capacity to fundamentally diminish the temperature of the cutting zone. In this manner, cryogenic machining is a greener and more secure procedure contrasted with customary machining methods.

10.3 Cryogenic cooling and engineering materials

This section focuses on the effectiveness of cryogenic cooling when machining various engineering materials such as nickel-based alloys, SS, and titanium alloys.

10.3.1 Nickel-based alloys

In a comprehensive investigation conducted by Kaynak (Kaynak, 2014), Inconel 718, which is one of the nickel-based alloys used in aircraft engines, was machined in different mediums (dry, MQL, cryogenic) under different cutting conditions. Cutting force, tool wear, tool life, and chip morphology were the machinability indicators considered. The experimental set was determined at different depth of cut, feed rate, and cutting speeds. To adjust the MQL and cryogenic medium, it was obtained by adjusting the pipe required for the passage of liquid nitrogen gas or MQL oil in parallel with the cutting tool. UNIST Coolube 2210EP metalworking oils are preferred for MQL and sprayed with a flow rate of 60 mL/h at 0.4 MPa air pressure. Liquid nitrogen gas was preferred as the cryogenic medium, and the machining process was carried out with a mass flow rate of about 10 g/s under 1.5 MPa pressure. When comparing the results in different mediums in terms of shear force, it is stated that cryogenic cooling and MQL, in general, contributed to improving shear performance when applied in appropriate shear parameters. In this study, according to the results of experiments performed at low cutting speed, all cutting forces were measured highest in dry medium and lowest in MQL medium. In the cryogenic cooling medium, it formed larger and thicker chips compared to dry and MQL. Although the increase in cutting speed decreased the spiral chip diameter, the use of lubrication/cooling systems did not change the chip breakability. It was observed that cutting tools and geometries, especially chip grooves, had a significant effect on the prevention/limitation of cryogenic gas penetrating the tool–chip contact layer, and play an effective role in reducing efficiency. Apparently the pace of event and movement of flank and cavity wear under chosen conditions were decreased by machining in the cryogenic medium. In experiments carried out at high cutting speeds, the best surface quality was obtained in this cryogenic medium where tool wear and thermal distortions were less. Some surface deformities were seen on the outside of the examples handled in a dry situation and at low cutting pace, and the most exceedingly awful surface was acquired. At the point when the consequences of trials completed in cooling/lubrication mediums were analyzed, it was seen that the surface quality expanded. Considering the results obtained, proper machining is carried out in the machining of the Inconel 718 using cryogenic cooling at high cutting speed. However, it is stated that it is recommended to use MQL machining at a lower cutting speed to improve the machining performance of this material.

Another important study (Dai, Zhuang, Pu, Zhang, & Ding, 2020), Inconel 718 was machined under three different mediums (dry, cryogenic precooling the workpiece and cryogenic cooling the cutting zone) and three different cutting speeds (40, 60, and 80 m/min). To create a cryogenic medium, nitrogen (LN_2) was applied to the cutting zone or the workpiece under a tension of 1 bar and a stream pace of 0.3 L/min. To analyze the impacts of cooling strategies, FEA and experimental analysis were carried out. The most important criteria in material machining applications are hardness, tool wear, surface quality, and fatigue behavior. Inconel 718 superalloy is delicate to immoderate plastic distortion during machining because of high cutting temperature and load. The dislocation formation inside the crystal structure of the material also causes the strain strengthening of the material along with the plastic deformation that occurs in the cutting process. The high cutting heat released in the cutting zone causes the surface hardness to increase as a result of the strain strengthening created by the plastic deformation. As a result, when analyzed for microhardness, its values on the treated region were fundamentally higher than the crude material because of deformation during cutting tasks. On account of the cryogenic held up technology utilized in turning, the surface quality of Inconel 718 part improved. Machining in a dry environment tends to create milder and rugged surfaces, because the surface is not cooled compared to machining in a cryogenic medium. The cooling condition significantly affects the machine influenced zone and also the level of work hardening. The depth of the work hardening layer achieved to around 60 microns without coolant, around 70 microns with precooling with LN_2, and around 80 microns with LN_2 that cools the cutting zone. The surface yield quality of the workpiece was anticipated by Meyer's law with the deliberate microhardness and cross length of the recess. The yield strength of the LN_2 cooled surface was greater than the dry section of this research, which improved the functional performance of the components.

Inconel 625, which is in the group of hard-to-work superalloys, has extensive applications in marine and aviation, due to its high corrosion protection and abrasion resistance. It is a group of materials that attract attention by researchers, due to their hard-to-process characteristics. Narkhede et al. (2020) conducted a comparison of Inconel 625 machining performance under dry and cryogenic cutting mediums with an industrial point of view by investigating tool—chip interface temperature, cutting forces, surface quality, and tool life. Cutting speed, depth of cut, and feed rate were varied at three levels, and Taguchi's L27 was used to design machining experiments. Liquid nitrogen was used to create a cryogenic medium, and then, this gas was sprayed to the cutting zone of the workpiece. In these 27 experiments, the parameters that reached the best surface roughness value were determined and 0.621-micron roughness value was reached for dry medium. In experiments using cryogenic coolant, it was measured as 0.42 less than 30% of this value. It is also seen that there was a significant decrease in the measured force values. As a result of the experiments measured by a thermal camera in both dry and cryogenic mediums, it was determined that the dry medium was 60°C and the cryogenic medium was 17°C. They stated that due to this temperature difference, lesser tool wear obtained in cryogenic cooling-based machining and hence longer tool life. Overall, it was observed that the use of LN_2-based cryogenic cooling increased the turning performance.

10.3.2 Stainless steel

Precipitation hardened (PH) SS are used in industries where mechanical properties such as high strength and corrosion resistance, outstanding wear protection, good malleability, and weldability are critical, such as in aerospace, defense, marine, petrochemical, and nuclear power plants (Arulkirubakaran, Balasubramanian, Raju, Palanisamy, & Manikandan, 2019; Junior, Diniz, & Teixeira Filho, 2009; Kochmański & Nowacki, 2006; Palanisamy & Senthil, 2018; Sivaiah & Chakradhar, 2017).

In an important study carried out by Sivaiah and Chakradhar (Sivaiah & Chakradhar, 2017), turning experiments were conducted in five different feed rates with two different media (wet and cryogenic) with constant cutting speed and depth. An emulsion type cutting liquid was utilized in turning process in flood medium. Liquid nitrogen gas was used in cryogenic machining. A compressor assembly was prepared to compress a TA 55 model cryocan and liquid nitrogen for storage and to transfer the liquid nitrogen to the tool workpiece contact area with a 2 mm diameter nozzle, available up to 3 bar pressure in the tank. The effects of machining parameters during turning the workpiece (17-4 PH SS) in cryogenic and wet medium on chip morphology, surface quality, tool side wear, tangential cutting force, feed force, thrust force, and cutting temperature were investigated and compared. High cutting temperatures are achieved when hard-to-machine materials are processed. Ordinary cutting fluids are utilized to conquer these cutting temperatures because it is undesirable to rise to high temperatures during cutting, but these fluids are unacceptable for health and environmental sustainability. As a results of the analyzes, a major difference was obtained in cryogenic machining, like, surface roughness, cutting force, tool wear, and cutting temperature 35%, 16%, 53%, and 78%, respectively. When the morphology of the chips obtained is examined, it can be seen that the cryogenic machining considerably increased chip breakability and suitable chip forms were produced compared to the chips obtained in the flood medium. Significant improvements have been observed in cryogenic machining, indicating enhancements in the surface and subproperties of the item, just as that the item is created in high quality and increased its manufacturability.

In another study by Sivaiah and Chakradhar (Sivaiah & Chakradhar, 2019), the turning operations of the 17-4 deposited hardened SS workpiece with the help of a tool holder developed by themselves were constructed the conventional cryogenic medium. Because, 17-4 PH SS material is a difficult-to-machining material, the traditional cutting of this material poses difficulties for manufacturers due to many restrictions on the cutting tool, more adhesion wear, and more use of conventional cooling than environmentally friendly arrangements. This challenge encourages manufacturers to seek different machining methods. In this study, the new tool holder, designed as an alternative to the conventional cryogenic method, aims to achieve cryogenic cooling in a new way by simultaneously providing a jet flow to both the cutting tip and side surfaces of the cryogenic cooling water (liquid nitrogen). In this study, the impact of 17-4 PH SS turning parameters processed in various conditions on tool wear, cutting temperature, microhardness, and surface integrity (surface roughness and surface topography) were explored. In both cryogenic cooling trends, experiments were carried out by providing a predefined LN_2 rate of flow 0.45 kg/min. Five different level experiments of feed rate, depth of cut, and speed parameters were carried out using the "one-factor method each time." According to the analysis made after the experiments, it is expressed that the proposed new cryogenic cooling trend fundamentally improved the turning performance attributes on the cryogenic cooling with an outer nozzle approach at the degrees of the all cutting speed, depth of cut, and feed rate, respectively, during the turning of workpiece. Compared with an external nozzle approach and the new cryogenic cooling trend proposed with cryogenic cooling, it was calculated that surface roughness values decreased by 35%, surface wear rate by 29%, and cutting temperature by 61%. Considering these results, it is stated that the recommended cooling trend 17-4 PH SS significantly increases efficiency when machining. The proposed cooling method has been observed to particularly improve the subatomic properties, thereby increasing the fatigue strength of the product obtained.

Khanna and Shah (Khanna & Shah, 2020) conducted turning of 15-5 PH SS in various mediums (dry, wet, MQL, liquid CO_2 supported cryogenic) with constant speed, depth of cut, and feed. The effects of cooling/lubrication on surface roughness (R_a), power consumption, and microhardness measurement, tool tip wear on SEM images, and chip morphology was investigated. In this study, using a 6 mm diameter cooling pipe for wet machining of 15-5 PH SS, a 5% concentration of mineral oil—water was applied to the surface between the insert and the sawdust at a flow rate of 3.75 L/min. By using easter-based polyol synthetic oil with antioxidant additives as MQL liquid, it is sprayed on the chip surface from 20 mm nozzle at 6 bar pressure at 14 mL/h. For cryonic machining, it is provided to apply on the surface between the tip and the sawdust with the help of a 2 mm diameter nozzle from liquid CO_2 thermoplastic pipe. Crater wear was measured as 14 microns in cryogenic machining, 19 microns in wet machining, and 89 microns in MQL. When the results were analyzed from the standpoint of tool life, it was observed that cryogenic machining increases the tool life by 44% contrast to the flood medium and 68% according to the MQL medium. In SEM micrographs, although there is no wear on the chip side surface for cryogenic machining, crack progression in dry machining, cutting edge breakage in wet machining, tool abrasion is detected by peeling the tool end coating in MQL machining. The average chip thickness obtained after cryogenic processing with CO_2 gas is 0.256 mm. Chips of thickness 0.285 mm in dry machining, 0.301 mm in wet machining, and 0.365 mm in MQL machining were obtained. The reason why the chip thickness obtained in dry machining is so close to the chip thickness obtained after cryogenic machining is that BUE occurs at the tool tip during cutting and acts as a different tip at the tool tip. As a result, the cutting force decreases, thus the power consumption decreases. Since the thickness of the chip is directly related to the amount of plastic deformation and the amount of plastic deformation is directly related to the cutting force, the chip thickness is directly proportional to the cutting force. It also has the lowest chip thickness, since it consumes less power than other coolant strategies during cryogenic processing. Since the most power is consumed after MQL-assisted machining, it is calculated that the chip thickness is highest. This shows that a suitable result has been obtained with the literature. In wet machining, the R_a values of the part were lowest. The second best R_a value was also reached in cryogenic machining. The worst R_a value was obtained on the surface machined in dry medium. Low tool wear is the most critical factor affecting R_a. It is believed that the optimum surface quality is accomplished in wet machining because of the emulsion which is constantly sprayed on the cutting tool tip interface with sawdust. In this way, the workpiece surface will be smoother since the tool tip will wear less. During cryogenic machining, the additional uniform and divided discrete style of sawdust was created throughout the machining, whereas the extremely fractured sawdust was ascertained for the dry machining. The smallest spiral sawdust was observed in progression with cryogenic machining.

10.3.3 Titanium alloys

Titanium (Ti) and its alloys are utilized in numerous sectors from the petrochemical industry to the biomedical industry due to their lightness, good corrosion resistance, high strength, long service life, etc. (Gupta and Laubscher, 2017). Titanium is included in the group of lightweight metals due to its density of 4.51 g/cm^3 (Mia et al., 2019). They are classified as pure Ti, α-Ti alloys, α-β Ti alloys, and β-Ti alloys (García-Martínez, Miguel, Martínez-Martínez, Carmen Manjabacas, & Coello, 2019). Difficulties are encountered in the machining of Ti and its alloys (Gupta & Laubscher, 2017). Due to low thermal conductivity, these alloys react chemically with the cutting tool at high temperatures, and due to the relatively small interaction area among the cutting tool/chip, the temperature at the tool tip reaches a very high value. For these reasons, Ti alloys are viewed as hard to machining materials. Their machinability is poor when machining with traditional cutting fluids using flood cooling. Cryogenic or MQL-assisted machining methods are considered to be an alternative to these cooling/lubrication systems, and studies are being carried out worldwide. The cryogenic machining of Ti alloys has increased eight times in the last decade compared to the first decade of 2000. This shows that the importance of cryogenic machining has been increasing in recent years. The LN_2 and CO_2 gases used in the cryogenic machining method stand out with their environmental friendliness as well as providing excellent cooling to the material.

Zhao et al. (2020) conducted a study on the surface quality (surface roughness, microhardness, microstructure, and residual stress) of a machined Ti grade 5 alloy after cryogenic milling. The results obtained are then compared to dry machining to provide sufficient data support to investigate the cryogenic milling mechanisms and industrial applications of the titanium alloy. Milling operations were applied at different cutting speeds to examine the effect of different jet temperatures (dry 20°C, −40°C, −80°C, −120°C, −160°C, −192°C) on the cutting force and surface integrity. The desired temperatures were obtained by adjusting the amount of N_2 in the LN_2 gas, and the temperatures were measured with the help of an infrared thermometer. A cryogenic N_2 jet with 0.6 MPa outlet pressure or LN_2 jet nozzle with 0.1 MPa outlet pressure was adjusted to be sprayed from the surface of the workpiece and tool tip to provide a cooling effect during cutting. When examining how cutting forces vary with temperature and cutting speed, it was observed that force values decreased as temperature decreased at constant cutting speeds. In addition, it was found that the force values increased when the cutting speed was changed where the temperature was determined as constant. The cutting speeds selected in the machining of a material have an important effect on the mechanism of the chip formation. With increasing cutting speed in the machining of traditional materials, chip morphology turns from a continuous structure to a discontinuous structure. In the machining, with the effect of plastic deformation, the chips move toward the front surface of the cutting tool, and when the certain tension level is reached, it causes breakage in the sliding band with the effect of shearing. Under these conditions, there is generally a continuous/chip-toothed chip formation in the machining of titanium. The surface roughness of the material reduces as the temperature of the LN_2 gas sprayed on the workpiece surface decreases, keeping the cutting speed constant. The fundamental purpose behind this is the temperature released during cutting is reduced and thrown out. When the surface quality of the materials processed in the dry environment and the cryogenic environment is compared, it is seen that the surface roughness of the material processed in the cryogenic environment decreases by approximately 24% and 34%. As a result of examining the surface quality in case of changing cutting speeds where the temperature is constant, when cutting speeds increased from 60 to 90 m/min, there was a significant increase in surface roughness. A slight increase was observed during the increase from 90 to 120 m/min. The explanation behind this is believed to be because of the raise of cutting power and the expansion of cutting heat moved to the workpiece as the cutting speed increments. The hardness of the surface treated with cryogenic cooling was found higher. Ti alloys react with oxygen and nitrogen during the cutting process. As a result, it is known that TiO_2 and TiN structures are formed, which greatly increase microhardness on the surface. It is also known that the heat released during cutting reduces the hardness of the surface. Because of the fact that the low thermal conductivity of Ti, this heat concentrates in one region, leading to a decrease in microhardness. In addition, friction of the sawdust on the surface may cause grain thinning on the surface, causing the microhardness to differ between the lower and upper parts of the surface. Regarding the microstructure, both cooling options have resulted in purification and bending to a depth of 10∼15 μm under the treated surface layers. This is mainly due to the solid thermomechanical connection between the chips and the tool part. In any case, the impacts of cutting speeds and cooling mediums on grain bending and thinning in the machined surface layer were not taken note. The cryogenic cooling medium performed better than dry milling to give the treated surface higher pressure residual stress. Additionally, higher shear rate levels not only increase the greatness of compressive residual stress but also increase the depth of the layer to which compressive stress is applied.

Shokrani, Dhokia, and Newman (2016) investigated the effects of cutting fluids on the surface integrity (roughness and microhardness) and microstructure of the material used to increase machinability and achieve higher productivity in machining operations. The main reason for doing this research was the increase in costs with the governments' regulations regarding the disposal and use of these fluids due to the harm caused by cutting fluids to health, environment, and people. As a result, manufacturers and researchers have started to produce projects to eliminate cutting fluids in machining operations. In this study, milling operations were carried out by determining the Ti6Al4V alloy workpiece in three different mediums (dry, wet, and cryogenic) and three different cutting speeds, feed rate, and cutting depth parameters. Experiments were conducted based on Taguchi L9 technique. Liquid nitrogen (LN_2) was used at −196°C at 1 bar pressure and 20 kg/h mass flow rate for cryogenic machining experiments. For this reason, a 0.8 mm cut and a 300:1 bandwidth and a 4 mm evaluation length were used for each measurement. Average surface roughness value was calculated by taking different measurements from each sample. Also, variance analysis was performed by taking the surface roughness values obtained. In addition to the average surface roughness, a cross-section of the machined block was prepared for the subsurface microhardness test. Sections of machined blocks were taken and microhardness tests were performed by applying 100 g load for 15 s at 10 micron intervals. The microhardness value was calculated by measuring the diagonals in the Vickers pyramid image created on the surface. Besides, scanning electron microscope was used to detect surface imperfections during cutting. It was found that, although the average roughness value decreases when the cutting speed increases from 30 to 115 m/min, roughness value does not change much when it increases from 115 to 200 m/min. The roughness value increases as the rate of feed increases (from 0.03 to 0.1 mm/tooth). Likewise, as the depth of cut increases (1−5 mm) the roughness value increases. It was found that the least roughness value was obtained in cryogenic medium. Cryogenic cooling has greatly improved the surface roughness in milling processes. The average roughness values of the samples obtained in cryogenic milling were 18% and 21% less compared to wet and dry milling, respectively. The lowest average roughness 0.58 micron was measured in cryogenic machining. The test parameters obtained are as follows: cutting speed is 115 m/min, the feed rate is 0.03 mm/tooth, and 5 mm depth of cut, which is 30% lower than wet machining and 40% lower than dry machining. Investigation of surface microhardness of machined samples showed that the highest increase in microhardness occurred in cryogenic machining. It is concluded that in machining Ti6Al4V alloy, cryogenic cooling significantly improved the surface integrity compared to dry and wet conditions.

Huang et al. (2020) machined Ti6Al4V under various environmentally friendly cooling conditions, that is, cryogenic gas (CG), MQL, and MQL + cryogenic. In this study, the oil flow rate used in the MQL method was 0.2 mL/min, and the preferred nitrogen gas in the cryogenic gas method was used to cool via a nozzle with a diameter of approximately 2 mm, which was quite small under the pressure of 0.2 MPa and at a flow rate of 54.2 L/min. In addition, copper pipes were used in the section from the tank to the nozzle and the coldness of the nitrogen gas was controlled by thermocouples. The experiments were carried out in different cooling systems and constant turning parameters. The roughness values were calculated utilizing a highly sensitive optical surface profilometer based on a Bruker Contour GT-X8 white light interferometer. The dominant factor affecting surface formation during the turning of the Ti6Al4V at low spindle speed was elastic recovery. Unlike the conventional cut with a relatively large cutting depth, each CG + MQL and CG methods resulted in higher roughness than the MQL method. This can be referred to elastic recovery caused by low temperatures from CG and a significant reduction in minimum uncut chip thickness. The surface hardness values of the Ti6Al4V alloy turned at different spindle speeds were measured, and it was found that the hardness increased up to a certain speed, and then, it decreased. It is stated that the reason is strain hardening and thermal softening. Similar results were found in every cooling system. At fixed spindle speeds, the hardest surface was determined to be the least hardened surface in MQL in CG system. However, the difference between the hardest surface and the softest surface was not much. The cutting forces were compared at different spindle speeds, and the minimum force value was obtained at the highest spindle speed. When compared between cooling systems, it was seen that CG achieved better results at all speed values. While the difference is high at low spindle speeds, similar cutting force values are obtained at high speeds. SEM images were taken after performing a 50 m workpiece cutting by selecting a tool tip to investigate the effect of tool wear on cooling systems. As a result of the examinations, "built-up-edge" was observed on the rake surface in all cooling systems. It was observed that the workpiece was less worn in the MQL system. This study identified CG + MQL as the best strategy due to obtaining the best surface quality and positive contribution to the environment and human health.

10.4 Summary

This chapter reported various aspects of cryogenic cooling-based machining. The investigations conducted by researchers found this technique effective and better as compared to other cooling and lubrication techniques such as flood cooling, MQL, and dry machining etc. in terms of better work surface quality, longer tool life, lower machining cost, lesser environmental footprints, and safer operations. Hybridization of cryogenic and MQL has also found to be of special interest and effective results. Some future attempts are required in the direction of minimizing the cost of cryogenic systems, investigating more cryogenic gases and their combinations, machining some more typical materials such as composites, and polymers under cryogenic environment, and extending sustainability studies etc.

References

Akincioğlu, S., Gökkaya, H., & Uygur, İ. (2015). A review of cryogenic treatment on cutting tools. *The International Journal of Advanced Manufacturing Technology*, *78*(9–12), 1609–1627.

Anand, R., Haq, M. I. U., & Raina, A. (2020). *Bio-based nano-lubricants for sustainable manufacturing. Nanomaterials and environmental biotechnology* (pp. 333–380). Springer.

Arulkirubakaran, D., Balasubramanian, K., Raju, R., Palanisamy, D., & Manikandan, N. (2019). Machinability studies on precipitation hardened stainless steel using cryo-treated textured carbide inserts. *Journal of Scientific and Industrial Research*, *78*, 216–222.

Balaji, V., Ravi, S., & Chandran, P. N. (2018). Optimization on cryogenic Co_2 machining parameters of AISI D2 steel using taguchi based grey relational approach and TOPSIS. *International Journal of Engineering & Technology*, 7(3.12), 885.

Balasubramanian, S., Gupta, M. K., & Singh, K. (2012). Cryogenics and its application with reference to spice grinding: A review. *Critical Reviews in Food Science and Nutrition*, *52*(9), 781–794.

Baldissera, P., & Delprete, C. (2008). Deep cryogenic treatment: A bibliographic review. *The Open Mechanical Engineering Journal*, *2*(1).

Bayraktar, S. (2020). *Cryogenic cooling-based sustainable machining. High speed machining* (pp. 223–241). Elsevier.

Bermingham, M., Kirsch, J., Sun, S., Palanisamy, S., & Dargusch, M. S. (2011). New observations on tool life, cutting forces and chip morphology in cryogenic machining Ti-6Al-4V. *International Journal of Machine Tools and Manufacture*, *51*(6), 500–511.

Bordin, A., Sartori, S., Bruschi, S., & Ghiotti, A. (2017). Experimental investigation on the feasibility of dry and cryogenic machining as sustainable strategies when turning Ti6Al4V produced by additive manufacturing. *Journal of Cleaner Production*, *142*, 4142–4151.

Boswell, B., Islam, M. N., Davies, I. J., Ginting, Y. R., & Ong, A. K. (2017). A review identifying the effectiveness of minimum quantity lubrication (MQL) during conventional machining. *The International Journal of Advanced Manufacturing Technology*, *92* (1–4), 321–340.

Bruschi, S., & Ghiotti, A. (2018). *Cryogenic machining of titanium alloys. Machining of light alloys: Aluminum, titanium, and magnesium* (p. 179). Taylor & Francis Goup, CRC Press.
Çetindağ, H. A., Çiçek, A., & Uçak, N. (2020). The effects of CryoMQL conditions on tool wear and surface integrity in hard turning of AISI 52100 bearing steel. *Journal of Manufacturing Processes, 56*, 463–473.
Dai, X., Zhuang, K., Pu, D., Zhang, W., & Ding, H. (2020). An investigation of the work hardening behavior in interrupted cutting Inconel 718 under cryogenic conditions. *Materials, 13*(9), 2202.
Deshpande, Y. V., Andhare, A. B., & Padole, P. M. (2018). How cryogenic techniques help in machining of nickel alloys? A review. *Machining Science and Technology, 22*(4), 543–584.
Dinesh, S., Senthilkumar, V., & Asokan, P. (2017). Experimental studies on the cryogenic machining of biodegradable ZK60 Mg alloy using micro-textured tools. *Materials and Manufacturing Processes, 32*(9), 979–987.
Fratila, D. (2013). *Sustainable manufacturing through environmentally-friendly machining. Green manufacturing processes and systems* (pp. 1–21). Springer.
García-Martínez, E., Miguel, V., Martínez-Martínez, A., Carmen Manjabacas, M., & Coello, J. (2019). Sustainable lubrication methods for the machining of titanium alloys: An overview. *Materials, 12*(23), 3852.
Ghosh, S., & Rao, P. V. (2015). Application of sustainable techniques in metal cutting for enhanced machinability: A review. *Journal of Cleaner Production, 100*, 17–34.
Gill, S. S., & Singh, H. (2013). *Cryogenic treatment of materials: cutting tools and polymers. Polymers at cryogenic temperatures* (pp. 245–273). Springer.
Gill, S. S., Singh, R., Singh, H., & Singh, J. (2009). Wear behaviour of cryogenically treated tungsten carbide inserts under dry and wet turning conditions. *International Journal of Machine Tools and Manufacture, 49*(3–4), 256–260.
Gupta, K. (2020). *Materials forming, machining and post processing*. Springer.
Gupta, K., & Laubscher, R. F. (2017). Sustainable machining of titanium alloys: A critical review. *Proceedings of the Institution of Mechanical Engineers, Part B: Journal of Engineering Manufacture, 231*(14), 2543–2560.
Hegab, H., Darras, B., & Kishawy, H. (2018). Towards sustainability assessment of machining processes. *Journal of Cleaner Production, 170*, 694–703.
Hsien, W. L. Y. (2015). *Towards green lubrication in machining*. Springer.
Huang, P., Li, H., Zhu, W.-L., Wang, H., Zhang, G., Wu, X., … Zhu, Z. (2020). Effects of eco-friendly cooling strategy on machining performance in micro-scale diamond turning of Ti–6Al–4V. *Journal of Cleaner Production, 243*, 118526.
Isakson, S. (2018). *Cryogenic machining of Ti-6Al-4V*. Department of Industrial and Materials Science, Chalmers University of Technology.
Jawahir, I., & Balaji, A. (2000). Predictive modeling and optimization of turning operations with complex grooved cutting tools for curled chip formation and chip breaking. *Machining Science and Technology, 4*(3), 399–443.
Jawahir, I. S., Attia, H., Biermann, D., Duflou, J., Klocke, F., Meyer, D., … Umbrello, D. (2016). Cryogenic manufacturing processes. *CIRP Annals, 65*(2), 713–736.
Jerold, B. D., & Kumar, M. P. (2012). Experimental comparison of carbon-dioxide and liquid nitrogen cryogenic coolants in turning of AISI 1045 steel. *Cryogenics, 52*(10), 569–574.
Josyula, S. K., & Narala, S. K. R. (2018). Performance enhancement of cryogenic machining and its effect on tool wear during turning of Al-Ticp composites. *Machining Science and Technology, 22*(2), 225–248.
Junior, A. B., Diniz, A. E., & Teixeira Filho, F. (2009). Tool wear and tool life in end milling of 15–5 PH stainless steel under different cooling and lubrication conditions. *The International Journal of Advanced Manufacturing Technology, 43*(7–8), 756.

Kalia, S. (2010). Cryogenic processing: a study of materials at low temperatures. *Journal of Low Temperature Physics, 158*(5–6), 934–945.

Kalsi, N. S., Sehgal, R., & Sharma, V. S. (2010). Cryogenic treatment of tool materials: A review. *Materials and Manufacturing Processes, 25*(10), 1077–1100.

Kaushal, A., Vardhan, A., Tiwari, A. C., & Saluja, S. K. (2016). Study the effect of cryogenic cooling on orthogonal machining process. *American Journal of Engineering Research (AJER), 5*, 96–104.

Kaynak, Y. (2014). Evaluation of machining performance in cryogenic machining of Inconel 718 and comparison with dry and MQL machining. *The International Journal of Advanced Manufacturing Technology, 72*(5–8), 919–933.

Kaynak, Y., & Gharibi, A. (2018). Progressive tool wear in cryogenic machining: the effect of liquid nitrogen and carbon dioxide. *Journal of Manufacturing and Materials Processing, 2*(2), 31.

Khanna, N., & Agrawal, C. (2020). *Titanium machining using indigenously developed sustainable cryogenic machining facility. Materials Forming, Machining and Post Processing* (pp. 183–205). Springer.

Khanna, N., & Shah, P. (2020). Comparative analysis of dry, flood, MQL and cryogenic CO2 techniques during the machining of 15-5-PH SS alloy. *Tribology International, 146*, 106196.

Kochmański, P., & Nowacki, J. (2006). Activated gas nitriding of 17-4 PH stainless steel. *Surface and Coatings Technology, 200*(22–23), 6558–6562.

Li, S., Xiao, M., Ye, G., Zhao, K., & Yang, M. (2018). Effects of deep cryogenic treatment on microstructural evolution and alloy phases precipitation of a new low carbon martensitic stainless bearing steel during aging. *Materials Science and Engineering: A, 732*, 167–177.

Liwen, H., Zhaocheng, W., Minjie, W., & Riguang, M. (2019). Research on mechanism of effect of cooling temperature on chip morphology of difficult-to-cut alloys. *Tool Engineering, 9*, 5.

Lu, T., Kudaravalli, R., & Georgiou, G. (2018). Cryogenic machining through the spindle and tool for improved machining process performance and sustainability: Pt. I, system design. *Procedia Manufacturing, 21*, 266–272.

Mia, M., Kumar Gupta, M., Adolfo Lozano, J., Carou, D., Yu Pimenov, D., Królczyk, G., ... Ranjan Dhar, N. (2019). Multi-objective optimization and life cycle assessment of eco-friendly cryogenic N2 assisted turning of Ti-6Al-4V. *Journal of Cleaner Production, 210*, 121–133.

Morkavuk, S., Köklü, U., Bağcı, M., & Gemi, L. (2018). Cryogenic machining of carbon fiber reinforced plastic (CFRP) composites and the effects of cryogenic treatment on tensile properties: A comparative study. *Composites Part B: Engineering, 147*, 1–11.

Muhamad, S. S., Ghani, J. A., & Haron, C. H. C. (2018). A review on future implementation of cryogenic machining in manufacturing industry. *Progress in Industrial Ecology, an International Journal, 12*(3), 260–283.

Nalbant, M., & Yildiz, Y. (2011). Effect of cryogenic cooling in milling process of AISI 304 stainless steel. *Transactions of Nonferrous Metals Society of China, 21*(1), 72–79.

Narkhede, G. B., Chinchanikar, S., Vadgeri, S. S., & Rathod, B. S. (2020). Comparative evaluation of machining performance of inconel 625 under dry and cryogenic cutting conditions. *IOP Conference Series: Materials Science and Engineering, 810*(1), 012036.

Özbek, N. A., Çİçek, A., Gülesİn, M., & Özbek, O. (2016). Application of deep cryogenic treatment to uncoated tungsten carbide inserts in the turning of AISI 304 stainless steel. *Metallurgical and Materials Transactions A, 47*(12), 6270–6280.

Palanisamy, D., & Senthil, P. (2018). A comparative study on machinability of cryo-treated and peak aged 15Cr-5Ni precipitation hardened stainless steel. *Measurement, 116*, 162–169.

Pervaiz, S., Kannan, S., & Kishawy, H. A. (2018). An extensive review of the water consumption and cutting fluid based sustainability concerns in the metal cutting sector. *Journal of Cleaner Production*, *197*, 134–153.

Pu, Z., Outeiro, J. C., Batista, A. C., Dillon, O. W., Puleo, D. A., & Jawahir, I. S. (2012). Enhanced surface integrity of AZ31B Mg alloy by cryogenic machining towards improved functional performance of machined components. *International Journal of Machine Tools and Manufacture*, *56*, 17–27.

Pusavec, F., Courbon, C., Rech, J., & Kopac, J. (2014). *An investigation of the effect of nitrogen phase on cryogenic machining performance and a case study on machining of inconel 718 alloy. International Manufacturing Science and Engineering Conference*. American Society of Mechanical Engineers.

Revuru, R. S., Posinasetti, N. R., Venkata Ramana, V. S. N., & Amrita, M. (2017). Application of cutting fluids in machining of titanium alloys—A review. *The International Journal of Advanced Manufacturing Technology*, *91*(5–8), 2477–2498.

Schoop, J., Ambrosy, F., Zanger, F., Schulze, V., Balk, T. J., & Jawahir, I. S. (2016). Cryogenic machining of porous tungsten for enhanced surface integrity. *Journal of Materials Processing Technology*, *229*, 614–621.

Senevirathne, S., Fernando M. 2012. Effect of cryogenic cooling on machining performance on hard to cut metals—A literature review. In: Proc. National Engineering Conference.

Senthilkumar, D., Rajendran, I., Pellizzari, M., & Siiriainen, J. (2011). Influence of shallow and deep cryogenic treatment on the residual state of stress of 4140 steel. *Journal of Materials Processing Technology*, *211*(3), 396–401.

Sert, A., & Celik, O. N. (2019). Characterization of the mechanism of cryogenic treatment on the microstructural changes in tungsten carbide cutting tools. *Materials Characterization*, *150*, 1–7.

Shokrani, A., Dhokia, V., Muñoz-Escalona, P., & Newman, S. T. (2013). State-of-the-art cryogenic machining and processing. *International Journal of Computer Integrated Manufacturing*, *26*(7), 616–648.

Shokrani, A., Dhokia, V., & Newman, S. T. (2012). Environmentally conscious machining of difficult-to-machine materials with regard to cutting fluids. *International Journal of Machine Tools and Manufacture*, *57*, 83–101.

Shokrani, A., Dhokia, V., & Newman, S. T. (2016). Investigation of the effects of cryogenic machining on surface integrity in CNC end milling of Ti–6Al–4V titanium alloy. *Journal of Manufacturing Processes*, *21*, 172–179.

Shukla, S., & Bajpai, V. (2019). *Cryogenic Machining. Innovations in Manufacturing for Sustainability* (pp. 29–52). Springer.

Sivaiah, P., & Chakradhar, D. (2017). Influence of cryogenic coolant on turning performance characteristics: A comparison with wet machining. *Materials and Manufacturing Processes*, *32*(13), 1475–1485.

Sivaiah, P., & Chakradhar, D. (2019). The effectiveness of a novel cryogenic cooling approach on turning performance characteristics during machining of 17-4 PH stainless steel material. *Silicon*, *11*(1), 25–38.

Stachurski, W., Sawicki, J., Wójcik, R., & Nadolny, K. (2018). Influence of application of hybrid MQL-CCA method of applying coolant during hob cutter sharpening on cutting blade surface condition. *Journal of Cleaner Production*, *171*, 892–910.

Suhaimi, M. A., Park, K.-H., Yang, G.-D., Sharif, S., & Kim, D.-W. (2018). Effect of cryogenic high-speed milling of compacted graphite iron using indirect spray system. *The International Journal of Advanced Manufacturing Technology*, *99*(9–12), 2149–2157.

Suhaimi, M. A., Yang, G.-D., Park, K.-H., Juzaili Hisam, M., Sharif, S., & Kim, D.-W. (2018). Effect of cryogenic machining for titanium alloy based on indirect, internal and external spray system. *Procedia Manufacturing*, *17*, 158–165.

Tahir, W., & Ikramullah S. (2013). Optimization and analysis of cutting parameters using cryogenic media in machining of high strength alloy steel. Conference: International Conference on Manufacturing Research (ICMR2013).

Vukelic, D., Simunovic, K., Simunovic, G., Saric, T., Kanovic, Z., Budak, I., & Agarski, B. (2020). Evaluation of an environment-friendly turning process of Inconel 601 in dry conditions. *Journal of Cleaner Production*, 121919.

Wang, Y., Dai, M., Liu, K., Liu, J., Han, L., & Liu, H. (2020). Research on surface heat transfer mechanism of liquid nitrogen jet cooling in cryogenic machining. *Applied Thermal Engineering*, 115607.

Wolf, J., Investigation of Chip-form and Tool-wear in Turning of Hardened AF9628 Alloy Under Various Cooling and Lubrication Conditions. 2019.

Yildiz, Y., & Nalbant, M. (2008). A review of cryogenic cooling in machining processes. *International Journal of Machine Tools and Manufacture*, *48*(9), 947–964.

Zhao, W., Ren, F., Iqbal, A., Gong, L., He, N., & Xu, Q. (2020). Effect of liquid nitrogen cooling on surface integrity in cryogenic milling of ti-6al-4 v titanium alloy. *The International Journal of Advanced Manufacturing Technology*, *106*(3), 1497–1508.

CHAPTER 11

Sustainability assessment in manufacturing: perspectives, challenges, and solutions

Manish Kumar[1] **and Monto Mani**[2]
[1]Centre for Product Design and Manufacturing, Indian Institute of Science, Bangalore, India
[2]Centre for Sustainable Technologies and Centre for Product Design and Manufacturing, Indian Institute of Science, Bangalore, India

Abbreviations

GRI	Global reporting initiative
LCA	Life-cycle assessment
SA	Sustainability assessment
SM	Sustainable manufacturing
WCED	World Commission on Environment and Development

11.1 Manufacturing and sustainability

Manufacturing can be defined as an activity to convert raw materials into usable products adopting various manufacturing processes (Groover, 2007). It has evolved from skill and labor intensive traditional manufacturing to highly energy-intensive automated manufacturing. It is a part of a much larger product life cycle that begins with design. At present, manufacturing of products involves international outsourcing that are used and disposed across in different geographies. Manufacturing consists of production machines and automation systems supported by information and communication technologies that require minimum human effort. Major economies rely on manufacturing for economic development since the industrial revolution. The location of manufacturing has changed over time depending on the availability of materials, cheap labor, and consumer markets (Gutowski, Allwood, Herrmann, & Sahni, 2013). Increased demand and complexity [at process, product, systems, organization, and market level (ElMaraghy, ElMaraghy, Tomiyama, & Monostori, 2012)], energy requirements to process a wide array of materials, and environmental emissions caused due to the processing have given rise to the

Sustainable Manufacturing
DOI: https://doi.org/10.1016/B978-0-12-818115-7.00013-4

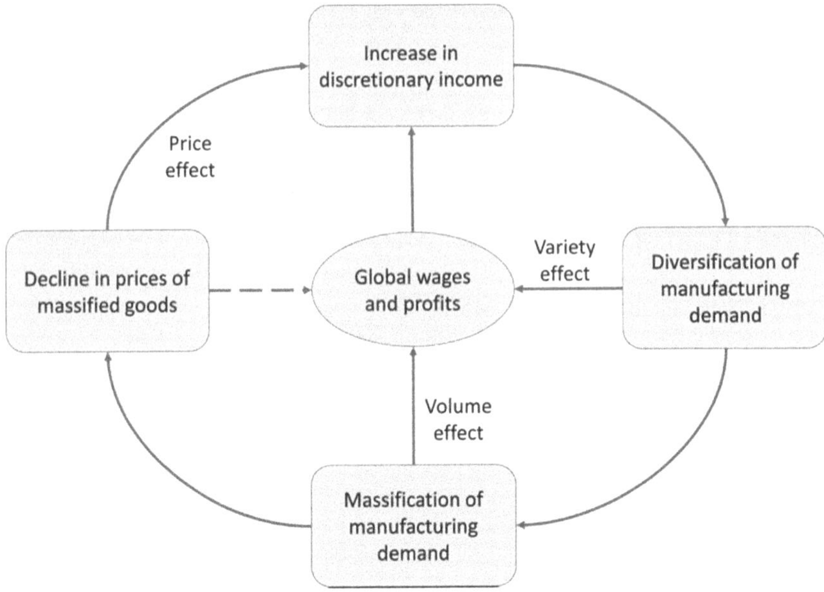

Figure 11.1 Global cycle of manufacturing, demand, and economy (UNIDO, 2018).

concept of material efficiency (Allwood, Ashby, Gutowski, & Worrell, 2013). Fig. 11.1 presents the virtuous cycle of industrial production and its effect on the demand of goods, global economy, and employment generation. Consumption or demand sustains manufacturing and leads to income generation by means of inclusive industrial development (UNIDO, 2018). Mass consumption of goods also impacts the environment by demanding resources and generating waste.

Industry is the second highest contributor to greenhouse gas (GHG) emissions in India (World Resource Institute, 2018) and shared 33% of global GHG emissions in 2014 (Rissman et al., 2020). Various strategies have evolved to operationalize material efficiency, such as, reduction of material usages in products and modular design of products for reuse, remanufacturing, recycle, etc., leading to concept of circular economy and smart manufacturing (World Resource Institute, 2018). Circular economy is a modern concept to move from open material and energy loops towards the closed loops by incorporating strategies such as reuse, repair, remanufacturing, and long-lasting designs (Geissdoerfer, Savaget, Bocken, & Hultink, 2017). Resource availability thus has become an important factor for manufacturing. Metal criticality assessment is a framework to estimate resource availability for various metals (Graedel et al.,

2012). Sustainability remains the central objective for advanced manufacturing developments (Esmaeilian, Behdad, & Wang, 2016), that is, to ensure the availability of resources required for manufacturing and to sustain a healthy environment and healthy lifestyle of communities. Rising concerns of sustainability followed by global initiatives such as Agenda 2030 have drawn attention of policy makers towards industrial sectors to develop guidelines for doing responsible business for sustainable development. Therefore, it is important to conduct sustainability assessments (SA) to address sustainability challenges and continue the virtuous cycle of manufacturing to fulfill the requirements of the global supply and demand and support the global economy. The key is to maintain a balance between impacts associated with the product life cycle and resource-use in the pursuit of sustainability.

In this chapter, Section 11.1 discusses manufacturing, its classification, sustainability perspectives, and SA. The concept of sustainable manufacturing (SM), existing tools for SA in manufacturing, their strengths and limitations, and application in industry have been discussed in Section 11.2. Section 11.3 presents challenges to implement SA tools and methodologies to overcome them. Section 11.4 summarizes the chapter.

11.1.1 Manufacturing: classification and levels

Two Latin words *manus* (hand) and *factus* (make), whose combination means "made by hand", were the origin of the word "manufacture" (Kalpakjian, 2001). Manufacturing is often used synonymously with production and is conceptually different from production. Production is the act of physically making the product from material constituents. In contrast, manufacturing encompasses several activities from material extraction to the selling of products and has a broader perspective as it includes entire considerations (such as product design, material sourcing, warehousing, supply chain, etc.) connected with material processing throughout the product life cycle (Herausgeber, 2004).

Manufacturing enterprises can be classified as micro-, small-, and medium-sized enterprises based on their turnover and number of employees. From a technology perspective, manufacturing firms can be classified as supplier-dominated, production-intensive, and science-based (Pavitt, 1984). Industrial classification systems vary from national to international standards, for example, Indian National Industrial Classification system categorizes manufacturing based on various products (such as food products, textiles, chemicals, metals, electronics, machinery, and etc.) (MoSPI, 2008). Global

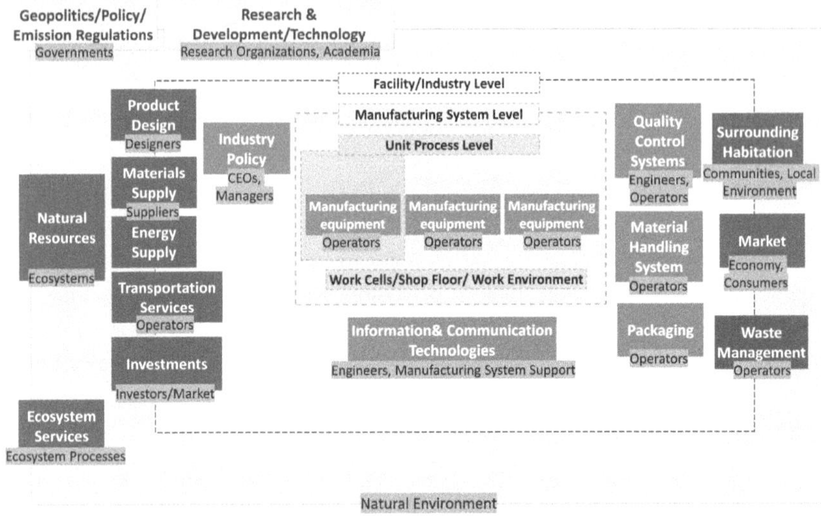

Figure 11.2 Manufacturing and associated stakeholders/entities.

Industrial Classification System comprises various industry sectors such as energy, materials, industrial, consumer discretionary, consumer staples, healthcare, etc. (MSCI 2020). North American Industry Classification System considers business involved with the mechanical, physical, and chemical transformation of substances and components, and assembly of component parts excluding construction under manufacturing classification (U.S. Office of Management and Budget). Also. manufacturing can be seen as operating at different levels, that is, unit manufacturing process, manufacturing system involving multiple machines, factory, multifactory, and global supply chains distributed across different geographies (Duflou et al., 2012). Fig. 11.2 shows various levels of manufacturing and various entities (highlighted in gray color) that affect its functionality. A manufacturing system can be characterized based on unit manufacturing operations and machines, layout of manufacturing system, level of automation involved, and variety of products (Groover, 2007). Decision makers at different levels of manufacturing face diverse operational challenges attributed to varying perspectives and system boundaries. Process parameters for individual processes are important for the machine operator. Sustainability indicators such as energy use, tool wear, waste generated, indoor (near-field) emissions, operator's safety and health, etc. are relevant at this level. Process parameters such as manufacturing precision determined during the production phase affects product performance and causes impacts in the use phase of product (Helu, Vijayaraghavan, & Dornfeld, 2011). Manufacturing system engineers focus on minimizing resource use (energy, water, and material) to produce a product with intended design parameters in a cost-effective way. Supply chain decisions have upstream (electricity generation and sourcing of various resources) and downstream impacts and require better data, tools, and connected global measurement systems for sustainability improvements (O'Rourke, 2014). Material sourcing, fair labor, and responsible business are some aspects imperative at the supply chain level. Social life-cycle assessments are important for assessing the conduct of industries at the supply chain level. Profit, procurement, and investment decisions-related indicators are important at the factory level. Further, materials selected in the design phase also have bearing on environmental impacts, human health and hence sustainability. Circularity indicators such as potential for reuse and recycle, end of life management are important to consider at the design stage. Moreover, the facility's location also plays an important role in estimating the sustainability performance of the manufacturing industry (Chen, Olhager, & Tang, 2014).

Manufacturing involves a wide variety of materials and geographically distributed production processes and activities. Lack of information sharing across product life-cycle stages and requirement of appropriate data for each stakeholder in the life-cycle stage are few challenges reported earlier in literature (Brundage et al., 2018). A good performance measure for a manufacturing system should be simple, proactive, and traceable at all levels throughout the manufacturing organization and should be able to be used by all decision-makers (Hon, 2005).

Implementing sustainability-oriented measures is vital at all levels of manufacturing to avoid suboptimization at individual level and therefore demands a holistic view. Systems-level innovation is important to guide process-level changes, and process-level interventions are important to fulfill systems-level sustainability requirements. Understanding the geographically distributed nature of manufacturing and its structure involving various levels is therefore crucial for operationalizing sustainability in manufacturing and effective decision-making.

11.1.2 Sustainability principles and perspectives

The term sustainability was popularized globally in the Brundtland commission's (formerly known as WCED) report "*Our Common Future*" in 1987, with a critical focus on securing environmental resources for future generations (General Assembly of the United Nations, 1987). The term has been debated immensely, and various definitions and interpretations have emerged. It has been researched from different point of views in history and has evolved with changing priorities of human beings such as maintenance of social order, stability, and rising concerns of limits to growth and environmental damages due to modern development and consumerism. Sustainability is considered to be associated with different capitals (natural, social, and economic). Natural capital consists of natural resources and natural services (also called ecosystem services). The weak sustainability perspective gives importance to economic capital, whereas strong sustainability focuses on conserving natural capital and intergenerational equity (Bell and Morse, 2012; Sala, Farioli, & Zamagni, 2013). (Glavič and Lukman, 2007) reviewed 51 sustainability terms and their definitions adopted by various authors and organizations. They observed most definitions fall under the environmental sciences domain, with an emphasis on the interconnections among various systems. The commonly asked questions are: what is to be sustained, in which space (spatial aspect), and for how long (temporal aspect) the sustainability has to be attained (Council, 1999; Bell and Morse, 2012).

Sustainability broadly deals with the ability of the system in its entity (earth) to support and sustain a healthy life; assessing sustainability requires holistic (integrated) consideration of human and environmental systems and information from diverse disciplines (Monto, Ganesh, & Varghese, 2005). Sustainability can be defined at the level of dynamic interactions between complex human systems and environmental systems and must include environmental, political, economic, and socio-cultural factors (Clayton and Radcliffe, 2018). Therefore, human beings are at the center of concerns for sustainable development. The term sustainability should always be referred to as an integrated pathway of development resulting from interactions of human and natural systems driven by context-dependent goals involving various actors and institutions (Clark & Harley, 2020). The sustainability of human species can only be defined and assessed at the level of dynamic interactions between human and environmental systems, with the anthropocentric concern of sustaining the environment and its services as a resource provider (a stand which is also challenged and debated). Human and natural systems are complex unbounded systems that involve multidimensional dynamic interactions. The sustainability of human species can only be determined by capturing complex interactions between human and environmental systems, requiring a systems approach and knowledge from wide range of disciplines (Clayton and Radcliffe, 2018). Systems thinking involves understanding system structure (i.e., interconnections among the entities and feedback mechanics) and its dynamic behavior (stock and flow, and nonlinear relationships) at different scales; to predict system behavior and produce desired outcomes by devising modifications (Arnold and Wade, 2015).

A plethora of terms have been coined to discern sustainability, with different approaches evolving from different domains, for example, *Green Chemistry* focuses on environmental and human health impacts due to chemical processes (Manley, Anastas, & Cue, 2008; Anastas, 1999), *Sustainable Production* (manufacturing used interchangeably) focuses on products and processes to minimize the use of energy and material, waste and emissions, and maximize profit and recyclability throughout life cycle (Lowell Centre for Sustainable Production, n.d.; US Department of Commerece, 2007), and *Industrial Ecology*, systemically studies integrated industrial and natural systems (Ehrenfeld, 2004; Ehrenfeld, 1997; Seager and Theis, 2002). *Circular Economy* focuses on closed-loop cycles to regenerate natural systems by maximizing recycling, reuse, and recovery of materials and energy (The Ellen MacArthur Foundation, n.d.; Stahel, 2016) and is now a prerequisite consideration for sustainability. The concept of planetary boundaries has been introduced to practicably define the limits that should not be transgressed to ensure sustainability (Rockström et al., 2009). Despite a variety of perspectives and definitions, there is no silver bullet to ensure sustainability. Therefore an integrated approach for context-specific time-scale and spatial boundaries needs to be followed to operationalize sustainability.

11.1.3 Sustainability assessment

SA is essentially an extended version of the existing general environmental assessment techniques such as impact assessment, environmental impact assessment, and strategic environment assessment (SEA) or theme-specific assessments such as social impact assessment (SIA), health impact assessment, etc. Ehrenfeld (1997; Seager and Theis 2002). There are multiple interpretations and terminologies used for SA, such as sustainability appraisal, integrated assessment, triple bottom line, etc.

SA approaches have been broadly classified as indicator/index oriented, product-related, and integrated assessments (Ness, Urbel-piirsalu, Anderberg, & Olsson, 2007). Details about various tools developed under the indicators category can be found in Singh, Murty, Gupta, & Dikshit (2012). Integrated assessments attempt to capture relationships between various aspects (environmental, social, and economic) of system entities and requires multidisciplinary knowledge (Hák, Moldan, & Dahl, 2012). Another set of classification categorizes SA tools into three categories: monetary, biophysical, and indicator-based (Gasparatos and Scolobig, 2012). Existing approaches follow either top-down (aggregated at national or regional level) or bottom-up approach (extensive data collection throughout the life cycle) (Gasparatos, El-Haram, & Horner, 2008). Future scenarios oriented SA tools can be categorized as procedural, analytical, combined scenario and assessment, strategic tools, and tools for aggregating impacts (Fauré, Arushanyan, Ekener, Miliutenko, & Finnveden, 2017).

Environmental Life-Cycle Assessment (E-LCA) or life-cycle assessment (LCA) framework has been accepted as ISO standards and widely practiced to estimate negative impacts associated with product or processes in different impact categories (International Organization for Standardization, 2006). Research in LCA has led to the development of consequential LCA to address the consequences of change in a market activity, regionalized LCA to address spatial (site-specific) challenges (Mutel et al., 2018; Margni et al., 2018), dynamic LCA to address temporal (short- and long-term) challenges, and work environment LCA (specifically for occupational settings). E-LCA covers a wide range of impact categories, with the latest recommendations suggesting the inclusion of new categories such as ecosystem services, socio-economic assets, and cultural and natural heritage (Verones et al., 2017). E-LCA has been successfully adopted widely for diverse sectors (manufacturing, human settlements, waste management, etc.) to support decision-making (Hellweg and Milà i Canals, 2014). Studies on exposure due to toxic materials throughout the supply chain have been evolving in recent LCA literature. More studies are being conducted to study near-field impacts in various product life-cycle stages (Huang, Ernstoff, Fantke, Csiszar, & Jolliet, 2017). Integration of LCA and risk assessment can be a potential solution to address site-specific impacts from a life-cycle perspective. Social LCA (S-LCA) complements E-LCA by considering both positive and negative social and socio-economic impacts occurring throughout the product life cycle (Benoît & Mazijn, 2009). Life-cycle sustainability assessment is a comprehensive framework to include impacts occurring in multiple dimensions (environmental, social, and economic) that addresses existing practical challenges in E-LCA by broadening and deepening the scope of assessment (Guinée et al., 2011).

SA studies must be comprehensive (by including thresholds and carrying capacity of the system), holistic in nature (capture interactions), include knowledge from various disciplines to ensure policy applications, and ensure stakeholder's participation (to include local context and cultural perspectives) throughout the assessment process. A systems approach is best suited to move from integrated assessment towards effective SA (Sala, Ciuffo, & Nijkamp, 2015). A SA method must have adequate scope (spatial and temporal), involve stakeholders, include environmental, social, and economic systems and interactions among their components, and be transparent and effective in collecting data and disclosing results (Pintér, Hardi, Martinuzzi, & Hall, 2012). Choosing appropriate time-scale over which sustainability has to be considered has been one of the difficult problems (Bond and Morrison-Saunders, 2011). Capacity of ecosystems to ensure availability of resources must be additionally considered at design stage along with assessing impacts across multiple spatial and temporal scales to ensure sustainability (Bakshi, Gutowski, & Sekulic, 2018). The involvement of (local) stakeholders throughout the assessment process is essential for operational SA (Taelman, Sanjuan-Delmás, Tonini, & Dewulf, 2019).

11.2 Sustainability assessment in manufacturing

This section briefly discusses the concept of SM, evolution of its definitions, and the sustainability purview of the definitions. Impacts associated with manufacturing decisions in the product life cycle have been discussed. A distinction between efficiency and effectiveness approach towards SA has been discussed using a classification of impacts (embodied and active). Insights from industrial practice studied by various authors is also presented. Definition of SM to ensure resource availability, the longevity of products within pollution bearing capacity of ecosystems and human beings has been proposed.

11.2.1 Sustainable manufacturing

SM is seen as either a life cycle, global closed-loop strategy or a paradigm, and has been defined using diverse terminology (Moldavska and Welo, 2017). There is no standard definition of SM to date. A widely cited definition proposed by US Department of Commerce covers aspects of minimizing resource-use and impacts and three dimensions of sustainability (US Department of Commerece, 2007). Definitions proposed by various authors have been summarized below (Table 11.1).

Table 11.1 Definitions of sustainable manufacturing (in chronological order).

Source	Definition	Year	Unique feature
Alting and Jøgensen (1993)	"Sustainable production means that products are designed, produced, distributed, used, and disposed with minimal (or none) environmental and occupational health damages, and with minimal use of resources (materials and energy)"	1993	Efficiency-oriented definiton (Pointing towards Effectiveness)
United Nations (2002)	"Sustainable consumption and production (SCP) is about the use of services and related products, which respond to basic needs and bring a better quality of life while minimizing the use of natural resources and toxic materials as well as the emissions of waste and pollutants over the life cycle of the service or product so as not to jeopardize the needs of further generations"	1994	Services, future generations
Veleva and Ellenbecker (2001)	"Sustainable production is the creation of goods and services using processes and systems that are: nonpolluting; conserving of energy and natural resources; economically viable; safe and healthful for workers; communities, and consumers; and socially and creatively rewarding for all working people"	2001	Zero pollution (effectiveness), emphasis on social dimension (creative rewarding)
US Department of Commerece (2007)	"Sustainable manufacturing includes things such as making products using less energy and materials, producing less waste, and using fewer hazardous materials as well as products that have greener attributes such as recyclability or lower energy use"	2008	Focus on some aspects of circular economy (recyclability)
OECD (2011)	"Sustainable manufacturing is all about minimizing the diverse business risks inherent in any manufacturing operation while maximizing the new opportunities that arise from improving your processes and products"	2011	Business perspective, inclined towards life-cycle management
Garetti and Taisch (2012)	"Sustainable Manufacturing can be defined as the ability to smartly use natural resources for manufacturing, by creating products and solutions that, thanks to new technology, regulatory measures and coherent social behaviors, are able to satisfy economic, environmental and social objectives, thus preserving the environment, while continuing to improve the quality of human life"	2012	Emphasis on social values (quality of human life)

(*Continued*)

Table 11.1 (Continued)

Source	Definition	Year	Unique feature
Nahkala (2013)	"In context of manufacturing, sustainability means the ability to produce specific product endlessly, considering the wide impacts of operations and the circulation of resources at the rate of production"	2013	Circular economy perspective
Jayal, Badurdeen, Dillon, and Jawahir (2010)	"Sustainable manufacturing at product, process and systems levels must: • demonstrate reduced negative environmental impact, • offer improved energy and resource efficiency, • generate minimum quantity of wastes, • provide operational safety, and • offer improved personnel health, while maintaining and/or improving the product and process quality with the overall life-cycle cost benefits"	2013	Levels of manufacturing (processes and systems)
Lee and Lee (2014)	"Sustainability in Manufacturing (SiM) is a measure of the manufacturing performance metrics for product design, process plan, and production system with respect to the environmental, economic, and social aspects of sustainability"	2014	Operational definition, sustainability metrics, measurement
Moldavska and Martinsen (2018)	"Sustainable Manufacturing can be defined as a complex behavior pattern to which any manufacturing organization should tend to evolve. This behavior pattern is defined by the criteria for SM such as to improve operational effectiveness, improve professional knowledge and competence of employees, increase the wealth of the society, and reduce discrimination"	2018	Operational effectiveness, more social metrics (such as discrimination, wealth)

Clearly, all definitions point towards the application of life-cycle approach in the multiple dimensions of sustainability. The prime focus has been on the relative improvements and efficient use of resources (e.g., by minimizing material use and waste avoidance) and minimizing various impacts due to pollutant emissions or use of toxic materials in the product life cycle. Only a few definitions point towards avoiding impacts (rather than minimizing) and effectively utilizing the available resources in a circular economy to establish a synergetic relationship between economy and ecology. There has been a limiting effort in connecting life cycle impacts with the absolute planetary limits.

Product life-cycle impacts can be classified as embodied impacts (embodied in the realization of a product, are mostly efficiency-based) and active impacts (occurring beyond manufacturing boundaries in use and pos-tuse phase are mostly effectiveness-based) (Kumar and Mani, 2019). Previous studies in the automotive manufacturing sector have shown that SA has not focused on the consumption and social dimensions (Pallaro, Subramanian, Abdulrahman, & Liu, 2015). Human health impacts due to exposure to chemicals in products through various exposure pathways are known as near-field impacts (Jolliet, Ernstoff, Csiszar, & Fantke, 2015). Absolute pollution and exposure limits need to be defined to avoid such impacts. Active impacts such as exposure to various chemicals because of product use, chemicals leaching to groundwater from discarded products at the end of life, etc. are difficult to quantify because of inherent uncertainty such as data, system boundary, dynamics of environment, and toxicity and its dispersal. Near-field impacts depend on chemical properties, product characteristics, frequency and duration of use, and product use circumstances/conditions. The distinction between two categories (active and embodied) is important because embodied impacts have been predominantly studied in the literature of SA in manufacturing (Kumar and Mani, 2018). Few product-specific studies have been conducted on active impacts of products in the past 5 years (impacts occurring due to product use), for example, shampoo (Huang et al., 2018) and building materials (Ernstoff et al., 2016). The concept of planetary boundaries could be helpful in defining the pollution limits for various impact categories at national scale (O'Neill, Fanning, Lamb, & Steinberger, 2018), for example, Swiss eco-factors ecological scarcity method assesses product life-cycle emissions and resource-use relative to the country's political targets (Frischknecht and Büsser Knöpfel, 2013). To follow an effective SA approach, one needs to holistically consider multiple levels of manufacturing to avoid suboptimization. Efficiency focuses at minimizing available resources, whereas effectiveness focuses at best utilization of available resources (Duflou et al., 2012). An effective sustainable design simultaneously addresses environmental, social, and economic challenges, and undesirable consequences (such as rebound effect) (Skerlos, 2015). Effective implementation of SM practices requires integration among different decision-makers and stakeholders [referred to as organizational integration (Moldavska and Welo, 2017)], in addition to currently focused integration among sustainability dimensions at strategic level (Zarte, Pechmann, & Nunes, 2019). There is a need to include worker's skill and creativity indicators in manufacturing classification to enhance craftsmanship and ensure long-term manufactuirng sustainability, as employment is one of the prime functions of the manufacturing in the economy and society (Campana and Cimatti, 2015). A systems-based approach could be helpful to develop a holistic SA system (Moldavska and Welo, 2016). Methodologies need to be developed to move from efficiency to effectiveness-based methods in product and chemical life cycles (Fantke and Illner, 2019).

Therefore, based on literature, it is evident that the following factors are important to ensure sustainability in manufacturing: (1) availability of resources (materials, energy, and water) or substitutes for the resources required for manufacturing (e.g., recycled materials); (2) impacts occurring (in various dimensions) due to various activities (processes) and products remain within the pollution (safety) limits. Increasing demand, complexity, and reduced life cycle of products could result in scarcity of certain materials (Graedel, Harper, Nassar, & Reck, 2015), which requires improved design decisions for end-of-life recovery (Ciacci, Reck, Nassar, & Graedel, 2015), for example, (Graedel et al., 2015) have estimated the criticality of 62 metals widely used in the electronics industry. Therefore, for manufacturing to be sustainable, the manufacturing activities (throughout the life cycle) should avoid extracting/demand resources that exceed the ecosystems capacity and limit impacts (in all relevant impact categories) within the pollution bearing capacity of entities (e.g., humans, environment) involved in the activities. Techno-ecological synergy framework is an approach to design synergy among technical and ecological systems for long-term sustainability (Bakshi, Ziv, & Lepech, 2015). This method accounts for supply from ecosystems as a novel contribution. Future studies need to include more circularity indicators for improved resource-effectiveness.

11.2.2 Sustainability assessment in manufacturing: tools and techniques

SAs initially started in chemical manufacturing industries in the early 2000s, followed by metal industries (predominantly aerospace and automotive) and later by electronics industries. The focus of assessment has been at various levels of manufacturing from individual parts or products, manufacturing processes, manufacturing systems, supply chains, and factory/enterprise levels. This section briefly discusses various methodologies implemented in manufacturing industries for SA at each level of manufacturing. SA in manufacturing can be broadly categorized into two categories (Moldavska and Welo, 2019):

- Indicator-based frameworks, for example, Global Reporting Initiative (GRI) and Organization for Economic Cooperation and Development toolkits, and
- Life-cycle-oriented tools and methods such as LCA, S-LCA, life-cycle costing, etc.

Most SAs in manufacturing rely on indicator-based approaches, with LCA as a tool within the broader framework. LCA has been used in combination with various simulation and multi-criteria decision tools. Different sets of environment, economic, and social indicators are used to assess manufacturing sustainability (Huang and Badurdeen, 2017; Huang and Badurdeen, 2018). SM indicators repository, developed by the National Institute of Standards and Technology, is a categorization of SA indicators related to manufacturing. It includes technological advancement and performance management indicators in addition to commonly used economic, environmental, and social sustainability dimensions (Joung, Carrell, Sarkar, & Feng, 2013). Materials use, energy use, air emissions in environmental dimension, cost indicators in economic dimension, and workers health and local community in social dimension are few indicators frequently used for SA in manufacturing (Ahmad, Wong, & Rajoo, 2019). Environmental dimensions are predominantly studied, majorly focused on resource-use such as energy, material, and water-use, GHG emissions, waste generated, etc. (Haapala et al., 2013; Despeisse, Mbaye, Ball, & Levers, 2012).

Manufacturing process-level SA studies have been conducted for various processes, for example, machining (Hegab, Darras, & Kishawy, 2018; Lu and Jawahir, 2015) and additive manufacturing (Doran, Smullin, & Haapala, 2016). Various indicators representing process parameters (cutting speed, feed, etc.), tool parameters (tool material, rake angle, etc.), and auxiliary materials (coolants, lubricants, etc.) have been estimated using mathematical modeling and thermodynamic analysis. Efforts for science-based modeling for sustainability characterization of manufacturing processes (Mani, Madan, Lee, Lyons, & Gupta, 2014), led to the development of ASTM standards for SM, and are currently under development (ASTM, 2019). Process-level studies primarily focus on environmental dimensions of sustainability and the technical aspects of manufacturing processes. Cooperative effort on process emissions in manufacturing-unit process life-cycle inventory (CO2PE!-UPLCI) framework includes sub-processes, for example, start-up and standby machining modes along with process parameters (Kellens, Dewulf, Overcash, Hauschild, & Duflou, 2012). Few studies focus on multiple dimensions (three dimensions of sustainability and technical aspects of manufacturing processes) following multicriteria decision models (Kluczek, 2016) and user experience aspects of manufacturing processes (Peruzzini and Pellicciari, 2017). Also, a few assessment studies of comparative nature (e.g., additive vs subtractive machining) have been conducted primarily to guide sustainable process selection (Priarone and Ingarao, 2017).

Manufacturing systems level SA in manufacturing includes a wide variety of metrics and indicators focusing on multiple sustainability dimensions. Existing methods widely used in manufacturing such as value stream mapping, have been incorporated (Faulkner and Badurdeen, 2014). Huang and Badurdeen (2018) presented a metrics-based sustainability evaluation method for manufacturing systems, production line, and factor/plant level, by including additional clusters to process-level assessment, for example, worker at line level and end of life management and stakeholder engagement at plant level (Huang and Badurdeen, 2018). Organization level impacts can be assessed using hybrid approaches such as AVOID (Finogenova, Bach, Berger, & Finkbeiner, 2019) by including upstream and downstream indirect impacts (causing due to indirect activities such as business trips) in addition to direct impacts. Sustainable supply chain management assessment focuses on impacts occurring across the supply chain, but requires integration among sustainability dimensions and more industry-specific studies (Rajeev, Pati, Padhi, & Govindan, 2017). Decision support for SM has been more focused at policy level implementation without reflecting effectiveness at operational level decisions such as process selection and manufacturing system design (Zarte et al., 2019). It is evident that a sufficient amount of work has been completed to identify and evaluate indicators for different levels of manufacturing.

Recent developments have led to an improved understanding of systems and their dynamic behavior by studying integration among sustainability dimensions and organizational levels using a systems approach. Zhang, Calvo-Amodio, & Haapala (2013) developed a systems-based conceptual model applicable at all levels for proactive decision-making related to sustainability in manufacturing. Onat, Kucukvar, Tatari, & Egilmez, 2016 have attempted to capture dynamic interactions between environmental, social, and economic indicators using causal loop diagrams (system dynamics approach). It is also important to address impacts at a larger scale beyond manufacturing boundaries and focus on concerns occurring in use and end of life phases (Herrmann, Hauschild, Gutowski, & Lifset, 2014). Consumer-oriented indicators (such as near-field impacts) receive less attention in current SA studies (Ahmad et al., 2019). Occupational health impacts in a manufacturing environment in existing SA studies are limited to impacts due to noise levels and acute injuries. Kijko et al. (2016) have proposed a methodology compatible with LCA to estimate potential occupational impacts due to chemical exposure throughout the supply chain of a product (Kijko et al., 2016).

Most studies on SA in manufacturing pay little attention to impacts associated with the upstream and downstream activities beyond manufacturing, such as resource availability, chemical exposure due to product, impacts of disposal, recycling, reuse, etc. The systems boundaries considered for assessment have been confined to the manufacturing stage, that is, either cradle-to-gate or gate-to-gate. Very few studies have considered the availability of resources, the capacity of ecosystem to supply resources, and manufacturing behavior in an integrated framework. SA in manufacturing ideally requires:

- Stock estimation of resources required for manufacturing, as well as the capacity of ecosystems to supply those resources, and
- Ensure impacts throughout the life cycle of a product are restricted within absolute boundaries (within pollution bearing capacity of local and global ecosystem and entities such as human beings).

Information from multiple sources would be required to ensure the amount of resources and meet the demand for manufactured goods/products. It could be beneficial to include market forecasts to estimate consumption patterns. Circular economy indicators could help in effectively utilizing the resources by extending the product use and reducing extraction. Indicators to understand the long-term impacts of exposure to materials and chemicals would be more effective than currently used efficiency improvement indicators. Following such approaches collectively would help in shifting towards effectiveness in SA.

11.2.3 Industry perspective on sustainability assessment in manufacturing

Due to rising sustainability concerns, many industries have started publishing sustainability reports on the impacts of product value chains. Industries report sustainability by using various index and ratings. GRI is a widely followed sustainability reporting standard in the industry which covers a broad range of environmental, social, and economic topics. Few enterprises have developed their own SA tools for internal practice, for example, Ford's Product Sustainability Index, Nike's SM Sourcing Index, General Motors Sustainability Metrics, Walmart Sustainability Index, etc. Few research articles have addressed the requirement of filling industry-academia gaps for increased adoption of existing SA tools in manufacturing industries. Belkhir, Bernard, & Abdelgadir (2017) reviewed sustainability reports of 40 companies and found no correlation between GRI reporting and absolute sustainability improvement. SAs in industry are mostly focused on sustainability practices (such as energy efficiency, use of renewable energy), instead of measuring sustainability performance (Moldavska and Welo, 2019). Small- and medium-scale industries limit their focus to economic dimensions and ignore social and environmental dimensions (Trianni, Cagno, Neri, & Howard, 2019; Das, Rangarajan, & Dutta, 2020). SIA in industrial product design practice remains widely focused at health and safety category and requires development of industry specific metrics and tools (Pack, Rose Phipps, Mattson, & Dahlin, 2019). Industrial sustainability practices are majorly driven by regulations and market pressures centered around environmental dimensions, and lack a proactive approach towards sustainability (Alayón, Säfsten, & Johansson, 2017). A study on small- and medium-scale German and Italian manufacturing industries shows a significant gap in adopting sustainability indicators due to lack of geography and application-specific indicators for individual industries (Trianni et al., 2019). Reluctance to share industry data adds to challenges to people outside the industry in assessing sustainability. Industries rely on standards for implementing new practices; therefore, the development of standards for SM is much needed (Smullin, Haapala, Mani, & Morris, 2016).

11.3 Challenges and solutions for effective sustainability assessment

The challenges in operationalizing sustainability in manufacturing can be broadly categorized as:

- Methodological basis or underlying logic (theoretical) challenges, and
- Application/translation or operationalization (practical) challenges.

Methodologically, effectiveness-based SA in manufacturing is a challenging task as it requires well-defined spatial and temporal system boundaries to ensure circularity in product life cycle, stakeholders involved, accurate site-specific data, data sources located at geographically distant locations, and absolute pollution limits (both local and global) for various impacts categories. Lack of data on various materials used in the product is also one of the most difficult challenges. Modern manufacturing paradigms and involved technologies such as smart manufacturing, big data analytics, and blockchains might fill some of the existing data gaps. Site-specific background concentration of pollutants is also necessary to accurately measure the impacts of various emissions. Site-specific assessments require data on environmental conditions in the specific geographies. Adopting a standardized method and predefined indicators could give us a false image of actual impacts, which calls for the need for regionalized (site-specific) assessments in combination with the global (site-generic) assessments. Challenge remains to maintain a balance between site-specific and site-generic assessments. Identifying effectiveness indicators is an important step, as case-specific indicator development sometimes requires domain-specific knowledge. It is important to identify hotspots in the value chain and trace various impacts back to the decisions taken in early life-cycle phases. Therefore, communicating assessment results to various decision-makers as per their relevant requirements is also crucial for effective SA. Furthermore, validating SA studies is an additional challenge.

Adopting a comprehensive framework is not feasible for all scale of manufacturing industries as conducting SAs is time and resource intensive and requires certain proficiency in assessment tools. The hierarchical disconnect among different decision-makers (managers, engineers, designers, etc.) could be a reason of ineffectiveness in the sustainability practices, for example, engineers and designers consider health and safety impacts widely for a product. In contrast, the executives and managers look at a holistic view of social impacts of a product (Pack et al., 2019). There is a requirement of an integrated approach towards the development of an effective SA method for manufacturing. Recent studies on chemical exposure, indoor pollution, and human toxicity suggest an urgent need to focus on active impacts in near-field environment, for example, health concerns due to prolonged exposure to artificial LED light (Nash et al., 2019), indoor pollution (Ham, 2019), and underestimated effects of BPA on human health (Gerona, Vom, Saal, & Hunt, 2019), etc. In the far-field, the impacts are required to be benchmarked against planetary boundaries. Efforts are required to determine absolute boundaries for various impact categories at national scales. The manufacturing community can learn from other domains that have successfully demonstrated absolute performance assessments, for example, construction and buildings (Brejnrod, Kalbar, Petersen, & Birkved, 2017). In addition, it is necessary to conduct more studies on circular economy scenarios to validate the relationship between circular economy and effective sustainability. Specific measures for different decision-makers involved in the product life cycle are required to be integrated on a single platform by incorporating systems thinking knowledge. Knowledge-based product lifecycle management (PLM) tools might help integrate product life cycle knowledge for improved resource optimization and SA (Belkadi, Bernard, & Laroche, 2015).

11.4 Summary

This chapter has shed light on the current state and challenges in operationalizing sustainability assessment in manufacturing. The importance of integrating information throughout the levels of manufacturing has been discussed. Including all sustainability dimensions and their interrelationships is also equally important. The assessment focus must expand beyond manufacturing boundaries to address active impacts and include circular economy principles and indicators. More studies are required to be conducted to assess site-specific impacts for a wide variety of products. Conducting proactive assessments for upcoming products and technologies could avoid unforeseen impacts and additional environmental burden. Collective efforts in the aforementioned directions might eventually help to shift towards effectiveness in SA in manufacturing.

References

Ahmad, S., Wong, K. Y., & Rajoo, S. (2019). Sustainability indicators for manufacturing sectors: A literature survey and maturity analysis from the triple-bottom line perspective. *Journal of Manufacturing Technology Management*, *30*(2), 312–334.

Alayón, C., Säfsten, K., & Johansson, G. (2017). Conceptual sustainable production principles in practice: Do they reflect what companies do? *Journal of Cleaner Production*, *141*, 693–701.

Allwood, J. M., Ashby, M. F., Gutowski, T. G., & Worrell, E. (2013). Material efficiency: Providing material services with less material production. *Philosophical Transactions of the Royal Society A: Mathematical, Physical and Engineering Sciences*, *371*, 1986.

Alting, D. L., & Jøgensen, D. J. (1993). The life cycle concept as a basis for sustainable industrial production. *CIRP Annals*, *42*(1), 163–167.

Anastas, P. T. (1999). Green chemistry and the role of analytical methodology development. *Critical Reviews in Analytical Chemistry*, *29*(3), 167–175.

Arnold, R. D., & Wade, J. P. (2015). A definition of systems thinking: a systems approach. *Procedia Computer Science*, *44*, 669–678.

ASTM, (2019). Subcommittee E60.13 on Sustainable Manufacturing. [Online]. Available from https://www.astm.org/COMMIT/SUBCOMMIT/E6013.htm. (Accessed 02 January 2019).

Bakshi, B. R., Gutowski, T. G., & Sekulic, D. P. (2018). Claiming sustainability: Requirements and challenges. *ACS Sustainable Chemistry & Engineering*, *6*(3), 3632–3639.

Bakshi, B. R., Ziv, G., & Lepech, M. D. (2015). Techno-ecological synergy: A framework for sustainable engineering. *Environmental Science & Technology*, *49*(3), 1752–1760.

Belkadi, F., Bernard, A., & Laroche, F. (2015). Knowledge based and PLM facilities for sustainability perspective in manufacturing: A global approach. *Procedia CIRP*, *29*, 203–208.

Belkhir, L., Bernard, S., & Abdelgadir, S. (2017). Does GRI reporting impact environmental sustainability? A cross-industry analysis of CO_2 emissions performance between GRI-reporting and non-reporting companies. *Management of Environmental Quality*, *28* (2), 138–155.

Bell, S., & Morse, S. (2012). *Sustainability indicators: Measuring the immeasurable?* Routledge.

Benoît, C., & Mazijn, B. (Eds.), (2009). *Guidelines for Social Life Cycle*. Assessment of Products.

Bond, A. J., & Morrison-Saunders, A. (2011). Re-evaluating sustainability assessment: Aligning the vision and the practice. *Environmental Impact Assessment Review*, *31*(1), 1–7.

Brejnrod, K. N., Kalbar, P., Petersen, S., & Birkved, M. (2017). The absolute environmental performance of buildings. *Building and Environment*, *119*, 87–98.

Brundage, M. P., et al. (2018). Analyzing environmental sustainability methods for use earlier in the product lifecycle. *Journal of Cleaner Production*, *187*, 877–892.

Campana, G., & Cimatti, B. (2015). Measures and methods for a new taxonomy in manufacturing enterprises. *Procedia CIRP*, *26*, 287–292.

Chen, L., Olhager, J., & Tang, O. (2014). Manufacturing facility location and sustainability: A literature review and research agenda. *International Journal of Production Economics*, *149*, 154–163.

Ciacci, L., Reck, B. K., Nassar, N. T., & Graedel, T. E. (2015). Lost by design. *Environmental Science & Technology*, *49*(16), 9443–9451.

Clark, W. C., & Harley, A. G. (2020). Sustainability Science: Toward a Synthesis. *Annual Review of Environment and Resources*, *45*, 331–386. Available from https://doi.org/10.1146/annurev-environ-012420-043621.

Clayton, T., & Radcliffe, N. (2018). *Sustainability: A systems approach*. Routledge.

Council, N. R. (1999). *Our common journey: A transition toward sustainability*. Washington, DC: The National Academies Press.

Das, M., Rangarajan, K., & Dutta, G. (2020). Corporate sustainability in SMEs: An Asian perspective. *Journal of International Business Studies*.

Despeisse, M., Mbaye, F., Ball, P. D., & Levers, A. (2012). The emergence of sustainable manufacturing practices. *Production Planning & Control*, *23*(5), 354–376.

Doran, M.P., Smullin, M.M., & Haapala, K.R., (2016). An approach to compare sustainability performance of additive and subtractive manufacturing during process planning.

In: ASME 2016 International Design Engineering Technical Conferences and Computers and Information in Engineering Conference, p. V004T05A047.

Duflou, J. R., et al. (2012). Towards energy and resource efficient manufacturing: A processes and systems approach. *CIRP Annals Manufacturing Technology, 61*(2), 587–609.

Ehrenfeld, J. (2004). Industrial ecology: A new field or only a metaphor? *Journal of Cleaner Production, 12*(8–10), 825–831.

Ehrenfeld, J. R. (1997). Industrial ecology: A framework for product and process design. *Journal of Cleaner Production, 5*(1–2), 87–95.

ElMaraghy, W., ElMaraghy, H., Tomiyama, T., & Monostori, L. (2012). Complexity in engineering design and manufacturing. *CIRP Annals, 61*(2), 793–814.

Ernstoff, A. S., Fantke, P., Csiszar, S. A., Henderson, A. D., Chung, S., & Jolliet, O. (2016). Multi-pathway exposure modeling of chemicals in cosmetics with application to shampoo. *Environment International, 92*, 87–96.

Esmaeilian, B., Behdad, S., & Wang, B. (2016). The evolution and future of manufacturing: A review. *Journal of Manufacturing Systems, 39*, 79–100.

Fantke, P., & Illner, N. (2019). Goods that are good enough: Introducing an absolute sustainability perspective for managing chemicals in consumer products. *Current Opinion in Green and Sustainable Chemistry, 15*, 91–97.

Faulkner, W., & Badurdeen, F. (2014). Sustainable Value Stream Mapping (Sus-VSM): methodology to visualize and assess manufacturing sustainability performance. *Journal of Cleaner Production, 85*, 8–18.

Fauré, E., Arushanyan, Y., Ekener, E., Miliutenko, S., & Finnveden, G. (2017). Methods for assessing future scenarios from a sustainability perspective,". *European Journal of Futures Research, 5*(1), 17.

Finogenova, N., Bach, V., Berger, M., & Finkbeiner, M. (2019). Hybrid approach for the evaluation of organizational indirect impacts (AVOID): combining product-related, process-based, and monetary-based methods. *International Journal of Life Cycle Assessment, 24*(6), 1058–1074.

Frischknecht, R., & Büsser Knöpfel, S. (2013). Swiss Eco-Factors 2013 according to the Ecological Scarcity Method. *Federal Office for the Environment, 256.*

Garetti, M., & Taisch, M. (2012). Sustainable manufacturing: Trends and research challenges. *Production Planning & Control, 23*(2–3), 83–104.

Gasparatos, A., El-Haram, M., & Horner, M. (2008). A critical review of reductionist approaches for assessing the progress towards sustainability. *Environmental Impact Assessment Review, 28*(4), 286–311.

Gasparatos, A., & Scolobig, A. (2012). Choosing the most appropriate sustainability assessment tool. *Ecological Economics, 80*(0), 1–7.

Geissdoerfer, M., Savaget, P., Bocken, N. M. P., & Hultink, E. J. (2017). The circular economy—A new sustainability paradigm? *Journal of Cleaner Production, 143*, 757–768.

General Assembly of the United Nations, (1987). Report of the world commission on environment and development. In: 96th plenary meeting, p. 11.

Gerona, R., Vom Saal, F. S., & Hunt, P. A. (2019). BPA: have flawed analytical techniques compromised risk assessments? *Lancet Diabetes & Endocrinology, 8*, 2019–2021.

Glavič, P., & Lukman, R. (2007). Review of sustainability terms and their definitions. *Journal of Cleaner Production, 15*(18), 1875–1885.

Graedel, T. E., et al. (2012). Methodology of metal criticality determination. *Environmental Science & Technology, 46*(2), 1063–1070.

Graedel, T. E., Harper, E. M., Nassar, N. T., Nuss, P., Reck, B. K., & Turner, B. L. (2015). Criticality of metals and metalloids. *Proceedings of the National Academy of Sciences of the United States of America, 112*(14), 4257–4262.

Graedel, T. E., Harper, E. M., Nassar, N. T., & Reck, B. K. (2015). On the materials basis of modern society. *Proceedings of the National Academy of Sciences of the United States of America, 112*(20), 6295.

Groover, M. P. (2007). *Fundamentals of modern manufacturing: Materials processes, and systems.* John Wiley & Sons.

Guinée, J. B., et al. (2011). Life cycle assessment: past, present, and future. *Environmental Science & Technology, 45*, 90–96, ACS Publications.

Gutowski, T. G., Allwood, J. M., Herrmann, C., & Sahni, S. (2013). A global assessment of manufacturing: economic development, energy use, carbon emissions, and the potential for energy efficiency and materials recycling. *Annual Review of Environment and Resources, 38*, 81–106.

Haapala, K. R., et al. (2013). A review of engineering research in sustainable manufacturing. *Journal of Manufacturing Science and Engineering, 135*(4), 041013.

Hák, T., Moldan, B., & Dahl, A. L. (2012). *Sustainability indicators: A scientific assessment* (67). Island Press.

Ham, B. (2019). Indoor chemical pollution impacts often remain invisible. *Science (80-), 366*(6469), 1084–1085.

Hegab, H. A., Darras, B., & Kishawy, H. A. (2018). Towards sustainability assessment of machining processes. *Journal of Cleaner Production, 170*, 694–703.

Hellweg, S., & Milà i Canals, L. (2014). Emerging approaches, challenges and opportunities in life cycle assessment. *Science (80-), 344*(6188), 1109–1113.

Helu, M., Vijayaraghavan, A., & Dornfeld, D. (2011). Evaluating the relationship between use phase environmental impacts and manufacturing process precision. *CIRP Annals, 60*(1), 49–52.

Herausgeber, (2004). CIRP Dictionary of Production Engineering Vol. 3 Manufacturing Systems.

Herrmann, C., Hauschild, M., Gutowski, T., & Lifset, R. (2014). "Life cycle engineering and sustainable manufacturing. *Journal of Industrial Ecology, 18*(4), 471–477.

Hon, B. (2005). Performance and evaluation of manufacturing systems. *CIRP Annals Manufacturing Technology, 54*(2), 139–154.

Huang A. and F. Badurdeen, (2017). Sustainable manufacturing performance evaluation at the enterprise level: Index- and value-based methods BT—Sustainable manufacturing performance evaluation at the enterprise level: Index- and value-based methods.

Huang, A., & Badurdeen, F. (2018). Metrics-based approach to evaluate sustainable manufacturing performance at the production line and plant levels. *Journal of Cleaner Production, 192*, 462–476.

Huang, L., Anastas, N., Egeghy, P., Vallero, D. A., Jolliet, O., & Bare, J. (2018). Integrating exposure to chemicals in building materials during use stage. *International Journal of Life Cycle Assessment*, 1–18.

Huang, L., Ernstoff, A., Fantke, P., Csiszar, S. A., & Jolliet, O. (2017). A review of models for near-field exposure pathways of chemicals in consumer products. *Science of the Total Environment, 574*, 1182–1208.

International Organization for Standardization. (2006). ISO 14040-environmental management—Life cycle assessment—Principles and framework. *International Organization for Standardization, 3*, 20.

Jayal, A. D., Badurdeen, F., Dillon, O. W., & Jawahir, I. S. (2010). Sustainable manufacturing: Modeling and optimization challenges at the product, process and system levels. *CIRP Journal of Manufacturing Science and Technology, 2*(3), 144–152.

Jolliet, O., Ernstoff, A. S., Csiszar, S. A., & Fantke, P. (2015). *Defining product intake fraction to quantify and compare exposure to consumer products.* ACS Publications.

Joung, C. B., Carrell, J., Sarkar, P., & Feng, S. C. (2013). Categorization of indicators for sustainable manufacturing. *Ecological Indicators, 24*, 148–157.

Kalpakjian, S. (2001). *Manufacturing engineering and technology*. India: Pearson Education.
Kellens K., W. Dewulf, M. Overcash, M.Z. Hauschild, and J.R. Duflou, (2012). Methodology for systematic analysis and improvement of manufacturing unit process life-cycle inventory (UPLCI)—CO_2 PE! initiative (cooperative effort on process emissions in manufacturing). Part 1: Methodology description. pp. 69—78.
Kijko, G., Jolliet, O., & Margni, M. (2016). Occupational health impacts due to exposure to organic chemicals over an entire product life cycle. *Environmental Science & Technology*, *50*(23), 13105—13114.
Kluczek, A. (2016). Application of multi-criteria approach for sustainability assessment of manufacturing processes. *Management and Production Engineering Review*, 7(3), 62—78.
Kumar M., & Mani M., (2018). Integrating active impacts in sustainability assessment in product (manufacturing) life cycle. In: 6th International Conference on Social Life Cycle Assessment (S-LCA), Pescara (Italy).
Kumar, M., & Mani, M. (2019). A systems-based sustainability assessment framework to capture active impacts in product life cycle/manufacturing. *Procedia Manufacturing*, *33*, 647—654.
Lee, J. Y., & Lee, Y. T. (2014). A framework for a research inventory of sustainability assessment in manufacturing. *Journal of Cleaner Production*, *79*, 207—218.
Lowell Centre for Sustainable Production, (n.d.). Principles of Sustainable Production. [Online]. Available from https://www.uml.edu/Research/Lowell-Center/About/Sustainable-Production-Defined.aspx.
Lu, T., & Jawahir, I. S. (2015). "Metrics-based sustainability evaluation of cryogenic machining. *Procedia CIRP*, *29*, 520—525.
Mani, M., Madan, J., Lee, J. H., Lyons, K. W., & Gupta, S. K. K. (2014). Sustainability characterisation for manufacturing processes. *International Journal of Production Research*, *52*(20), 5895—5912.
Manley, J. B., Anastas, P. T., & Cue, B. W. (2008). Frontiers in Green Chemistry: Meeting the grand challenges for sustainability in R&D and manufacturing. *Journal of Cleaner Production*, *16*(6), 743—750.
Margni, M., Mutel, C., Reinhard, J., Stolz, P., Van Zelm, R., & Vieira, M. (2018). *Regionalization in LCA: Current status in concepts, software and databases. 69th LCA forum*. Swiss Federal Institute of Technology.
Moldavska, A., & Martinsen, K. (2018). Defining sustainable manufacturing using a concept of attractor as a metaphor. *Procedia CIRP*, *67*, 93—97.
Moldavska, A., & Welo, T. (2016). Development of manufacturing sustainability assessment using systems thinking. *Sustainability*, *8*(1), 5.
Moldavska, A., & Welo, T. (2017). The concept of sustainable manufacturing and its definitions: A content-analysis based literature review. *Journal of Cleaner Production*, *166*, 744—755.
Moldavska, A., & Welo, T. (2019). A holistic approach to corporate sustainability assessment: Incorporating sustainable development goals into sustainable manufacturing performance evaluation. *Journal of Manufacturing Systems*, *50*, 53—68.
Monto, M., Ganesh, L. S., & Varghese, K. (2005). *Sustainability and human settlements: Fundamental issues, modeling and simulations*. Sage.
MoSPI. (2008). *National Industrial Classification 2008* (pp. 1—193). Central Statistical Organisation.
MSCI, (2020). The Global Industry Classification Standard (GICS®). [Online]. Available from https://www.msci.com/gics. (Accessed 01 January 2020).
Mutel, C., et al. (2018). Overview and recommendations for regionalized life cycle impact assessment. *International Journal of Life Cycle Assessment*, *24*, 856—865.
Nahkala, S. (2013). *Aligning product design methods and tools for sustainability. Re-engineering manufacturing for sustainability* (pp. 53—58). Springer.

Nash, T. R., et al. (2019). Daily blue-light exposure shortens lifespan and causes brain neurodegeneration in Drosophila. *npj Aging and Mechanisms of Disease*, *5*(1), 8.

Ness, B., Urbel-piirsalu, E., Anderberg, S., & Olsson, L. (2007). Categorising tools for sustainability assessment. *Ecological Economics*, *60*(3), 498–508.

OECD. (2011). *OECD sustainable manufacturing tookit—Seven steps to environmental excellence* (p. 54) US Department of Commerce.

Onat, N. C., Kucukvar, M., Tatari, O., & Egilmez, G. (2016). Integration of system dynamics approach toward deepening and broadening the life cycle sustainability assessment framework: a case for electric vehicles. *The International Journal of Life Cycle Assessment*, *21*(7), 1009–1034. Available from https://doi.org/10.1007/s11367-016-1070-4.

O'Neill, D. W., Fanning, A. L., Lamb, W. F., & Steinberger, J. K. (2018). A good life for all within planetary boundaries. *Nature Sustainability*, *1*(2), 88.

O'Rourke, D. (2014). The science of sustainable supply chains. *Science (80-)*, *344*(6188), 1124–1127.

Pack, A. T., Rose Phipps, E., Mattson, C. A., & Dahlin, E. C. (2019). Social impact in product design, an exploration of current industry practices. *Journal of Mechanical Design*, *142*(7).

Pallaro, E., Subramanian, N., Abdulrahman, M. D., & Liu, C. (2015). Sustainable production and consumption in the automotive sector: Integrated review framework and research directions. *Sustainable Production and Consumption*, *4*, 47–61.

Pavitt, K. (1984). Sectoral patterns of technical change: towards a taxonomy and a theory. *Research Policy*, *13*(6), 343–373.

Peruzzini, M., & Pellicciari, M. (2017). User experience evaluation model for sustainable manufacturing. *International Journal of Computer Integrated Manufacturing*, *31*, 1–19.

Pintér, L., Hardi, P., Martinuzzi, A., & Hall, J. (2012). Bellagio STAMP: Principles for sustainability assessment and measurement. *Ecological Indicators*, *17*, 20–28.

Priarone, P. C., & Ingarao, G. (2017). Towards criteria for sustainable process selection: on the modelling of pure subtractive versus additive/subtractive integrated manufacturing approaches. *Journal of Cleaner Production*, *144*, 57–68.

Rajeev, A., Pati, R. K., Padhi, S. S., & Govindan, K. (2017). Evolution of sustainability in supply chain management: A literature review. *Journal of Cleaner Production*, *162*, 299–314.

Rissman, J., et al. (2020). Technologies and policies to decarbonize global industry: Review and assessment of mitigation drivers through 2070. *Applied Energy*, *266*, 114848.

Rockström, J., et al. (2009). Planetary boundaries: Exploring the safe operating space for humanity. *Ecology and Society*, *14*(2).

Sala, S., Ciuffo, B., & Nijkamp, P. (2015). A systemic framework for sustainability assessment. *Ecological Economics*, *119*, 314–325.

Sala, S., Farioli, F., & Zamagni, A. (2013). Progress in sustainability science: Lessons learnt from current methodologies for sustainability assessment: Part 1. *International Journal of Life Cycle Assessment*, *18*(9), 1653–1672.

Seager, T. P., & Theis, T. L. (2002). A uniform definition and quantitative basis for industrial ecology. *Journal of Cleaner Production*, *10*(3), 225–235.

Singh, R. K., Murty, H. R., Gupta, S. K., & Dikshit, A. K. (2012). An overview of sustainability assessment methodologies. *Ecological Indicators*, *15*(1), 281–299.

Skerlos, S. J. (2015). Promoting effectiveness in sustainable design. *Procedia CIRP*, *29*, 13–18.

Smullin M.M., Haapala K.R., Mani M., & Morris K.C., (2016). Using industry focus groups and literature review to identify challenges in sustainable assessment theory and practice. In: ASME 2016 International Design Engineering Technical Conferences and Computers and Information in Engineering Conference, 2016, p. V004T05A048.

Stahel, W. R. (2016). The circular economy. *Nature*, *531*(7595), 435–438.

Taelman, S., Sanjuan-Delmás, D., Tonini, D., & Dewulf, J. (2019). An operational framework for sustainability assessment including local to global impacts: Focus on waste management systems. *Resources, Conservation & Recycling: X*, *2*, 100005.

The Ellen MacArthur Foundation, (n.d.). What is a circular economy? [Online]. Available from https://www.ellenmacarthurfoundation.org/circular-economy/concept.

Trianni, A., Cagno, E., Neri, A., & Howard, M. (2019). Measuring industrial sustainability performance: Empirical evidence from Italian and German manufacturing small and medium enterprises. *Journal of Cleaner Production*, *229*, 1355–1376.

United Nations, (2002). Sustainable Consumption and Production, [Online]. Available from https://sustainabledevelopment.un.org/topics/sustainableconsumptionandproduction. (Accessed 19 June 2018).

UNIDO. (2018). *Industrial development report—Demand for manufacturing: Driving inclusive and sustainable industrial development*. Vienna: UNIDO.

US Department of Commerece, (2007). [Online]. Available from https://www.trade.gov/green/documents/introduction-to-sustainable-manufacturing.pptx. (Accessed 05 January 2018).

Available from https://www.census.gov/eos/www/naics/. (2017).

Veleva, V., & Ellenbecker, M. (2001). Indicators of sustainable production: Framework and methodology. *Journal of Cleaner Production*, *9*(6), 519–549.

Verones, F., et al. (2017). LCIA framework and cross-cutting issues guidance within the UNEP-SETAC Life Cycle Initiative. *Journal of Cleaner Production*, *161*, 957–967.

World Resource Institute, (2018). By the Numbers: New Emissions Data Quantify India's Climate Challenge, [Online]. Available from https://www.wri.org/blog/2018/08/numbers-new-emissions-data-quantify-indias-climate-challenge. (Accessed 27 December 2019).

Zarte, M., Pechmann, A., & Nunes, I. L. (2019). Decision support systems for sustainable manufacturing surrounding the product and production life cycle—A literature review. *Journal of Cleaner Production*, *219*, 336–349.

Zhang, H., Calvo-Amodio, J., & Haapala, K. R. (2013). A conceptual model for assisting sustainable manufacturing through system dynamics. *Journal of Manufacturing Systems*, *32*(4).

CHAPTER 12

REMOVED: Life cycle engineering: past, present, and future

Anbesh Jamwal, Rajeev Agrawal and Monica Sharma
Department of Mechanical Engineering, Malaviya National Institute of Technology, J.L.N. Marg, Jaipur, India

This chapter has been removed due to multiple instances of misattributed information, improper citations and inaccuracies regarding the sources and definitions related to life cycle engineering.

Please see Elsevier policy on Book/Chapter removal (https://www.elsevier.com/about/policies/book-chapter-removal).

Sustainable Manufacturing
DOI: https://doi.org/10.1016/B978-0-12-818115-7.00012-2

CHAPTER 13

Plastic recycling and regeneration

Jilse Sebastian and Lijo Paul
Department of Mechanical Engineering, St. Joseph's College of Engineering and Technology Palai, Choondacherry, India

13.1 Introduction

Plastic material has been introduced as replacement for many existing materials including steel, glass, and even concrete for various industrial applications. Plastics are used in the most of the domestic as well as industrial applications as they are light weight, low cost, and advanced technical properties. The name *Plastics* was originated from "*plastikos*" which means capable of being shaped or molded (Woldemar d'Ambrières, 2019). Plastic is a polymeric material made from the polymers. Synthetic and natural polymers are used to make plastics which can be made into various product through molding. This has resulted in high-production rate in last 60 years. Majority of the plastics are used for making disposable packing items and other short span items which are thrown to environment very fast. The present consumption of plastics is unsustainable. In this chapter we mainly focus on types of plastics, effect of plastic wastes on environment, methods of plastic recycling in land, water, and air, and recycle products from plastics including fuels.

Plastics are nonbiodegradable polymers which are used to make most of the products used in day-to-day life. Based on processing methods, generally plastics are classified into two categories namely thermoplastic and thermosetting plastics. Thermoplastic can be reshaped in multiple times through heating, whereas thermosetting plastics can be solidified to permanent shapes due to irreversible polymerization. In thermoplastic, linear-molecular chains are responsible for softening with heat and hardening on cooling (Grigore, 2017; Pascault, Sautereau, Verdu, & Williams, 2002). In thermoset plastics, irreversible polymerization occurs and is cured with heat from chemical reactions.

Sustainable Manufacturing
DOI: https://doi.org/10.1016/B978-0-12-818115-7.00016-X

Thermoplastic polymers are classified into three types as given below (Woldemar d'Ambrières, 2019):

1. Crystalline thermoplastics: These plastics are translucent with molecular chains with regular arrangements. They have higher mechanical impact resistance which includes polypropylene (PP), low-density polyethylene (LDPE), and high-density polyethylene (HDPE).
2. Amorphous thermoplastics: These plastics are transparent with random molecular arrangements which include poly vinyl chloride (PVC), polymethylmethacrylate (PMMA), and acrylonitrile butadiene styrene.
3. Semicrystalline polymers: These plastics have combined properties of crystalline and amorphous polymers which includes polyester polybutylene terephthalate and polyamide imide.

13.2 Plastic pollution and environment

A tremendous increase in the production and application of plastics has seen constantly over the pay few decades reaching more than 300 million metric tons (Woldemar d'Ambrières, 2019). But the crucial problem from this fact is that only 9% of plastic is properly recycled and the rest is used for landfill or dumped in the wild including oceans since 1950 (Woldemar d'Ambrières, 2019). Every year more than 8 million metric tons of plastic wastes have been swept along ocean beds and rivers which take more than 100 years for decomposition resulting in threat to marine life (Hopewell, Dvorak, & Kosior, 2009). Human activities are the main reasons for plastic pollution as they dispose it to environment without proper recycling and disposal of the same. Some major causes for plastic pollution include:

- Dumping of plastic waste on landfills and soil
- Air pollution through plastic burning
- Plastic toy wastes
- Usage and dispose of disposable plastic cups, bottles, knives, fork, spoon, containers, etc.
- Lack of proper recycling and reusing of plastic products

It is projected that 6–12 million metric tons of plastic waste gets disposed into oceans (Woldemar d'Ambrières, 2019). Many developed nations restrict plastic waste accumulation through 30% recycling, whereas it is the worst situation with 0% recycling in developing nations (Woldemar d'Ambrières, 2019). Currently, most plastic waste goes to landfill and does not get decomposed for years (Woldemar d'Ambrières, 2019). More than becoming a critical problem, recycling of plastics is a herculean task for most of the developing nations and its management varies from country to country. Currently, most of the plastic waste is thrown to environment one way or another. Mostly 50% of plastics have single-use disposable applications and 20%–25% for long-term infrastructure and the remainder for durable consumer applications (Amin & Amin, 2011).

13.3 Plastic recycling

From the environmental and socio-economic standpoints, plastic recycling in economic ways is the best solution to save the environment (Woldemar d'Ambrières, 2019). Recycling is the most effective solution to plastic pollution and stands as the most dynamic area in plastic industry today (Brazel & Rosen, 2012; Yang et al., 2012). Recycling process reduce the oil usage, carbon dioxide emission, and dumping of plastic wastes for landfills. Advancements in technologies and methods available for collection, sorting, and processing of plastic have resulted in improved ways for plastic recycling. With cooperation of industry, society, and government, there exist lot of scope to recycle plastics rather than using them for landfills and provide significant socio-economic gains.

At present, mechanical recycling process is used at a wider scale for treatment of plastic solid waste (PSW). The recycling includes removal of organic residue through washing, followed by shredding, melting, and remolding of the polymer and is mostly mixed with virgin plastics to produce selected properties in manufacturing (Ribeiro et al., 2016). Composite material with biodegradable plant fibers are easy to recycle than common type of polymer composites (Ashori, 2008). Most of the thermoset plastics with cross-linked polymers are not recyclable in mechanical methods (Elo & Sundin, 2009). Setting up a system to recycle plastic waste allows a local industry to emerge and recover value from the recycled material.

13.3.1 Recycling stages

The major steps followed in plastic recycling include:

1. Collection: Plastics are collected from customers and industries for recycling.
2. Sorting: Plastic that was collected is separated further based on its make and type and transferred to shredding machine.
3. Washing: Plastic wastes are washed after sorting in order to remove impurities and to improve quality of recycled products.

4. Shredding: Washed plastic waste is fed to shredding units through conveyor belts where plastic are teared into small pellets.
5. Identification and classification of plastic: Proper quality test are conducted on the pellets through various techniques.
6. Extruding: In this process, shredded plastic pellets are melted and extruded into filaments which can be used for making different recycled products.

In this section, different case studies that were carried out for effective utilization of plastics are discussed further.

13.3.2 Case study 1: plastic ropes from disposable water Poly Ethylene Terephthalate (PET) bottles

Conventional way of recycling PET bottles is by molding, which emits harmful gases. There is a wide variety of recycling units exist. But many of them use heavy and large equipment. Production cost for such units are also high. So a better solution for a compact recycling unit is suggested in the current work. A machine which can be used to recycle the unwanted PET bottles is suggested. The plastic bottles have converted into ropes by stripping and reheating process in this case study. The parts are created and assembled, and their applicability was checked. Proper spinning and cutting operation have been ensured in the machine (Leng, Padhan, & Sreeram, 2018).

The width of the machine is chosen as 50 cm based on the easy on carry as shown in Fig. 13.1. The machine is run by two different DC motors. The motors were chosen for the proper working of the spinning, cutting, and collecting operations. The schematic sketch of the same is

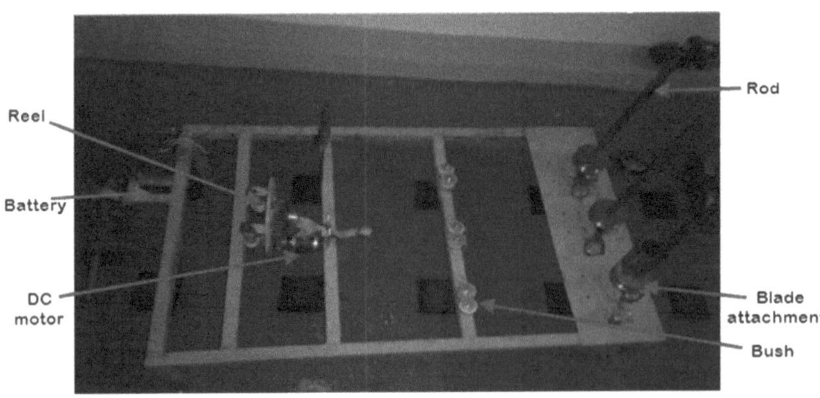

Figure 13.1 Experimental setup for production of plastic ropes from PET bottles.

given in Fig. 13.2. One of the motors had high torque and low speed, and other had high speed and low torque. Motors are attached to the frame by using suitable bearings. Plastic roller wheels are chosen because of their ability to withstand deformation. It is light weight which reduces the weight that is acting on the structure. Moreover the wheel provides adequate grip to the strip which peeled from the bottle. Metallic drum with bundle are made of Japan metal sheet that is rigid and sturdy. Gear drives are provided to transfer the power from the motor to the metallic drum (Jamdar, Kathalewar, Dubey, & Sabnis, 2017). Three bottles are arranged at equal intervals on a separate pillar for their smooth peeling.

Strip bundles are attached with the drum, and rotate along with it. Collecting wheels are attached to the machine at the extreme end side

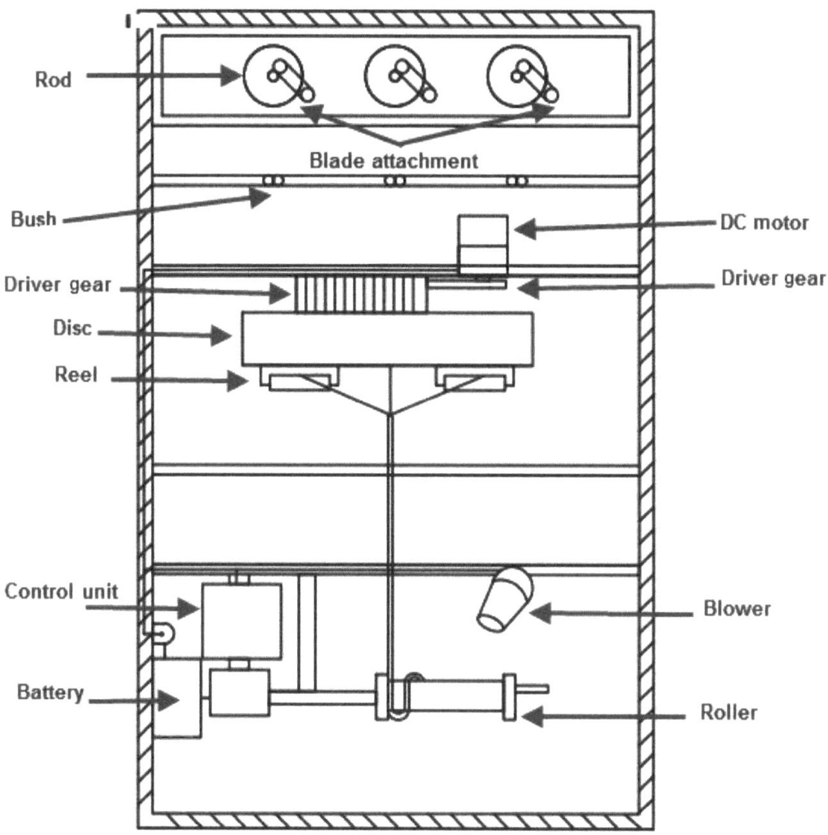

Figure 13.2 Schematic of experimental setup for production of plastic ropes from PET bottles.

based on the sketch. Separate low speed motor is used to drive the collecting wheel. The machine is hand held, and a handle is given that is attached to the frame and helps in easy movement with minimum effort.

13.3.2.1 Fabrication

The machine is made in different steps. Different parts that were made separately are assembled together. These steps are briefed as given below:

- Pulley of known weight is applied for the even distribution of load over the bottle in order to prevent the upward motion of bottle while cutting operation.
- Cutting blade is attached to the root space of the bottle with the help of bolt and nut arrangement by welding operation.
- Plastic rollers are attached in order to prevent the bending of the peeled strip.
- Gears are attached for the proper power transmission to the disk.
- Finally, the parts are connected and diagnosed. The frame is covered with the metal sheets. Collecting wheels are also attached to the frame at the extreme end (Welle, 2013).

13.3.2.2 Working

The aim of the plastic bottle recycling machine is to produce the plastic rope from the plastic bottle. So that at the bottom, the plastic bottle is cut out by using the stainless steel blade. Then, the bottle is placed over the bottle holder, and the strip is peeling out from the bottle. The peeled out strip is further allowed to pass the rollers in order to avoid the tilting of the strip coming out from the plastic bottle. Then, the strip twisted with other strips comes from the bundles, and a gradual heating is done by using the blower. The heating is essential in order to the avoid the unwinding tendency of the plastic strips and make it as single unit, and finally, the rope is stored over the rotating storage wheel.

13.3.2.3 Testing

Analysis of the plastic rope obtained from the plastic bottle recycling machine is performed on the spring tensile testing machine. This is carried out by fixing the rope on a fixed support and another end to a moving support due to loading. It is obtained in such a way that a single strip has a load carrying capacity of about 80 N, and the overall rope has a load carrying capacity of about 180 N. The plastic rope obtained is very strong for household application, and it is flexible and durable. The rope has a tensile strength of 180−200 N, and the single strip plastic is about 60 N. The plastic rope produced from this machine is about 12−15 m long, and it can be used in various household applications such as hanging cloths for drying purpose and tying together etc.

13.3.3 Case study 2: 3D-printer filament from waste plastics

In this work, a device is proposed with a plastic extruder and a 3D printer which is used to convert the waste plastic into commercially needed products. In this work, plastic are shredded and passed through an extruder as given in Fig. 13.3 to produce 3D printing filaments. Later these filaments are used in fused deposition modeling (FDM) machine to manufacture 3D products (Raju & Sharma, 2014). Tables 13.1 and 13.2 provide the specifications of the extrusion machine and 3D printer respectively used in the current study.

Major steps involved in the current work are:

1. Collection of shredded HDPE plastic
2. Extruded into filament using extruder machine
3. Fed into FDM machine
4. 3D products are produced

Collection of shredded HDPE plastic

- Shredded HDPE plastic is collected from local market at a cost of Indian National Rupees 90 per kg
- Different colored bottles can be chosen for different colored filament

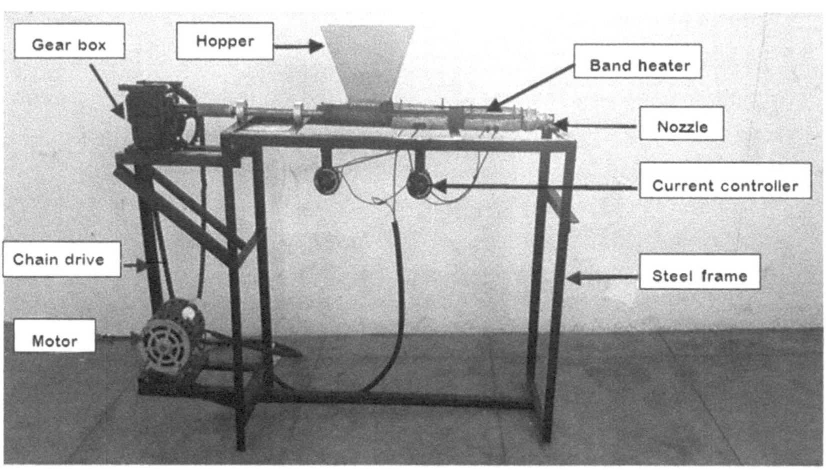

Figure 13.3 Extruder machine.

Table 13.1 Specification of extruder machine.

Parts	Specification
Motor	0.25 hp, 1440 rpm
Chain	Cycle chain
Gear box	1:30
Screw shaft	Length-600 mm, pitch depth-5 mm, pitch length-10 mm, outer diameter-48 mm
Barrel cylinder	600 mm
Hopper	1 mm thickness
Band heater	480°C−500°C
Current controller	0−200
Nozzle	1.75−3 mm
3D printer	Annet A8

Table 13.2 Specifications of 3D printer.

Model name	Annet A8
Nozzle dia	0.4 mm
Printing speed	10−120 mm/s
Filament dia	1.75 mm
Build size	220 × 220 × 240 mm
File format	STL, G-code, OBJ
Working condition	10°C−30°C

13.3.3.1 Extruder machine

- In plastic extrusion, plastic flakes are melted down, filtered for uniform consistency, pushed through a die cut for the final product shape, and then, cooled. It is a complicated process that takes many forms.
  The main procedure for working of extruder is as follows:
- Initially, the shredded plastic is fed into the hopper of the extruder along with additives or colorants (Ragaert, Delva, & Van Geem, 2017).
- The beads from the hopper moves to the feed throat which passes them to large spinning screw working within a horizontal barrel.
- The beads while passing through crew get heated to the melting point. Plastic on reaching the end of the screw gets properly mixed and becomes a semi solid.

- At the end of the screw, the extruded plastic flows across a screen and a breaker plate, which serve two functions. The screen removes any contaminants or inconsistencies in the plastic, and the breaker plate changes the motion of the plastic from rotational to longitudinal.
- The plastic coming out of the extruder is rolled in form of fine filament. Consistency in this flow is critical to achieving an end product with integrity.

13.3.3.2 Extruded filament

- HDPE filament is extruded from the extrusion machine by feeding HDPE plastic flakes as shown in Fig. 13.4.
- The diameter of the filament can be varied by varying the diameter of the hole drilled in the nozzle (Fig. 13.5). In this project, filament diameters are between 1.75 and 3 mm

13.3.3.3 3D printer

•3D printing is the processes in which material is joined or solidified under computer control to create a three-dimensional object. In this project, an FDM 3D printer to print 3D objects with the filament is used.

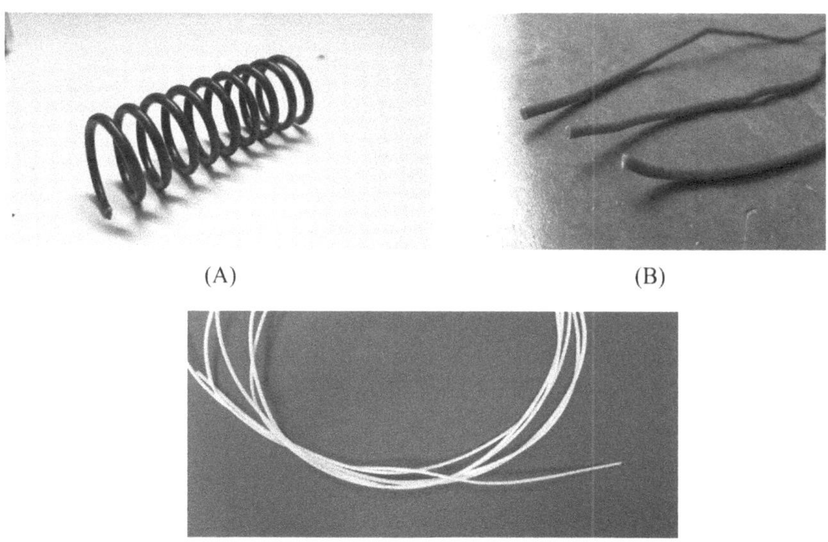

(A) (B) (C)

Figure 13.4 Extruded HDPE filaments. *HDFE*, high-density polyethylene.

Figure 13.5 Nozzle.

Figure 13.6 3D-printed products.

13.3.3.4 3D-printed objects

Using HDPE filament, 3D-printed objects (Fig. 13.6) are made using Annet A8 3D printer.

13.3.3.5 Test results

- From analysis, it is noted that HDPE plastics are perfect alternatives for currently used plastics because of their intermediate melting point which make them use in current 3D printers.

- It has a tensile strength close to that of currently used filaments.
- The cost of making 1 kg of filament from this project is in between 100 and 150 Indian National Rupees.

13.3.3.6 Highlights of the work

- 3D-printer filament is successfully produced.
- Reduction in the cost 3D-printer filament.
- 3D-printed products are made using the filament.
- Promotes recycling of HDPE plastics in useful way.

The proposed filament making process utilizes the waste plastics in a very smart way. The proposed design decreases the cost of 3D-printer filament with the extrusion machine. This technique nicely blends differently colored plastics together and outputs a homogeneous and smooth color.

13.4 Plastic waste to energy conversion

Previous sections have detailed about the various methods to recycle plastic waste materials to produce new and relevant products. Even though the idea of recycling is good, it is not practical to go for recycling of entire plastic waste produced. In addition, the present techniques of recycling are lengthy, costly, and energy intensive. Simultaneously, energy demand of the world is increasing day-by-day. Combining together, the idea to convert plastic waste to energy is impressing. In addition, the distressing depletion of natural energy sources enhances the technology developments for waste to energy production, especially nonbiodegradable plastic waste to energy conversion.

Conversion of plastic waste to energy can be categorized in two subdivisions such as direct method and indirect method. In the first method, the exothermic heat recovered from the chemical reaction of plastic material is directly used for power production, That is, the heat energy recovered can be utilized for conversion of water to steam or for gas heating so that steam or gas turbines can be operated for power production. In the later method, plastic materials can be converted into potential fuels which can then be used in conventional combustion systems for power production. Some of these techniques are discussed in the forthcoming sections.

13.4.1 Plastic pyrolysis for liquid fuel production

The process of pyrolysis (thermal cracking) of plastic waste materials comes under the category of indirection energy conversion technique.

This is one of the finest techniques for fuel production from plastic waste as it can provide both liquid and gaseous fuels having high-energy values. Since plastics are derived from petroleum, its molecular structure consist of complex polymers which make it long-chain organic compound. Pyrolysis is an endothermic reaction in which these complex long-chain molecules undergo thermochemical degradation to form units of short-chain molecules (compounds). During pyrolysis of plastic waste, the long-chain polymers are broken down into products of all the three phases solid (char), liquid, and gas in the absence of oxygen. However, the proportion of solid tar yield is proven to be less than 10% by mass of plastic waste (Sharuddin, Abnisa, Daud, & Aroua, 2018). The liquid phase pyrolysis products have properties similar to diesel or petrol as the plastics are originally derived from petroleum itself. The gaseous phase product (syngas or synthetic gas) also has good calorific value which can be used as gaseous fuel.

Pyrolysis of plastic to produce liquid fuel is studied by many researchers since the process can yield large fraction of liquid oil around 80 wt.%. The converted liquid fuel can be directly used in industrial furnaces, boilers, combustion chambers of steam/gas turbine units, and diesel engines without much treatments (Sharuddin et al., 2018). The process parameters of pyrolysis can be controlled for optimization of product yield. The major parameters that influence the yield of liquid fuel production are temperature, chamber pressure, type of reactors, residence time, and type and rate of fluidizing gas and catalyst used.

The major controlling parameter in pyrolysis reaction is temperature. The polymeric chain cracking reaction rate mainly depends on temperature of reaction. Temperature of thermal degradation depends on the chemical structure plastics. For common plastics such as PET, HDPE, LDPE, PP, and PS, thermal degradation starts at about 350°C. Nevertheless, for PVC, the degradation starts at lower temperature of 220°C. Furthermore, the operating temperature have strong influence on the product preference. If liquid product is preferred, the operating temperature should in the range of 300°C–500°C, whereas for gaseous or char product, higher temperature more than 500°C is suggested which is applicable for almost all plastics (Sharuddin et al., 2018; Sharuddin, Abnisa, Daud, & Aroua, 2016).

Most plastic pyrolysis experiments conducted in laboratory scale are performed in batch or semibatch reactors. These reactors are preferable as the operating parameter control is easy. However, catalytic pyrolysis is not advisable for these reactors as there is a chance of coke formation on the catalyst outer surface which may degrade the effectiveness of the catalyst. For catalytic plastic pyrolysis process, continuous-flow fluidized bed reactors are best preferred as the catalyst can be reused many times (Olazar et al., 2009).

Murata et al. mentioned that higher pressure increases the gaseous product yield but only apparent at high temperatures. Pressure also affected the carbon number distribution of the liquid product by shifting to the lower molecular weight side. Besides, the rate of double-bond formation decreased when pressure increased (Murata, Sato, & Sakata, 2004; Sharuddin et al., 2016). At lower temperature, influence of pressure on residence time is higher. Residence time is the average time period that particle lies inside the reactor which affects the product distribution. Mastral et al. found that higher residence time rises primary product conversion product which leads to the formation of more thermally stable products such as light hydrocarbons and noncondensable gas. However, the residence time is not much significant while carrying out the experiments at higher temperatures (Mastral, Esperanza, Berrueco, Juste, & Ceamanos, 2003; Sharuddin et al., 2016).

Use of catalyst in pyrolysis reaction is to increase the rate of chemical reaction and to improve the hydrocarbon distribution so that the obtained pyrolysis liquid should have similar properties to the conventional fuels of petrol and diesel. The presence of catalyst reduces the reaction temperature and hence increases the rate of reaction. In addition, reduction in reaction temperature reduces the energy input requirement for the pyrolysis reaction. Widely used catalysts in plastic pyrolysis are zeolites and silica-alumina catalysts (Sharuddin et al., 2018). It has been proven that the usage of zeolite catalyst in real municipal plastic waste may also help to reduce the impurities in the oil produced (Miskolczi, Angyal, Bartha, & Valkai, 2009). Silica-alumina catalyst is an amorphous acid type catalyst. It is reported that the final end product of plastic pyrolysis is highly influence by acidity of catalyst. According to the experiments by Sakata et al. (1997), it was found that the catalyst with lower acidity produced higher amount of liquid oil (up to 74.3 wt.%). Recently, Ngernyen et al. has used low cost binder-free pelletized bentonite clay as catalyst for pyrolysis of four waste plastics of resins PS, PP, LDPE, and HDPE. They were able to produce liquid based fuels with higher calorific values and having less viscosity with different types of plastic wastes (Budsaereechai, Hunt, & Ngernyen, 2019).

The authors have studied the liquid oil yield of PET, PS, HDPE, LDPE, and combinations of HDPE and LDPE using a mini-scale batch pyrolysis reactor unit. The pyrolysis process unit for liquid fuel production installed at St. Joseph's College of Engineering and Technology, Palai, India, is shown in Fig. 13.7. The reactor is installed with two stage condenser units. Since the condensation of plastic fumes is arranged in two stages, the gaseous product outlet is almost nil.

Upon analysis of results, it was found that plastic category PS (up to 79 wt/wt.%) provides maximum liquid yield, whereas PET category, the least (32 wt/wt.%) (Vijayakumar & Sebastian, 2018). Pyrolysis reaction is not desirable for PVC resin category as it produce Hydrogen Chloride fumes which is highly toxic. However, the combination of HDPE and LDPE has produced a competitive liquid yield of 73% when the combination was 60:40, respectively (Vijayakumar, 2019).

13.4.2 Gasification of plastic waste

Gasification of plastic waste materials is categorized under indirect waste to energy conversion technique as the output syngas can be stored or can be fed to combustion systems which operates on gaseous fuels. It is an attractive technology due to the adaptability of its final merchantable products such as electricity, steam, dimethyl ether, liquid phase fuels, methanol, ethanol, and other chemicals. Gasification of municipal solid waste (MSW) is an easy technology, as it can use both biodegradable and nonbiodegradable materials simultaneously without the hectic work of mechanical separation.

Figure 13.7 Pyrolysis reactor unit for liquid fuel production.

An operational temperature of 600–800°C is required for gasification plastic wastes. Fluidized gasification systems are widely in use nowadays even though a few research outputs are available with fixed bed gasification (both updraft and downdraft). Air or oxygen or air plus steam is used as a gasification agent. Air gasification process is considered as most reliable process for making combustible syngas for energy extraction. The primary product of gasification is a gaseous mixture of carbon monoxide and hydrogen, with minor percentages of other gaseous hydrocarbons. This gas mixture can be used as a substitute for natural gas for energy production or as feedstock for the production of chemicals in chemical industries. Generally 20%–40% of stoichiometric combustion air only needed for gasification of the PSW (Brems, Dewil, Baeyens, & Zhang, 2013). During gasification, the hydrocarbon feedstock (waste plastics), essentially oxidizes in a controlled manner to generate the depolymerization heat which is endothermic.

Gasification process can jointly valorize plastics with other feedstock materials such as biomass and other biodegradable components in MSW. The composition of the gas output will certainly depend on both composition of feedstock materials as well as the gasifying agent. It is reported that air gasification of waste plastics produces syngas having heating value 6–8 MJ/m^3 (Lopez et al., 2018), whereas steam gasification can produce syngas having heating value above 15 MJ/m^3 even at nitrogen free environment (Wilk & Hofbauer, 2013). However, it is essential to have an efficient gas cleaning system installed with gasification unit as high tar content reported in waste plastics gasification than those reported in biomass gasification (Erkiaga, Lopez, Amutio, Bilbao, & Olazar, 2013).

Gasification of plastic waste is potential and emerging technology in the present days. The major advantage of this technique is that both biodegradable and nonbiodegradable waste matter can be directly fed to the gasification system hence to avoid the hectic task of plastic components segregation from MSW. Even though the technology of gasification with biomass and coal is almost matured, only limited studies have been done with gasification plastic matter in MSW. Considering the importance, a detailed discussion of PSW gasification technology is made in this chapter.

13.4.3 Reactors for plastic gasification

The gasifying units established at present are designed to feed biomass or coal. The chamber as well as accessories design are highly depend on the nature of feedstock. It is essential to have the following capabilities for the gasifying units for effective gasification of PSW.

The gasifying unit should be able to (Sansaniwal, Pal, Rosen, & Tyagi, 2017):

1. Undergo fast plastic depolymerization which indeed required high heat transfer rates in chamber.
2. Avoid operational problems related to the sticky nature of plastics.
3. Provide appropriate residence time to favor tar cracking.
4. Provide accessories to feed catalyst in situ.

13.4.3.1 Fixed bed reactors

Fixed bed gasification reactors are simple in design and operation and requires limited investment. A wide variety of designs are available for fixed bed reactors according to the use and compatibility to the installation space. Ahmed et al. have used fixed bed reactor for gasification of polystyrene at temperatures ranging between 700°C and 900°C and found that syngas yield increases with temperature (Arena, Zaccariello, & Mastellone, 2009). Later on, they have successfully done cogasification of polyethylene with wood chips in steam gasification mode (Koppatz, Pfeifer, & Hofbauer, 2011). Similarly, He et al. have studied plastic gasification in continuous feed mode using Ni/Al_2O_3 as reforming catalyst in situ (Ahmed & Gupta, 2009). It was found that the use of catalyst enhanced the reaction rate and so the feed rate could be increased. Nevertheless, the main encounters for fixed bed gasifiers are difficulty in scaling up, poor heat transfer rate, limited gas—solid contact, and operational difficulty in continuous regime. Therefore studies on gasification of waste plastics or their cogasification with coal and biomass are limited with fixed bed reactors.

13.4.3.2 Fluidized bed reactors

Fluidized bed gasification reactors are the most commonly used gasifiers for PSW gasification. Bubbling fluidized bed and circulating fluidized bed methods are the traditionally used types (Basu, 2010). Schematic diagrams of three types of fluidized bed gasification systems are shown in Fig. 13.8. It can be seen that the bubbling bed system is simple, but with limited particle size and nonuniformity in fluidization. In circulating bed type, the fluidization is much improved which leads to higher gas conversion rate with low tar yields. Dual fluidized bed is a much extended version of circulating fluidized bed with very high fluidity and have the wide flexibility

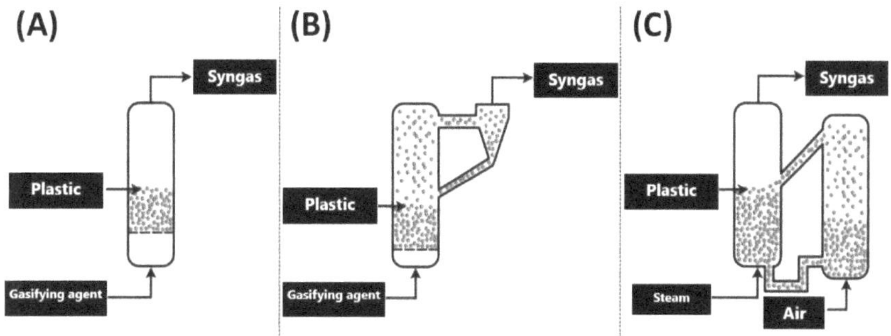

Figure 13.8 Different types of fluidized beds for gasification (A), bubbling fluidized bed (B), and circulating fluidized bed (C). *From Dual fluidized bed (Lopez, G., Artetxe, M., Amutio, M., Alvarez, J., Bilbao, J., & Olazar, M., (2018). Recent advances in the gasification of waste plastics. A critical overview. Renewable and Sustainable* Energy, *82, 576–596), with kind permission of Elsevier.*

in feedstock particle size. However, installation of large-scale models of the latter two types requires high investment cost. Therefore it can be seen that maximum number of plastic gasification studies have been reported in bubbling fluidized reactors. In comparison with fixed bed reactors, bubbling bed reactors have high heat and mass transfer rates, exceptional gas–solid contact and good control of temperature during gasification process (Molino, Chianese, & Musmarra, 2016). Generally, air (or oxygen) is used as the gasifying agent for fluidized bed gasification of plastic wastes. Arena et al. conducted study on a pilot plant bubbling fluidized bed gasifier with continuous feed of up to 90 kg/h and found a much lower tar reduction in the valorized content when applied with natural olivine as an in situ tar reduction agent (Erkiaga, Lopez, Amutio, Bilbao, & Olazar, 2014). However, there are reports from other researchers who has applied plastic feed rates in the 1–4 kg/h gasifiers operating in continuous regime (Lopez et al., 2018).

13.4.3.3 Spouted bed reactor

Spouted bed reactors were earlier used for drying of biomass, plastic wastes, and tires. They were rarely but successfully used for pyrolysis of biomass and gasification of low-grade coal. However, Erkiaga et al. have done experimental studies with a conical spouted bed reactor (CSBR) operating with continuous biomass (forest pinewood waste) feed (Erkiaga et al., 2014). It was found that CSBR features cyclic and vigorous particle movement which avoids bed defluidisation due to the agglomeration.

Heat transfer rate between phases in the reactor is quiet high so that wide feedstock particle size distribution is possible. In addition, gas flow rate control is very good which allows operating in a wide steam/feedstock ratio with a short residence time. These features make this type of reactor much suitable for gasification of biomass as well as cogasification with PSW. The main limitation of this method of gasification is the short residence of the volatiles, which hinders tar cracking reactions. Nevertheless, Lopez et al. have done steam gasification of different mixtures of biomass and HDPE plastics and found that the tar content in the gaseous product only 9.74 g N/m^3 which is even lower than biomass sole gasification (Lopez, Erkiaga, Amutio, Bilbao, & Olazar, 2015).

13.4.4 Cogasification

A major benefit of gasification of MSW over pyrolysis is its ability to use heterogeneous mixture feed, that is, both biodegradable as well as nonbiodegradable waste matter (PSW) can be fed together so that successful gasification is possible. This heterogeneous feedstock gasification process is termed as cogasification. In most cases, waste plastics have cogasified with biomass. It can be coprocessed with coal too. Pinto et al. reported that addition of PE plastic waste with pine wastes has improved the release of H_2 in the gas output with the expense slight amount of CO. It was measured almost 98% of energy conversion during the experiments. It was identified that temperature was the most influenced parameter which determines the gases composition. The rise of temperature favored the formation of H_2 and decreased the formation of hydrocarbons, tar, and char (Miranda et al., 2002). Later on, the same team have done cogasification in an improved setup with gas cleaning unit. The gas cleaning unit consist of a cyclone separator and two reactors with equipped with catalyst bed. Ni-based catalysts were used for the effective reduction of tar content in syngas (Pinto et al., 2009).

13.4.5 Plasma pyrolysis and gasification

Plasma pyrolysis of plastic waste can be kept under the scheme of indirect method of plastic waste to energy conversion. The technology involves the presence of a plasma arc for the cracking of feedstock matter. It is an emerging and potential technique for plastic waste conversion, as the chance of partial cracked products in the reactor outlet is practically nil. Hence, technique is advisable for processing of mixed waste matters which include hospital waste, synthetic fiber wastes, medicinal wastes, etc. In most cases, the output of plasma reactors are gaseous products which can also designated as syngas. However, there are plasma pyrolysis reactors having additional fractional condensation units for liquefaction of these gaseous output to convenient liquid fractions.

Plasma can be considered as a high-temperature "thermal flame" which can disintegrate almost all kind of organic and inorganic compounds included in MSW. For pyrolysis reaction, thermal plasma is utilized while nonequilibrium plasma (cold plasma) find application in many electrical appliances. The plasma generation in reactors are made usually through the gas excitement by the application of heavy electrical discharge between the end of two electrodes. The plasma jet produced at one end of the electrode extends to the next electrode through sufficient high velocity gas flow. The inelastic collisions of excited electrons in the region further ionizes the gas field and maintains a stable plasma flame.

13.4.5.1 Plasma pyrolysis

A series of reactions take place in plasma pyrolysis process in which the most common ones are eliminations and rearrangements. Depending upon the operating conditions, other reactions such as oxidation, reduction, substitution, or addition may also happen during the process. Like earlier discussed pyrolysis reactions, plasma pyrolysis too takes place in oxygen starvation condition. Plasma pyrolysis uses extremely high temperatures (2000–10,000 K) of plasma arc to disintegrate waste material into simple molecules almost completely. Hot plasmas are principally adequate for destruction of toxic molecules which may produce during normal pyrolysis process. In addition, plasma pyrolysis has the advantage of reduction in volume of feedstock matter, after releasing ultraviolet radiations from it. The major advantage of plasma pyrolysis its very short period (0.01–0.5 s) to complete pyrolysis reaction (Dave & Joshi, 2010). Time for complete waste transformation depends on waste and temperature.

Pragnesh and Asim have done plasma pyrolysis of PE and PP plastics and reported that presence of PE increases alkane content. They also reported that the plasma pyrolysis of PS plastics with earlier mixture yields higher aromatic content in end product. In addition, the presence of PP favors alkene formation. Consequently, it can be concluded that the presence of both PS and PP increase the octane rating of end product (Dave & Joshi, 2010). Bockhorn et al. investigated the thermal degradation of PE as well as PS at 430°C–480°C for reaction duration ranging between

20 min and 6 h. PE decomposes into a large number of paraffinic and olefinic compounds without a residue, whereas PP decomposes into a large number of aliphatic compounds without a residue (Bockhorn, Hornung, Hornung, & Schawaller, 1999). This again proves the elimination of polymer residue through plasma pyrolysis process.

A project of Petroleum Conservation Research Associations in association with Central Mechanical Engineering Research Institute of India had successfully implemented a plastic waste plasma pyrolysis unit with own developed plasma generation technology which uses carbon and graphite electrode arrangement in a tripod orientation. The length of plasma arc was adjustable with leadscrew attachment. The syngas developed was able to fed for an internal combustion (IC) engine generator unit after passing through a gas cleaning unit which consist of water scrubbed and cyclone separator units (Dubey, 2019). The schematic representation of plasma arc pyrolysis gasifier unit is shown in Fig. 13.9 (Puncochara et al., 2012).

Industrial scale operation of plasma pyrolysis units is already established in which most of them are using readily available plasma torches for

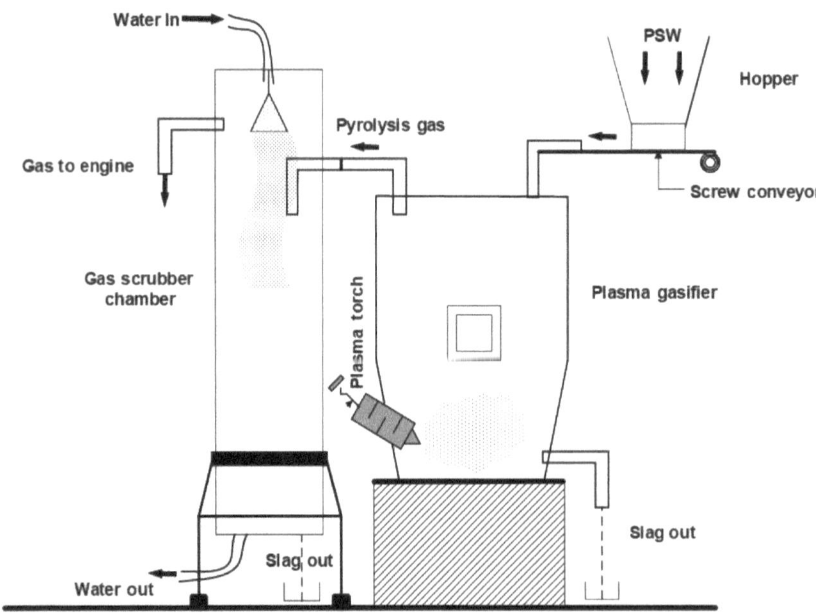

Figure 13.9 Schematic representation of plasma arc pyrolysis gasifier unit *From Puncochara, M., Rujb, B., & Chatterjeeb, P.K., (2012). Development of process for disposal of plastic waste using plasma pyrolysis technology and option for energy recovery.* Procedia Engineering, *42, 420–430, with kind permission of Elsevier.*

plasma generation. There are several manufacturers around the world who can provide portable and mobile plasma reactors for industrial waste treatment purposes.

13.4.5.2 Plasma gasification

Park et al. had used a V-type Plasma torch for steam plasma gasification system of PE at 1200°C. PE was decomposed to provide 65% of H_2 and 30% of CO at steam—PE ratio of 1.2 per kg of PE. The gasification was done in cogasification mode using wood chips and polyolefin sheets as feedstock materials. It has been found that injection of high-temperature steam/air mixture into pyrolysis gas can effectively decompose tar components in pyrolysis gas into CO and H_2, resulting in an almost tar-free, clean gas product. They had successfully demonstrated a dual fueled diesel engine combined with gasification of wastes to achieve 30% engine thermal efficiency (Hyun Seo Park, Kim, & Kim, 2005).

Tadeusz et al. reported the process of plasma gasification of polyethylene plastics with cellulose (from hospital waste). It was reported that gasification of 1 kg of this feedstock (proportion of paper, plastic and water at 6:3:1, respectively), 1 kWh electric energy should be consumed. However, upon analysis, the energy content of CO and H_2 in the output gas concentration, it was estimated that electric energy over 3 kWh can be produced by the combustion of syngas which prove economic efficiency of the technology (Mączka, Śliwka, & Wnukowski, 2013). The following reaction steps illustrate plasma pyrolysis process done by Tadeusz et al.:

$$\text{Cellulose } C_6H_{10}O_5 + \text{plasma} \rightarrow CH_4 + 2(CO) + 3(H_2O) + 3C \rightarrow n(CO) + mH_2$$

$$\text{Polyethylene}(CH_2-CH_2)_n + \text{plasma} + H_2O \rightarrow x(CH_4) + z(CO) + \text{radicals} \rightarrow n(CO) + mH_2$$

A block diagram representing steps of plasma gasification plastic wastes with contaminated biomass to produce liquid fuels is shown in Fig. 13.10.

It can be concluded that the use of plasma techniques instead of traditional combustion produces syngas containing very limited quantity of tar as well as partial combustion products. It can be seen that most of the plasma pyrolysis reactors are working in continuous mode, whereas batch production mode is applied in electrically/flame heated pyrolysis systems. The utilization of higher energy input for plasma pyrolysis reaction is justified by the improved energy potential of the output syngas. Higher

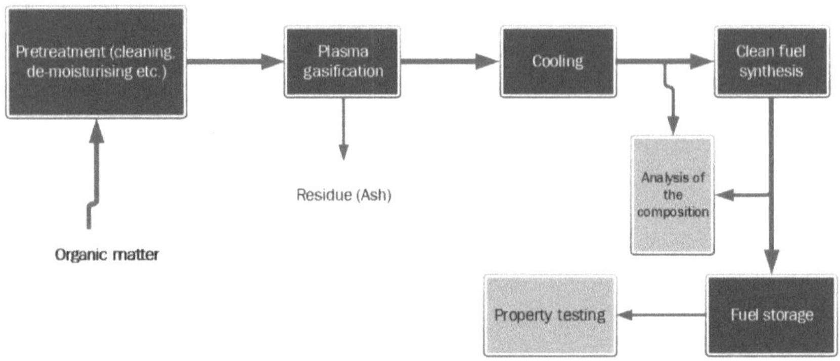

Figure 13.10 Process steps in plasma gasification (Mączka et al., 2013).

concentration of H_2 in the syngas is a value addition for the attractiveness of plasma gasification.

13.4.6 Power production from incinerated heat

Incineration is the most established and convenient technique for waste removal. Though incineration is widely used in both industrial and household solid waste disposal process, energy recovery from the same is limited only to large-scale incineration units. Incineration is the process of controlled burning of waste at sufficient aerobic conditions above 500°C. This results in liberation of heat energy with visible products as gases and ash. Net energy yield depends upon the density and composition of the waste. Relative percentage of moisture and inert materials, combustion temperature, size and shape of the constituents, and design of the combustion system are the main influencing parameters of incineration technology. In practice, about 65%–80% of the energy content of the organic matter can be recovered as heat energy, which can be utilized either for direct thermal applications or for power generation methods (Patil, Kulkarni, & Patil, 2014). Since the burning of the plastic waste happens in incinerator at elevated temperature in closed chamber, the exit of toxic dioxins is almost eliminated. The incineration plants which are associated with an electric power production unit where the energy for power production is attained by the waste heat recovery from incinerator. Hence, the method of energy recovery from incineration can be categorized under direct waste to energy conversion.

Solid waste incineration plants are designed, constructed and operated to achieve maximum combustion of the waste matter, so that exit emissions should be least harmful for environment. Generally, waste

It is seen that plastic recycling is one of the solutions to reduce dumping of disposable plastic in the environment. Some case studies conducted by authors presented in the chapter encourage further work in the field. Various methods for energy conversion and power generation from plastics can be focused as future work. Large-scale projects require strong and evidenced models as detailed in this chapter. Government and agency funding, societal support, institutional collaboration, and efforts from researchers are the prime factors to establish this field further.

Acknowledgment

The authors are gratefully acknowledging the Center for Engineering Research and Development (CERD), Government of Kerala, India for awarding funding for the projects as mentioned below:

- Student Project Scheme (No. KTU/RESEARCH 2/4597/2019) for conducting case study on 3D-printer filament from waste plastics for second author.
- Research SEED fund (No. KTU/RESEARCH 2/3894/2018) for installation of plastic pyrolysis chamber for the first author.

References

Ahmed, I. I., & Gupta, A. K. (2009). Hydrogen production from polystyrene pyrolysis and gasification: Characteristics and kinetics. *International Journal of Hydrogen Energy, 34* (15), 6253–6264.

Amin, S., & Amin, M. (2011). Thermoplastic elastomeric (TPE) materials and their use in outdoor electrical insulation. *Reviews on Advanced Materials Science, 29*, 15–30.

Arena, U., Zaccariello, L., & Mastellone, M. L. (2009). Tar removal during the fluidized bed gasification of plastic waste. *Waste Management (New York, N.Y.), 29*(2), 783–791.

Ashori, A. (2008). Wood-plastic composites as promising green-composites for automotive industries!. *Bioresource Technology, 99*, 4661–4667.

Basu, P., 2010. *Biomass Gasification and Pyrolysis.*

Bockhorn, H., Hornung, A., Hornung, U., & Schawaller, D. (1999). Kinetic study on the thermal degradation of polypropylene and polyethylene. *Journal of Analytical and Applied Pyrolysis, 48*(2), 93–109.

Brazel, C. S., & Rosen, S. L. (2012). *Fundamental Principles of Polymeric Materials.* Hoboken, NJ: John Wiley & Sons.

Brems, A., Dewil, R., Baeyens, J., & Zhang, R. (2013). Gasification of plastic waste as waste-to-energy or waste-to-syngas recovery route. *Natural Science, 5*(6), 695–704.

Budsaereechai, S., Hunt, A. J., & Ngernyen, Y. (2019). Catalytic pyrolysis of plastic waste for the production of liquid fuels for engines. *RSC Advances, 9*(10), 5844–5857.

Dave, P. N., & Joshi, A. K. (2010). Plasma pyrolysis and gasification of plastics waste—A review. *Journal of Scientific and Industrial Research (India)., 69*(3), 177–179.

Dubey, B., 2019. Plastic Waste Management—Landfilling, Other Applications. *YouTube*, May 06, 2019 [Video file]. Available: https://www.youtube.com/watch?v = vi1s8Iqedas &feature = emb_logo. (Accessed 10 May 2019).

Elo, K., & Sundin, E., 2009. Automation of Plastic Recycling—A case study, December 2009.

Erkiaga, A., Lopez, G., Amutio, M., Bilbao, J., & Olazar, M. (2013). Syngas from steam gasification of polyethylene in a conical spouted bed reactor. *Fuel*, *109*, 461–469.

Erkiaga, A., Lopez, G., Amutio, M., Bilbao, J., & Olazar, M. (2014). Influence of operating conditions on the steam gasification of biomass in a conical spouted bed reactor. *Chemical Engineering Journal*, *237*, 259–267.

Grigore, M. E. (2017). Methods of recycling, properties and applications of recycled thermoplastic polymers. *Recycling*, *2*, 24.

Hopewell, J., Dvorak, R., & Kosior, E. (2009). Plastics recycling: Challenges and opportunities. *Philosophical Transactions of the Royal Society B*, *364*.

Hyun Seo Park, S. J. K., Kim, C. G., & Kim, S. J. (2005). Characterization of PE gasification by steam plasma. *Journal of Industrial and Engineering Chemistry*, *12*(2), 216–223.

Jamdar, V., Kathalewar, M., Dubey, K. A., & Sabnis, A. (2017). Recycling of PET wastes using electron beam radiations and preparation of polyurethane coatings using recycled material. *Progress in Organic Coatings*, *107*, 54–63.

Koppatz, S., Pfeifer, C., & Hofbauer, H. (2011). Comparison of the performance behaviour of silica sand and olivine in a dual fluidised bed reactor system for steam gasification of biomass at pilot plant scale. *Chemical Engineering Journal*, *175*(1), 468–483.

Leng, Z., Padhan, R. K., & Sreeram, A. (2018). Production of a sustainable paving material through chemical recycling of waste PET into crumb rubber modified asphalt. *Journal of Cleaner Production*, *180*, 682–688.

Lopez, G., Artetxe, M., Amutio, M., Alvarez, J., Bilbao, J., & Olazar, M. (2018). Recent advances in the gasification of waste plastics. A critical overview. *Renewable and Sustainable Energy*, *82*, 576–596, May 2017.

Lopez, G., Erkiaga, A., Amutio, M., Bilbao, J., & Olazar, M. (2015). Effect of polyethylene co-feeding in the steam gasification of biomass in a conical spouted bed reactor. *Fuel*, *153*(x), 393–401.

Mączka, T., Śliwka, E., & Wnukowski, M. (2013). Plasma gasification of waste plastics. *Journal of Ecological Engineering*, *14*(1), 33–39.

Mastral, F. J., Esperanza, E., Berrueco, C., Juste, M., & Ceamanos, J. (2003). Fluidized bed thermal degradation products of HDPE in an inert atmosphere and in air-nitrogen mixtures. *Journal of Analytical and Applied Pyrolysis*, *70*(1), 1–17.

Ministry of New and Renewable Energy, 2016. MNRE Report to Lok Sabha on MSW.

Miranda, M., Gulyurtlu, I., Cabrita, I., Pinto, F., Franco, C., & Andre, R. N. (2002). Cogasification study of biomass mixed with plastic wastes. *Fuel*, *81*, 291–297.

Miskolczi, N., Angyal, A., Bartha, L., & Valkai, I. (2009). Fuels by pyrolysis of waste plastics from agricultural and packaging sectors in a pilot scale reactor. *Fuel Processing Technology*, *90*(7–8), 1032–1040.

Molino, A., Chianese, S., & Musmarra, D. (2016). Biomass gasification technology: The state of the art overview. *Journal of Energy Chemistry*, *25*(1), 10–25.

Murata, K., Sato, K., & Sakata, Y. (2004). Effect of pressure on thermal degradation of polyethylene. *Journal of Analytical and Applied Pyrolysis*, *71*(2), 569–589.

Olazar, M., Lopez, G., Amutio, M., Elordi, G., Aguado, R., & Bilbao, J. (2009). Influence of FCC catalyst steaming on HDPE pyrolysis product distribution. *Journal of Analytical and Applied Pyrolysis*, *85*(1–2), 359–365.

Pascault, J.-P., Sautereau, H., Verdu, J., & Williams, R. J. (2002). *Thermosetting Polymers* (Volume 64). Roca Raton, FL: CRC Press.

Patil, A. A., Kulkarni, A. A., & Patil, B. B. (2014). Waste to energy by incineration. *Journal of Computer Science and Technology*, *3*(6), 2278–3814.

Pinto, F., André, R. N., Franco, C., Lopes, H., Gulyurtlu, I., & Cabrita, I. (2009). Co-gasification of coal and wastes in a pilot-scale installation 1: Effect of catalysts in syngas treatment to achieve tar abatement. *Fuel*, *88*(12), 2392–2402.

Plants, W., 2018. Waste-to-Energy Plants. [Online]. Available: https://www.drishtiias.com/printpdf/waste-to-energy-plants.

Puncochara, M., Rujb, B., & Chatterjeeb, P. K. (2012). Development of process for disposal of plastic waste using plasma pyrolysis technology and option for energy recovery. *Procedia Engineering*, *42*, 420–430.

Ragaert, K., Delva, L., & Van Geem, K. (2017). Mechanical and chemical recycling of solid plastic waste. *Waste Management*, *69*, 24–58.

Raju, G., & Sharma, M. L. (2014). Optimization of plastic extrusion. *International Journal Advanced Mechanical Engineering*, *4*, ISSN 2250-3234 November 6.

Ribeiro, M. C. S., Fiúza, A., Ferreira, A., Dinis, M. D. L., Meira Castro, A. C., Meixedo, J. P., & Alvim, M. R. (2016). Recycling approach towards sustainability advance of composite materials' industry. *Recycling*, *1*, 178.

Sakata, Y., Azhar Uddin, M., Muto, A., Kanada, Y., Koizumi, K., & Murata, K. (1997). Catalytic degradation of polyethylene into fuel oil over mesoporous silica (KFS-16) catalyst. *Journal of Analytical and Applied Pyrolysis*, *43*(1), 15–25.

Sansaniwal, S. K., Pal, K., Rosen, M. A., & Tyagi, S. K. (2017). Recent advances in the development of biomass gasification technology: A comprehensive review. *Renewable and Sustainable Energy*, *72*, 363–384, December 2016.

Sharuddin, S. D. A., Abnisa, F., Daud, W. M. A. W., & Aroua, M. K. (2016). A review on pyrolysis of plastic wastes. *Energy Conversion and Management*, *115*, 308–326.

Sharuddin, S. D. A., Abnisa, F., Daud, W. M. A. W., & Aroua, M. K. (2018). Pyrolysis of plastic waste for liquid fuel production as prospective energy resource. *IOP Conference Series: Materials Science and Engineering*, *334*(1).

Urbancl, D., Zlak, J., Anicic, B., Trop, P., & Goricanec, D. (2016). The evaluation of heat production using municipal biomass co-incineration within a thermal power plant. *Energy*, *108*(1), 140–147.

Vijayakumar, A., 2019. MTech Dissertation, SJCET Palai, Kottayam.

Vijayakumar, A., & Sebastian, J. (2018). Pyrolysis process to produce fuel from different types of plastic—A review. *IOP Conference Series: Materials Science and Engineering*, *396*(1).

Wasilewski, Z. R. (2013). Energy recovery from waste plastics. *CHEMIK*, *67*(5), 435–445.

Welle, F. (2013). Is PET bottle-to-bottle recycling safe? Evaluation of post-consumer recycling processes according to the EFSA guidelines. *Resources, Conservation and Recycling*, *73*, 41–45.

Wilk, V., & Hofbauer, H. (2013). Conversion of mixed plastic wastes in a dual fluidized bed steam gasifier. *Fuel*, *107*, 787–799.

Woldemar d'Ambrières, 2019. Plastics recycling worldwide: Current overview and desirable changes, FieldActions Science Reports, 19.

Yang, Y., Boom, R., Irion, B., van Heerden, D.-J., Kuiper, P., & de Wit, H. (2012). Recycling of composite materials. *Chemical Engineering and Processing—Process Intensification*, *51*, 53–68.

CHAPTER 14

Sustainable supply chain management in manufacturing industries

E. Manavalan, M. Thanigai Arasu and Jayakrishna Kandasamy
School of Mechanical Engineering, VIT University, Vellore, India

14.1 Introduction

In the evolution of modern world, carbon footprint is one of the major issues that influences economy, environmental, and healthcare. The sustainable management process focuses in reducing the carbon footprint, improving environment, and facilitating green economic initiatives (Singh & Trivedi, 2016). The sustainable management process is also used in various sectors such as agriculture, automobile, and healthcare where supply chain plays a vital role (Fulconis, Pache, & Reynaud, 2019).

The sustainability is highly necessitated in providing precious quality life for future generation. Fig. 14.1 illustrates the typical sustainable development process. The objective of sustainable development is to fulfill the current requirements of society and, at the same time, to minimize the environment, social, and economic impacts of the product.

The sustainable development should establish a good relationship between economic profit, public, and planet by integrating with green product development, lean manufacturing, safety, healthcare, and closed-loop supply chain (Kleindorfer, Singhal, & Van Wassenhove, 2005; Matos and Hall, 2007; Oelze, Hoejmose, Habisch, & Millington, 2016).

In the context of sustainable development, organizations started focusing on implementing the sustainability in the products which they manufacture, process, and practices they follow across supply chain. Thus organizations are transforming their linear supply chains into circular supply chains (Kleindorfer et al., 2005; Stevens & Johnson, 2016). Circular supply chain helps to minimize the usage of new input resources and energy leakages to

Sustainable Manufacturing
DOI: https://doi.org/10.1016/B978-0-12-818115-7.00010-9

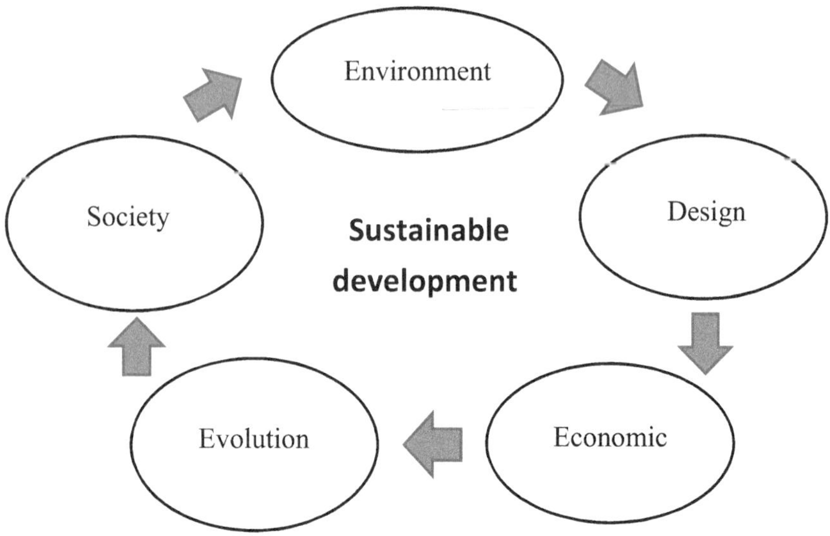

Figure 14.1 Sustainable development cycle.

eliminate the solid wastes and further use renewable energy (Gaustad, Krystofik, Bustamante, & Badami, 2018; Matos & Hall, 2007).

14.2 Supply chain transformation

The aim of supply chain transformation is to align the supply chain members' goals and maximize the overall operational efficiency. It is about providing a process which can improve the service level, make the product cost effective, and can maintain stock optimally. Organizations need to cope with global competition and frequent change in customer demand. Supply chains need to align in these challenging conditions. It is inevitable to manage both vertical and horizontal supply chain efficiently in terms of procuring goods, manufacturing products, and transporting it through distribution centers (O'Rourke, 2014). The supply chain transformation provides opportunities to maximize productivity, reduce total cost of ownership, and improve customer satisfaction (Tsoulfas & Pappis, 2008).

With this context, this section discusses the different types of supply chain and its impacts on the sustainability. Typically, the supply chain can be crafted in the following classifications namely, linear supply chain, circular supply chain, and shared supply chain (Reefke & Sundaram, 2018).

Linear supply chain focuses only on the forward flow, that is, procure raw materials from suppliers, make the item, and dispose it once the end-of-life (EOL) of the item is reached. Circular supply chain focuses both on forward and reverse supply chain. The objective of circular supply chain is to improve the sustainability of the organization. It focuses on minimizing the wastages in the process and product, and wherever necessary adapt reuse, repair, and recycle strategies. In a shared supply chain, multiple organizations get mutual benefits by sharing the same logistics to move the goods and use the same warehouses to store the items. This approach reduces overall operational costs for manufacturers and helps customers to get the items faster.

14.2.1 Linear supply chain

The linear supply chain is a forward product value chain in which the suppliers, producers, distribution service, and customers linked together for feed-forward material flow (Manavalan & Jayakrishna, 2019a). Basically, it follows the traditional supply chain process that is procure-manufacture-dispose model. Thus the linear supply chain forms a simple and easy one-way supply process flows from producers to the customers as illustrated in Fig. 14.2. It includes the process of raw material consuming, testing, manufacturing, and transportation consecutively with the reduction of supplier works and transportation charges.

The linear supply chain focuses on the sales and profits of an organization and not pays much attention to sustainability. The typical example of linear supply chain is "beverage manufacturing." Consider a beverage factory processing their soft drinks and filling up with a plastic bottle. After sales, the end-user consumes the drinks and disposes the plastic bottle in the form of solid waste. The focus on recycle of the plastic bottle is not focused in linear supply chain.

Figure 14.2 Linear supply chain.

14.2.2 Closed-loop or circular supply chain

Organizations are focused on improving cost-cutting measure in the challenging economic conditions. Closed-loop supply chain helps organizations to handle the reverse logistics properly leads to cost savings. The challenges in reverse logistics are the uncertainties over the quality, cost variation, and customer involvement in returning the product (Gnoni et al., n.d.). Thus it is important to study the reverse logistics to understand the challenges involved. The closed-loop supply chains are designed to bring the pollution-free environment and focuses on the sustainability of the organization (Govindan & Hasanagic, 2018). As illustrated in Fig. 14.3, the scraps are either recycled or remanufactured or repaired which forms a circular supply chain (https://www.soundonsound.com/techniques/refurbishing-your-old-equipment). After recovery, the grade of the products is tested and classified as follows: grade 1, fit for reuse; grade 2, remanufactured/repaired; and grade 3, scrapped and recycled (Gaustad et al., 2018; https://www.soundonsound.com/techniques/refurbishing-your-old-equipment). The grade classification allows the organizations to extend the EOL products, which can minimize the wastages and improve the sustainability. Following are few examples to extend the EOL of the products, sell as new (just cleaning again), sell by repacking as new, sell by reprocessing

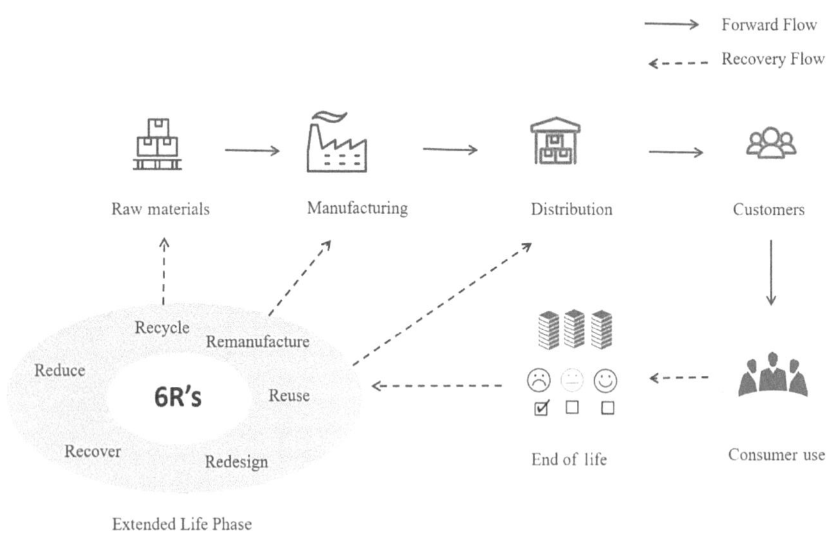

Figure 14.3 Circular supply chain.

and new packaging, sell by discounts and offers, repair and refurnishing compliments, resource recycling.

Washer manufacturers are the good example of this type. Consider a washer manufacturer produces metal washers. Typically, the process starts with procuring steel rods that are machined. During the manufacturing process, few items are scrapped. The scraps (generated waste) are sent to recycle factory (metal fabricators) for further processing. Likewise, the waste recovered from customer also recycled (Pan et al., 2015; Roos, 2014).

Closed-loop supply chain effectively minimizes the product waste, environmental pollution, and excess resources usage and energy consumption to regulate green economy (Genovese, Acquaye, Figueroa, & Koh, 2017).

At times, the process of setting effective waste treatment plant/process is slightly difficult and expensive. Sometimes the cost of remanufacturing is higher than the normal manufacturing processes.

14.2.3 Shared supply chain

As illustrated in Fig. 14.4, shared or integrated supply chain is a cost-effective model as the same transportation can be used by different manufacturers and the same distribution center can be used to store the items (Song & Gao, 2018; Zhang, Liu, Zhang, & Bai, 2015). The emerging technologies such as Internet of Things (IoT), global positioning system (GPS), and autonomous robots can help supply chain members to communicate efficiently and make the shared supply chain as a successful model (Geissdoerfer, Morioka, de Carvalho, & Evans, 2018; Stevens & Johnson, 2016). The advantages of shared supply chain are higher profit margin and flexibility.

14.2.3.1 Higher profit margin

Shared space or logistics can help manufacturers to reduce the operational costs which may result in higher profit, and also it helps the manufacturers

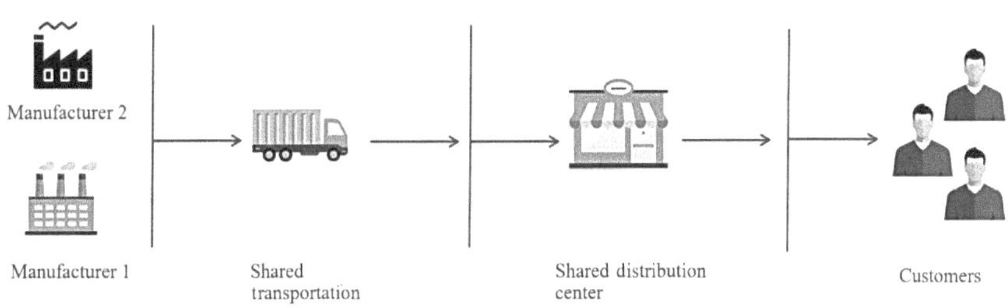

Figure 14.4 Shared supply chain.

to deliver the goods faster (Khan, Hussain, & Saber, 2016). In this model, it is achievable to make more deliveries at the same price.

14.2.3.2 Flexibility

Increase in flexibility is more important for integrated supply chain. The flexibility can be achieved, by integrating all of a business's functions. Shared supply chain network gives more flexibility to deliver the customer demand in seamless manner (https://www.inspirage.com/2016/09/top-4-benefits-integrated-supply-chain).

14.3 Goals of sustainable supply chain management

The major objective of supply chain management is to fulfill the customer demand quickly but in recent times, and it becomes harder to achieve it considering the complex supply chain network involvement. The goals of typical supply chain are to reduce the cost, improve the process, track asset consumption, and improve inventory visibility. Governments are keen on insisting supply chain companies to reduce carbon footprint and improve sustainability. In this context, the goals of sustainable supply chain (SSC) are to reduce the environmental impacts by minimizing the product and energy waste, improve social responsibility by improving workers' health and safety, increase revenue by selling more sustainable products which improves economic stability. It is the duty of an organization to balance its environmental, economic, and social practices. The recent study on sustainability provides more opportunity to review the flow of materials, streamline the processes, and improvise the operations (Su et al., 2016). In addition, it also reveals the importance of corporate social responsibility. Some of the areas to focus in sustainable supply chain management (SSCM) are to reduce unnecessary logistics movement which can save fuel consumption, to use bio-degradable items for packing, to find easier ways to dispose/collect waste products, to conserve water and energy wherever applicable, and to use more recycled raw materials than new one.

14.3.1 Sustainable supply chain management and its implementation

The key function of SSC is to facilitate manufacturing. Sustainable manufacturing is the integration of process and the products that are collaborated to make eco-friendly goods.

Manufacturing and supply chain are interrelated. To achieve SSC, it is necessary to gain sustainability in manufacturing as well. The three pillars of sustainable manufacturing are process, product, and environment.

14.3.1.1 Process

The idea is to transform the linear process (product to waste) into the circular process (product to waste to new resource) focused from raw material stage till it reaches the customer, and to bring back the EOL product to the manufacturing facility to recycle it. This is in the effort to promote green manufacturers and increase circular factories to make eco-friendly and reusable products that are highly sustainable (Lewandowski, 2016). In recent days, the manufacturers are focusing on the environmentally friendly design, process, product, and flexibility in the closed-loop supply chain process (He, 2017).

14.3.1.2 Product

The use of bio-degradable products has extensively minimized environmental effects, and the recycled products have reduced the wastes in closed-loop supply chain (Farooque, Zhang, Thurer, Qu, & Huisingh, 2019). The solid wastes are collected and reused as a resource for new product development (reprocessing and re-inventory) preventing pollutions and industrial wastes.

14.3.1.3 Environment

In recent times, organizations are realizing the importance of corporate social responsibilities and uses sustainable raw materials wherever possible to make the environment greener. The usage of eco-friendly products and minimizing the wastes in the process helps in improving the resource utilization, productivity, and economic profitability (Dubey et al., 2017).

14.4 Issues in sustainable supply chain management adoption and implementation

A recent study suggests that increased cost in implementing SSC is one of the key barriers for organizations (Luthra & Mangla, 2018a). Further, the researchers also reveals that the changing culture, bringing the sustainable development from the proof of concept to reality, uncertainty over the costs when there is unexpected surge the demand, lack of sustainability risk assessment are also key drivers which restricts organization to implement SSCM in an organization (Esfahbodi, Zhang, & Watson, 2016).

14.4.1 Government norms

Considering the environment and public welfare, the governments and various organizations published several norms and regulations to the industries to target for green society and eco-friendly environment. The following important points can be seen:

- Industrial environmentalism (1960–70): It enables the industries to concern the processing environment and to make their own regulation evaluating environmental impacts (https://www.uwosh.edu/facstaff/barnhill/ES-243/pp%20outline%20New%20Environmentalism.pdf).
- Regulatory environmentalism (1970–82): In 1970 these norms were created by the US Environmental protection agency. During this period, the industrial environment and the business process are developed considering the laws on environmental threats (http://www.thegreenmedium.com/blog/2015/9/1/the-green-mediums-first-contributor-matt-gwozd).
- Environmentalism as social responsibility (1982–88): At this stage, major nongovernment organizations were collaborated with government, due to the increase in their funding on environmental compliances (Afgan, 2010).
- Strategic environmentalism (1988–93): In this period, the industries are forced to improve their environment skills for environmental protectiveness (Afgan, 2010).
- Environmental management as an opportunity (1993–present): In this stage, green process has been implemented significantly to create higher advantages on the organization for developing their green products and green business world to accelerate some financial advantages like: low-cost utility, energy, and productivity which form the way for sustainable management (Afgan, 2010).

14.4.2 Environmental regulations pertaining to transportation

Logistics movement is one of the key aspects of supply chain. The organizations are responsible to follow the environmental regulations defined as per transportation policy so that it reduces fuel consumption and emissions and improve flexibility in logistics infrastructure (Ellram & Golicic, 2015). The primary factors which influence pollution are emission of pollutants and transportation of hazardous materials. Governments deploy stringent measures to regulate the movement of hazardous materials so that large number of fatal accidents can be avoided. Following are the environmental regulations which can be followed during transportation of materials, (1) symbols and safety signs of hazardous materials must be pasted both sides of the vehicle; transporting hazardous material wastes, (2) the driver must have some knowledge and skill to handle hazardous materials and also they must have personal protective tools and equipment, (3) speed of the vehicle should be monitored so that security of the cargo is maintained, (4) labor regulation should be implemented so that working hours of driver are properly maintained to avoid accidents because of fatigue, (5) proper maintenance of vehicle is required periodically so that carbon emission remains under control.

14.4.3 Barriers in sustainability adoption by stakeholders in the supply chain

Based on the literature, there are various barriers that hamper the implementation of SSC. But, all of them do not have major impacts so it is important to know the highly influencing factors which hinder the adoption of SSC by stakeholders. Following are the barriers in implementing SSC by stakeholders in the supply chain (Meixell & Luoma, 2015), (1) investment cost is high, (2) nonexistence of stringent government regulation, (3) lack of subject matter experts in sustainability within the organization, (4) lack of education on the benefits of reverse logistics, (5) hard to establish and monitor the suppliers to maintain the sustainability, (6) lack of environmental awareness, and (7) disposing hazardous wastes are expensive. Inspite of the barriers listed by the stakeholders, it is highly encouraged to adopt sustainability so that the organization can reap the benefits in long run. Moreover, the reduction in waste will result in cost reduction which may have led to generate more revenue and market share for the organization (Mani, Agrawal, & Sharma, 2016).

14.5 Tools to measure supply chain sustainability

The need of sustainability measurement is required for improving the supply chain operations performance, product quality, and environment friendly products. There are few important tools to measure the sustainability range, namely, country environmental analysis (CEA), sustainable business performance indicators (SBPI), and cost-benefit analysis (CBA). These tools can be explored for measuring sustainability in supply chain.

14.5.1 Country environmental analysis (CEA)—2001

CEA is an environmental analytic tool used to identify the environmental concerns and sustainable development activities inside the country to create a potential bridge between environment, economic, and organizational society by shaping the key input of earliest stages decision-making process on environmental development and sustainability. CEA provides programmatic decisions model that helps to manage risks/effects at the project level itself.

14.5.2 Sustainable business performance indicators (SBPI)—2000

SBPI is a systemized digital instrument developed in collaboration with measurement technology department at RISE (Research Institute of Sweden) to provide customized digital tracking data's and measuring sustainability performance. SBPI is fully a digital-based measurement method compiling the values of process, product development, product quality, sustainability, and compliances over a time period using comparable, repeatable, and objective measurement methods.

14.5.3 Cost-benefit analysis (CBA)—1936

CBA is a decision-making technique for fundamental decision-making, development projects, facility investments, environmental regulations, etc., by illustrating both the advantage and disadvantage of the project systematically to estimate all the costs and benefits comparing contemplated source of action with alternative sources of action. The result value of CBA may be positive or negative based on the sound of influence and input parameters.

Another approach in measuring SSC is achieved by measuring the factors that influences the three pillars of sustainability such as environment, social, and economy practices (Taticchi, Garengo, Nudurupati, Tonelli, & Pasqualino, 2015). The factors which influence the environmental sustainability are air, water, land, materials, energy, and association. These factors can be measured as follows: the indicators for the factor "air" are carbon footprint discharges during manufacturing and transportation, for the factor "water" are total water consumed versus water retreated, for the factor "land" are toxin items toxin items generated versus landfill unwanted items, for the factor "material" are percentage of hazardous materials used, percentage of recyclable content in the product, for the factor "energy" are total energy used and energy from natural resources, and for the factor "association" check if the organization is ISO 14000 certified.

The factors which influence the social sustainability are industry, customer usage, and association. These factors can be measured as follows: The indicators for the factor "industry" are salary and employee healthcare benefits, for the factor "customer usage" are product hazardousness, and healthcare benefits, for the factor "association" check if the organization is provided with safety compliance, education given on safety to employee.

The factors which influence the economic sustainability are economic performance, financial health, brand, and association. These factors can be measured as follows: The indicators for the factor "economic performance" are item failure rate and response time to fulfill an order, for the factor "financial health" are revenue versus profit, debt-free organization, for the factor "association" check if the organization is ISO 9000 certified.

14.6 Supplier selection for sustainability

The supply of sustainable raw materials contributes majorly in achieving sustainability in the supply chain. Thus it is imperative to select the right supplier who encourages green supply chain. The prospective supplier should be evaluated based on their environmental awareness and assurance on the environmentally friendly product supply. In traditional supply chain, the vendors' ratings are based on the performance, quality, and cost. Due to the environmental and government pressures, many organizations started to evaluate their suppliers based on the environmental regulation, pollution norms, and recycling procedure followed by the supplier. Some indicators to evaluate supplier in the context of sustainability are percentage of recyclable materials, percentage of reusable materials, percentage of hazardous and toxic materials used, percentage of hazardous and toxic waste generation, and environmental regulation compliances (Alikhani, Torabi, & Altay, 2019).

In order to maintain the green production inside the organization, supplier raw materials also should be green. Thus it is imperative to select supplier based on few other aspects such as efficient planning performed by supplier so that there is no supply disruption, meeting legal requirements of the organization, and prompt in the external communication.

14.6.1 Planning

It is organizations' responsibility to check whether the supplier is certified with ISO 14001 system. In this context, suppliers are evaluated based on their environmental impacts. At supplier end, following factors are validated, whether the solid wastes are bio-degradable, whether the resource usage and energy usages are effectively utilized. Also, suppliers are evaluated based on the following environmental factors namely water contamination, air pollution, land contamination and land degradation, health hazard, and depletion of natural resources

14.6.2 Legal requirements

Legal requirement is one of the aspects in organization while selecting a supplier. It is imperative for organization to evaluate the supplier based on the legal standards maintained by the supplier. Organization should check for the legal factors before selecting a supplier. The legal factors are international agreement certificate (while dealing with international supplier), corporate guidelines and policies, industry codes, and environmental legislation practices.

14.6.3 External communication

One of the important aspect of seamless production is to maintain sufficient raw material. It requires a proper communication with supplier and ensure that the supplier is meeting the delivery dates. Supplier should be evaluated on the goods supplied versus goods rejected, and this requires effective mangement of communication with supplier to know when a good will be rejected. From a supplier side, it is important to communicate the downstream members in the supply chain when there is change in tier-2 supplier for any reason.

14.7 Closing the loop with proper end-of-life disposal strategies

As described earlier, closing the loop means the reverse logistics of the supply chain. This section discusses the green supply chain and its impact on sustainability and waste management strategies.

14.7.1 Green supply chain and its impact on sustainability

During these days, organizations are playing a vital role in achieving and maintaining supply chain sustainability. Green supply chain is achieved by changing their process, and products aimed at reducing the environmental impacts and improving their performance over environmental, social and economic activities (Carrano, Pazour, Roy, & Thorn, 2015; Meng, Lou, Peng, & Prybutok, 2016). Before the green practices, the environmental dimension is considered only on the basis of sustainability concept "cleaner production," proposed by the UNEP (United Nations Environment Program) in 1989.

Cleaner production: The process of using the resources and energies effectively along with the elimination of toxic raw materials and reduction in toxicity of all emissions and solid wastes. The benefits that can be availed using cleaner production strategies are (1) it improves economic, social, and environmental achievements, (2) it improves the efficiency in low-cost conservation of energy and raw materials, (3) it improves ecosystem and compliance with market needs, (4) it provides cohesive working environments for employees, and (5) it creates good company image on public perspectives.

Fig. 14.5 illustrates the green supply chain procedure with the benefits achieved by each function in the supply chain community. The green supply chain encourages suppliers to use less toxic and sustainable materials. In production environment, green procedure can be implemented to

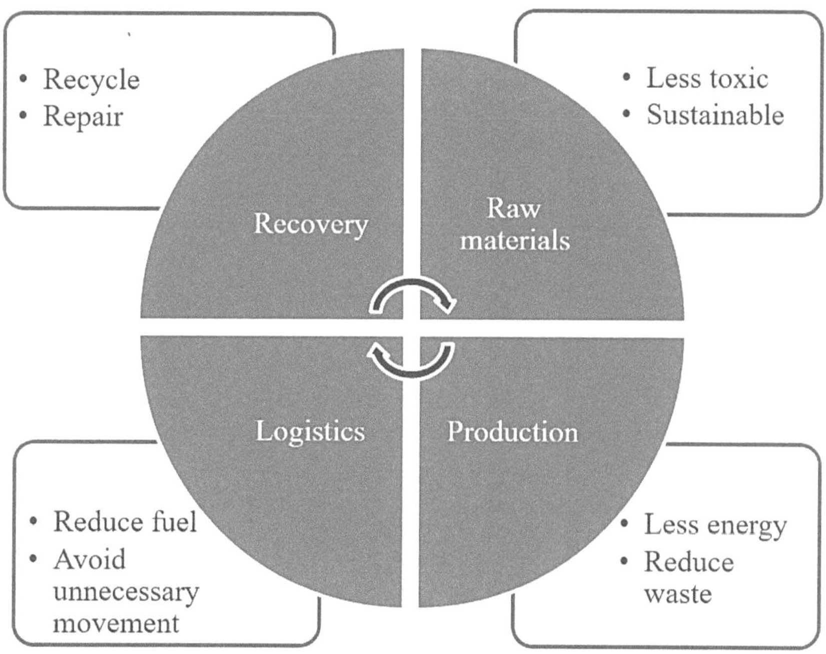

Figure 14.5 Green supply chain procedures.

reduce the energy usage or recover the energy by reusing it in another form. The logistics plays a key role in green supply chain, as it contributes considerable carbon footprints. Thus logistics management should be planned properly to avoid unnecessary movement of materials. In addition to green logistic practices, shared supply chain can also help to reduce fuel consumption. The reverse logistics will help to recover the used products and disposition it, as per 6R's principle (Bhardwaj, 2016; Chin, Tat, & Sulaiman, 2015).

The effectiveness of green practices, economic, and environmental sustainability of an organization is measured by the ISO 14001 certification. This certification stimulates the company to implement their facilities and verity of technologies to cut down the material and energy waste during production operation. The major benefits of implementing ISO 14001 are improved company image and reputation, increased customer loyalty and trust, improved motivation and staff morale, improved performance, profit, and opportunities, better identification and achievement of market advantages and operation efficiency, better adoption of preventive measures for environmental accidents.

14.7.2 Waste management strategies

The waste management and 6R process manages the wastes from inception to final disposal which includes waste collection, transportation, and waste treatment to regulate future production and operation process (Kanagaraj, Senthilvelan, Panda, & Kavitha, 2015; Tan et al., 2015). The wastes can be obtained in different three phases such as solid, liquid, and gas. Each phase requires different treatment and disposal methods (Lin & Tseng, 2016).

14.7.2.1 Reduce

"Reduce" means consuming or using fewer resources in the first phase. During purchase of the item, the buyer not only evaluate it on the basis of its durability and quality but also contemplates the reduction failure rate of the item thus wastes and also percentage of sustainable materials used in the item. For example, while purchase of an item, buy items where there is reduction in usage of toxic materials and buy refillable material to reduce the price

14.7.2.2 Reuse

It implies "using the materials in more than one form from its original form" instead of throwing away after each and every use. "Reuse" prevents the new resources required and extension of EOL of older resources which are usually classified as "wastes." For example, reuse paper and plastic bags rather than frequent disposal and also use cotton towels rather than paper towels, since cotton towels can be washed and reused; reuse bottles and glass jars as containers.

14.7.2.3 Recycle

Recycle is the process of converting the one form of material waste into another form of resource product using some physical and chemical processes to reduce energy usage and new resources being used to preventing the green ecosystem. For example, do not throw away the papers and plastic bags that are recyclable, waste generated during the production process is to be collected and recycled.

14.8 Role of contemporary technologies in ensuring supply chain sustainability in industry 4.0

In recent days, organization started evaluating the industry 4.0 (I4.0) requirements to transform into digital world. A detailed literature study has been done to know the role of I4.0 technologies such as IoT, Artificial Intelligence, Big data, and Cloud computing in SSC. SSCM collaborates with supply chain members to ensure that sustainability practices such as environment, social, and economic are efficiently managed. Thus it is a key to have a proper information system which supports I4.0 technologies. The I4.0 technologies allow the manufacturers to monitor the life cycle of product across the supply chain. One of the key challenges for manufacturer in supply chain is to ensure the product is delivered on time in good condition. IoT technology can be used to track the product visibility and condition of the product. By using this, product wastages can be reduced and the customer satisfaction in SSC can be improved. In reverse supply chain, IoT helps the organization to recover the EOL products, and further, it will be classified for recycle, repair or dispose. Reusable container can be tagged with rapid frequency identification (RFID) and can be handled using reverse logistics to reuse it. I4.0 will be more helpful in perishable supply chain where the shelf lives of the items are short. The proper monitoring of the food items with IoT can reduce the food waste and maintain the food fresh till it reaches the end user. RFID sensors can also be used to identify spoiled foods.

The process to implement SSC is difficult earlier as it was hard to share the real time information across information systems (Iyer, Germain, & Frankwick, 2004). As illustrated in Fig. 14.6, it is possible to monitor the information across the SSC with IoT, cyber-physical systems (CPSs), and cloud technologies. The goal of I4.0 is to make smart factory transforms information and manage machines autonomously (Luthra & Mangla, 2018b). CPS allows the machines to work autonomously. For example, equipment and devices in smart factory understands the routing process required to make a particular product variant. It also identifies the operating parameters and configurations required. The communication between upstream and downstream members in the supply chain is important to improve the visibility so that stakeholders know the complete details until EOL of the product. Cloud Computing can be helpful to process the information related available resources, stocks, resource allocations, shipment faster, and it can also provide accurate information. Thus it is important to study the influence of I4.0 on the SSC. The key factors integration between I4.0 and SSC are initial investment, management strategy, preparedness of the organization, technology awareness, empower the core team, involve subject matter experts, organization culture, transparent communication, projects and IT team involvement, and educate the end-users. Both I4.0 and SSC are growing gradually that realigns the current trends of manufacturing and logistics management. The factors which influence the integration of I4.0 and SSC can be further studied in detail by academicians and industrialist.

The focus on sustainability is gaining the required momentum; thus organizations are envisioning that I4.0 technologies offers tremendous opportunities realizing SSC. With the help of CPS, equipment can be

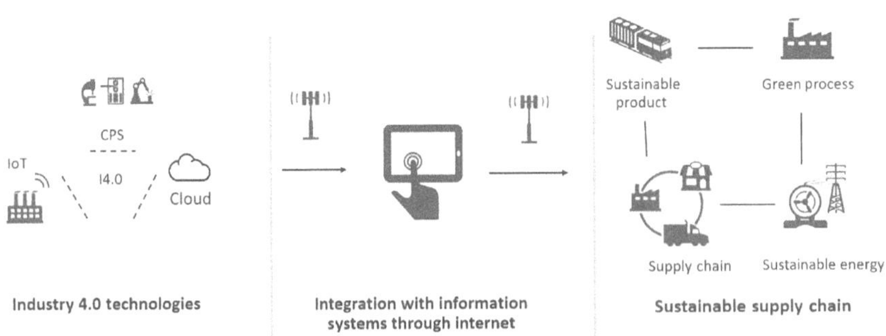

Figure 14.6 Integrated model for understanding the interaction between industry 4.0 and sustainable supply chain.

monitored and failures can be predicted well in advance so that production downtime can be minimized or avoided. Studies also reveal that technology can help small and medium enterprise to reuse the equipment instead of new purchase of the new equipment. Blockchain can help the organization to trace the original source of the product and helps organizations to sign digital contracts with the suppliers (Saberi, Kouhizadeh, Sarkis, & Shen, 2019).

14.8.1 Influence of industry 4.0 on sustainable supply chain

Digital technologies in industrial revolution are setting the tone for the future supply chain. I4.0 is the focal point for various sectors where digital disruption is happening (Almada-Lobo, 2015; Stock & Seliger, 2016). The innovative technologies in I4.0 helps organization to meet the customer expectations quickly, allows customizing the product in volume. The combination of technology with supply chain amplifies the value of sustainability which will redefine the industry in the next decade. IoT-embedded supply chain provides better visibility in the organization which is key factor in various discrete and process manufacturing industries. In addition, it is also helpful in the supply chain of semiconductor and pharmaceuticals industries where high quality and precision are very much required. It is highly encouraged for the organizations which are interested in deploying I4.0 technology, understand the technical infrastructure requirements and available IoT architecture before moving on with implementation. Initially the deployment of I4.0 is focused on productivity, and gradually, there is an interest by contemporary organization to look for sustainability with technology. It offers the foundation for bringing the new opportunities for organization to fulfill sustainability requirements. The review reveals that the SSC with IoT was not studied earlier, and it is considered as a research gap.

14.8.2 Current trends of internet of things in sustainable supply chain

IoT plays a vital role in supply chain as there is huge investment made by large organization in various sectors such as oil and gas, automobile, and food processing industries. In recent times, the accumulation real-time information is huge, and it cannot be processed by traditional information systems. Thus there is a need for big data to process the information faster and provide it to stakeholders to take important decisions quickly. In general, the success of IoT technology deployment depends on the I4.0 principles such as interoperability, scalability, and reliability. In addition, the technology should be affordable by many organizations. Various studies are also happening to make the IoT affordable and also to ensure the data is safe and secured. Basically, the crux of IoT is to connect the machines, devices with the digital systems. Thus this is applicable for widespread industries and viable option for many supply chain companies. I4.0 helps in reducing the human errors where workers work in hazardous environments. IoT embedded with artificial intelligence can help the buyers to know if there is any shipment delay supplier by the supplier. It will alert the stakeholders on the supply disruption and allows the buyer to take informed decision to get the supply from alternative supplier. During distribution of the product to the customers, the cargo can be monitored, and also, the driver fatigue can be tracked with the IoT. The condition-based monitoring system can be implemented by maintenance department to monitor the machines and equipment to reduce the unplanned downtime. In mining supply chain, drones can be used considering the workers safety (Manavalan & Jayakrishna, 2019b).

Even though the initial investment is required to deploy I4.0 technology, it really provides more value to the organization in terms of bringing more visibility of goods, tracking the shipment in real-time, ensuring the reverse logistics of supply chain is effective. I4.0 technology helps in monitoring product performance in real time, proper maintenance of equipment, and effective recovery of EOL products systematically (Maksimovic, 2018). Thus, I4.0 tools facilitate in achieving the overall goal of SSC.

14.8.3 Internet of things embedded supply chain management

IoT helps the organizations to overcome the challenges they face in supply chain operations. The key aspect of IoT is sharing the information on real-time seamlessly. Few studies show that the IoT is implemented in food supply chain with the combination of sensors and internet. IoT monitors the food supply chain on real-time and alerts the stakeholders when there is drop in prepacked food quality. IoT-embedded supply chain is in preliminary stage and studies shows that there are tremendous research opportunities exist for academicians and organization. Supply chain organizations are required to analyze the present systems and evaluate the I4.0 requirements to invest in technology to get the benefits in extended period. Further, the SSC insists the usage of renewable energy such as water, air, etc. The objective of I4.0 in SSC is to ensure that organizations are transforming themselves into digital organizations, minimize carbon footprint, and improve the sustainability requirements. The advantage of moving to I4.0 is not only to connect one factory, but also multiple factories can be interlinked with the help of cloud resulting in digital supply chain (Ben-Daya, Hassini, & Bahroun, 2019).

14.9 Managerial implications

The overall chapter examines the various aspects of supply chain such as linear to circular and circular to shared supply chain. It discusses the importance of SSC and the issues in adopting SSC. The review reveals that the high investment cost is the major barrier behind its adoption. In addition, the tools to measure the supply chain sustainability have been discussed. Organizations are interested in selecting green supplier so that they can support to make environment friendly products using sustainable products, establish circular supply chain, and recover EOL products with the overall goal of reducing the carbon footprints and improve overall supply chain operational efficiency. This chapter assures that organizations can get real-time information with the help of I4.0 technologies but the organization should be matured enough in supply chain to move towards digital technologies. In addition, the closing the loop with proper EOL disposal strategies are also discussed. This helps the supply chain community with the shared information and trust among the its members.

14.10 Conclusion

Supply chain is becoming complex with the wide range of products along with global competitive business environment. The key factors which affect the sustainability of supply chain are the wastages in product, leakage of energy, unnecessary movement of materials, worker health and safety, strictly adhere to government regulations. Due to government and environmental pressure organization are emphasized to align their supply chain to make it more sustainable. SSC targets to address the socio-economic-environmental issues, and organization are required to adjust themselves to transform their linear supply to circular supply chain. Organizations should also explore shared supply chain model as opportunity to be more cost effective, and it provides mutual gains. The contemporary organizations should encourage green practices and maintain better corporate social responsibility and protect the environment. The transformation of supply chain from linear to circular by implementing reverse logistics, circular to shared supply chain are summarized. The issues in SSC adoption and the barriers in sustainability adoption by stakeholders are discussed. The tools to measure the supply chain sustainability, factors to consider selecting the sustainable suppliers, explore the possibilities of closing the loop with EOL disposal strategies were elaborated in this chapter. The role of I4.0 technologies ensuring supply chain sustainability is discussed. This chapter reveals that organizations should focus on moving towards I4.0 technologies as it promises wide opportunities in achieving sustainability in supply chain. The I4.0 technology is going to be a key driver in reducing environmental risks and making eco-friendly products. Further it focuses on workers' health and safety with the overall goal of reducing carbon footprints.

14.11 Future research direction

The chapter summarizes that there are lot of benefits in adopting SSC embedded with I4.0 technologies. But it requires highly skilled people in the organization to implement it. Also, the chapter reveals that the investment on sustainability and technology seems to considerably high in initial phases, and hence, small-scale industries may find it challenging to deploy it. The proposed tools to measure sustainability of supply chain, I4.0 technologies are generic for any SSC organization. In future, researchers are encouraged to study specific industry and conduct cost benefit analysis which can give an overall view to the stakeholders to implement SSC embedded with I4.0 technologies.

References

Afgan, N. H. (2010). Sustainability paradigm: Intelligent energy system. *Sustainability.*, *2*(12), 3812–3830.

Alikhani, R., Torabi, S. A., & Altay, N. (2019). Strategic supplier selection under sustainability and risk criteria. *International Journal of Production Economics*, *208*, 69–82.

Almada-Lobo, F. (2015). The Industry 4.0 revolution and the future of Manufacturing Execution Systems (MES). *Journal of Innovation Management*, *3*(4), 16–21.

Ben-Daya, M., Hassini, E., & Bahroun, Z. (2019). Internet of things and supply chain management: A literature review. *International Journal of Production Research*, *57*(15-16), 4719–4742.

Bhardwaj, B. R. (2016). Role of green policy on sustainable supply chain management. *Benchmarking: An International Journal*, *23*.

Carrano, A. L., Pazour, J. A., Roy, D., & Thorn, B. K. (2015). Selection of pallet management strategies based on carbon emissions impact. *International Journal of Production Economics, 164*, 258–270.

Chin, T. A., Tat, H. H., & Sulaiman, Z. (2015). Green supply chain management, environmental collaboration and sustainability performance. *Procedia Cirp, 26*, 695–699.

Dubey, R., Gunasekaran, A., Papadopoulos, T., Childe, S. J., Shibin, K. T., & Wamba, S. F. (2017). Sustainable supply chain management: Framework and further research directions. *Journal of Cleaner Production, 142*, 1119–1130.

Ellram, L. M., & Golicic, S. L. (2015). Adopting environmental transportation practices. *Transportation Journal, 54*(1), 55–88.

Esfahbodi, A., Zhang, Y., & Watson, G. (2016). Sustainable supply chain management in emerging economies: Trade-offs between environmental and cost performance. *International Journal of Production Economics, 181*, 350–366.

Farooque, M., Zhang, A., Thurer, M., Qu, T., & Huisingh, D. (2019). Circular supply chain management: A definition and structured literature review. *Journal of Cleaner Production, 228*, 882–900.

Fulconis, F., Pache, G., & Reynaud, E. (2019). Frugal supply chains: A managerial and societal perspective. *Society and Business Review, 14*.

Gaustad, G., Krystofik, M., Bustamante, M., & Badami, K. (2018). Circular economy strategies for mitigating critical material supply issues. *Resources, Conservation and Recycling, 135*, 24–33.

Geissdoerfer, M., Morioka, S. N., de Carvalho, M. M., & Evans, S. (2018). Business models and supply chains for the circular economy. *Journal of Cleaner Production, 190*, 712–721.

Genovese, A., Acquaye, A. A., Figueroa, A., & Koh, S. L. (2017). Sustainable supply chain management and the transition towards a circular economy: Evidence and some applications. *Omega, 66*, 344–357.

Govindan, K., & Hasanagic, M. (2018). A systematic review on drivers, barriers, and practices towards circular economy: A supply chain perspective. *International Journal of Production Research, 56*(1-2), 278–311.

Gnoni, M. G., Tornese, F., Thorn, B. K., Carrano, A. L., & Pazour, J. A. (2018). A Measurement Tool for Circular Economy Practices: a Case Study in Pallet Supply Chains. 15th IMHRC Proceedings (Savannah, Georgia. USA – 2018). 12. https://digitalcommons.georgiasouthern.edu/pmhr_2018/12.

He, Y. (2017). Supply risk sharing in a closed-loop supply chain. *International Journal of Production Economics, 183*, 39–52.

http://www.thegreenmedium.com/blog/2015/9/1/the-green-mediums-first-contributor-matt-gwozd (accessed April, 2020).

https://www.inspirage.com/2016/09/top-4-benefits-integrated-supply-chain (accessed April, 2020).

https://www.soundonsound.com/techniques/refurbishing-your-old-equipment (accessed April, 2020).

https://www.uwosh.edu/facstaff/barnhill/ES-243/pp%20outline%20New%20Environmentalism.pdf (accessed April, 2020).

Iyer, K. N., Germain, R., & Frankwick, G. L. (2004). Supply chain B2B e-commerce and time-based delivery performance. *International Journal of Physical Distribution & Logistics Management, 34*(8), 645–661.

Kanagaraj, J., Senthilvelan, T., Panda, R. C., & Kavitha, S. (2015). Eco-friendly waste management strategies for greener environment towards sustainable development in leather industry: A comprehensive review. *Journal of Cleaner Production, 89*, 1–7.

Khan, M., Hussain, M., & Saber, H. M. (2016). Information sharing in a sustainable supply chain. *International Journal of Production Economics, 181*, 208–214.

Kleindorfer, P. R., Singhal, K., & Van Wassenhove, L. N. (2005). Sustainable operations management. *Production and Operations Management*, *14*(4), 482–492.
Lewandowski, M. (2016). Designing the business models for circular economy—Towards the conceptual framework. *Sustainability.*, *8*(1), 43.
Lin, Y. H., & Tseng, M. L. (2016). Assessing the competitive priorities within sustainable supply chain management under uncertainty. *Journal of Cleaner Production*, *112*, 2133–2141.
Luthra, S., & Mangla, S. K. (2018a). When strategies matter: adoption of sustainable supply chain management practices in an emerging economy's context. *Resources, Conservation and Recycling*, *138*, 194–206.
Luthra, S., & Mangla, S. K. (2018b). Evaluating challenges to industry 4.0 initiatives for supply chain sustainability in emerging economies. *Process Safety and Environmental Protection*, *117*, 168–179.
Maksimovic, M. (2018). *Greening the future: Green internet of things (G-IoT) as a key technological enabler of sustainable development. Internet of things and big data analytics toward next-generation intelligence* (pp. 283–313). Cham: Springer.
Manavalan, E., & Jayakrishna, K. (2019a). An Analysis on sustainable supply chain for circular economy. *Procedia Manufacturing*, *33*, 477–484.
Manavalan, E., & Jayakrishna, K. (2019b). A review of internet of things (IoT) embedded sustainable supply chain for industry 4.0 requirements. *Computers & Industrial Engineering*, *127*, 925–953.
Mani, V., Agrawal, R., & Sharma, V. (2016). Impediments to social sustainability adoption in the supply chain: An ISM and MICMAC analysis in Indian manufacturing industries. *Global Journal of Flexible Systems Management*, *17*(2), 135–156.
Matos, S., & Hall, J. (2007). Integrating sustainable development in the supply chain: The case of life cycle assessment in oil and gas and agricultural biotechnology. *Journal of Operations Management*, *25*(6), 1083–1102.
Meixell, M. J., & Luoma, P. (2015). Stakeholder pressure in sustainable supply chain management. *International Journal of Physical Distribution & Logistics Management*, *45*(1/2), 69.
Meng, K., Lou, P., Peng, X., & Prybutok, V. (2016). A hybrid approach for performance evaluation and optimized selection of recoverable end-of-life products in the reverse supply chain. *Computers & Industrial Engineering*, *98*, 171–184.
Oelze, N., Hoejmose, S. U., Habisch, A., & Millington, A. (2016). Sustainable development in supply chain management: The role of organizational learning for policy implementation. *Business Strategy and the Environment*, *25*(4), 241–260.
O'Rourke, D. (2014). The science of sustainable supply chains. *Science*, *344*(6188), 1124–1127.
Pan, S. Y., Du, M. A., Huang, I. T., Liu, I. H., Chang, E. E., & Chiang, P. C. (2015). Strategies on implementation of waste-to-energy (WTE) supply chain for circular economy system: A review. *Journal of Cleaner Production*, *108*, 409–421.
Reefke, H., & Sundaram, D. (2018). Sustainable supply chain management: Decision models for transformation and maturity. *Decision Support Systems*, *113*, 56–72.
Roos, G. (2014). Business model innovation to create and capture resource value in future circular material chains. *Resources*, *3*(1), 248–274.
Saberi, S., Kouhizadeh, M., Sarkis, J., & Shen, L. (2019). Blockchain technology and its relationships to sustainable supply chain management. *International Journal of Production Research*, *57*(7), 2117–2135.
Singh, A., & Trivedi, A. (2016). Sustainable green supply chain management: Trends and current practices. *Competitiveness Review*, *26*.
Song, H., & Gao, X. (2018). Green supply chain game model and analysis under revenue-sharing contract. *Journal of Cleaner Production*, *170*, 183–192.

Stevens, G. C., & Johnson, M. (2016). Integrating the supply chain... 25 years on. *International Journal of Physical Distribution & Logistics Management, 46*.

Stock, T., & Seliger, G. (2016). Opportunities of sustainable manufacturing in industry 4.0. *Procedia Cirp, 40*, 536–541.

Su, C. M., Horng, D. J., Tseng, M. L., Chiu, A. S., Wu, K. J., & Chen, H. P. (2016). Improving sustainable supply chain management using a novel hierarchical grey-DEMATEL approach. *Journal of Cleaner Production, 134*, 469–481.

Tan, S. T., Ho, W. S., Hashim, H., Lee, C. T., Taib, M. R., & Ho, C. S. (2015). Energy, economic and environmental (3E) analysis of waste-to-energy (WTE) strategies for municipal solid waste (MSW) management in Malaysia. *Energy Conversion and Management, 102*, 111–120.

Taticchi, P., Garengo, P., Nudurupati, S. S., Tonelli, F., & Pasqualino, R. (2015). A review of decision-support tools and performance measurement and sustainable supply chain management. *International Journal of Production Research, 53*(21), 6473–6494.

Tsoulfas, G. T., & Pappis, C. P. (2008). A model for supply chains environmental performance analysis and decision making. *Journal of Cleaner Production, 16*(15), 1647–1657.

Zhang, J., Liu, G., Zhang, Q., & Bai, Z. (2015). Coordinating a supply chain for deteriorating items with a revenue sharing and cooperative investment contract. *Omega, 56*, 37–49.

CHAPTER 15

Achieving sustainability in machining of cylindrical gears

Vishal Kharka and Neelesh Kumar Jain
Department of Mechanical Engineering, Indian Institute of Technology Indore, Simrol, India

15.1 Introduction

15.1.1 Sustainable manufacturing

In recent years, ecological issues such as wastes, effluents, emissions, health, and safety of workers in manufacturing industries have witnessed extensive attention, research, development, and innovation attempts to explore and establish sustainable manufacturing practices. The primary objective of the modern manufacturing processes now has been to significantly reduce their adverse environmental impacts by introducing various environment-friendly techniques keeping in view the societal and economic aspects (Gupta, Laubscher, Davim, & Jain, 2016; Pusavec, Kramar, Krajnik, & Kopac, 2010). This concept has led to the development and preference for "sustainability" in almost every field of manufacturing. Sustainable manufacturing is environment-friendly manufacturing of products by utilizing green strategies and techniques that are less harmful, safe, and economical (Pusavec, Krajnik, & Kopac, 2010). Its main focus is on developing and implementing energy-efficient, environment and operator friendly, and economical processes for manufacturing of the products commercially. As shown in Fig. 15.1A, Jawahir and Dillon (2007) identified six primary parameters that significantly affect and define sustainability of a manufacturing process. These elements serve as the metrics for evaluating sustainability and performance of a manufacturing process. Gupta et al. (2016) have mentioned that sustainability depends on three important factors: social, environmental, and economical (Fig. 15.1B).

Various environmental standards have been introduced as a part of sustainability drive to help organizations and industries minimize their environmental-unfriendly processes and operations. Constant emphasis has been laid on careful selection of manufacturing processes, shortening the process chain, monitoring the manufacturing process parameters, analyzing

Sustainable Manufacturing
DOI: https://doi.org/10.1016/B978-0-12-818115-7.00002-X

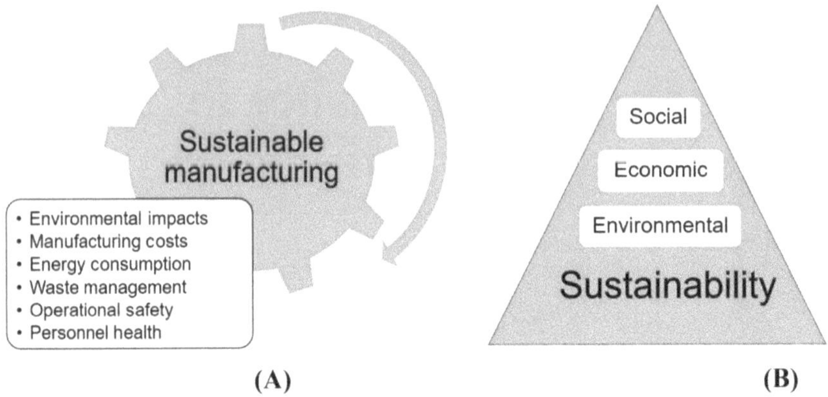

Figure 15.1 Concept of sustainability: (A) elements of sustainable manufacturing and (B) sustainability pyramid.

the product life cycle, and thereby developing an inclination towards environment-friendly products to follow the regulations concerning the environment.

15.1.2 Conventional gear manufacturing processes

Manufacturing of gears can be broadly categorized into three primary types according to nature of the processes: subtractive or material removal (i.e., machining), deformative, and additive (Gupta, Jain, & Laubscher, 2017). Each of the processes has been discussed briefly in Table 15.1 along with their characteristics and limitations.

15.1.2.1 Subtractive manufacturing processes for cylindrical gears

Subtractive or material removal or machining, involving cutting of gear teeth over a prepared blank, is the most commonly used manufacturing process for gears. Gear machining processes can be subdivided into two categories: (1) form-cutting and (2) generative processes. *Form-cutting* process uses a cutting tool having the same form as the space between two adjacent teeth on a gear. Cutting tool is basically a replica of the gear to be cut. Broaching, milling, shear cutting, and teeth cutting on shaper are the basic form-cutting type gear machining process for cylindrical gears. *Milling* is an important and extensively used process for manufacturing cylindrical gears using generally disk type cutters or ends mill cutters. It is mostly preferred for small volume production of low precision gears. Milled gears find applications where the microgeometry deviations are not

Table 15.1 Details of conventional gear manufacturing processes.

Nature	Process	Characteristics and limitations
Subtractive processes	*Form cutting processes*	
	Gear milling	• Small volume production of low precision gears • Higher deviations in gear microgeometry
	Gear broaching	• High-volume production • To manufacture internal gears, racks, splines, sector gears
	Gear cutting on shaper	• High-volume production, low-quality gears • Economical
	Shear cutting	• For high-volume production spur gears • Preferred for external gears
	Generative process	
	Gear hobbing	• To manufacture spur, helical, worm gears, and splines • Simple and economical but not preferred for high accuracy
	Gear shaping	• To manufacture almost all types of gears • Preferable for both external and internal gears
	Gear planning	• Less accurate than shaping and hobbing
Additive processes	Sand casting	• Yields low quality of gears and preferred for large-size gears
	Die casting	• Preferred for small sized gears having high quality • Expensive process
	Investment casting	• Improved quality, higher production rate, and cost-effective
	Powder metallurgy	• Improved gear quality • Economical process for mass production of different gears
	Injection molding	• Manufacturing nonmetallic and plastic gears • Yields good gear quality
Deformative processes	Stamping	• Economical process for high-volume production of sheet-metal gears having low quality
	Fine blanking	• For manufacturing only straight spur gears having good geometric properties
	Extrusion and cold drawing	• To manufacture spur gears, helical gears, worms, splines • Good surface integrity of gears
	Gear rolling	• High-quality gears with high-fatigue strength • Shorter process times
	Gear forging	• Low-quality gears thus requiring finishing operations

of a major concern. *Broaching* is a relatively easier and convenient process of gear machining that uses multitooth cutting tool called broach. Linear movement of the broach tool causes removal of the material from the gear blank. Teeth on the broach tool are designed in such a way that the depth of cut increases with each tooth as the broaching process progresses. However, process is expensive due to the high costs of broach tools and needs separate broach tools for different gears. It can be used to manufacture both internal and external gears. A shaper machine can also be used for large volume production of low-quality spur and helical gears. Each tooth is cut individually by linear movement of a tool moving parallel to center axis of the workpiece. *Shear cutting* process is similar to the broaching process and is primarily used for commercial manufacturing of spur gears.

Generative processes use a cutter in the form of gear or a rack cutter to generate gear teeth. Module of the cutter should be the same as that of the gear to be machined. Gear hobbing, gear shaping, and gear planning are generative processes for gear manufacturing. *Gear hobbing* is one of the most economical and widely used processes to manufacture cylindrical gears. Progressive cuts made by the rotating cutting tool, known as hob, generate the required gear teeth. However, high accuracies cannot be achieved by this process necessitating subsequent finishing processes to achieve the desired gear quality. *Gear shaping* process generates gear teeth by a toothed cutter moving in axial direction. It is used to manufacture both external and internal cylindrical gears. Fig. 15.2 depicts machining processes used to manufacture cylindrical gears.

15.1.2.2 Deformative manufacturing processes for cylindrical gears

Deformative processes use dies of respective shape and size to manufacture gears. Deformative processes for manufacturing cylindrical gears include stamping, fine blanking, forging, extrusion, rolling, and cold drawing. These processes offer certain advantages with respect to reduced process chain, no chip generation, increased strength. Most of these processes require finishing operations to obtain the preferred gear quality. Fig. 15.3 shows stamping, forging, and extrusion processes. *Stamping* is a sheet-metal operation in which a die is punched against the sheet blank to obtain the gear of desired size and shape. It is used to manufacture small thickness gears. *Fine blanking* is a derivative process of stamping process that uses additional tools and attachments to manufacture gears of better dimensional accuracy and precision than the stamping process. *Extrusion* is

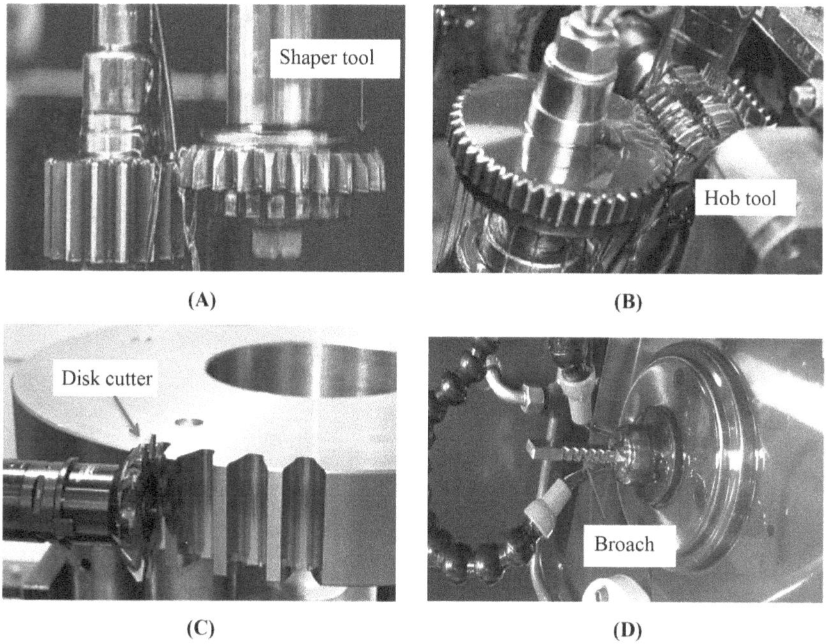

Figure 15.2 Machining processes to manufacture cylindrical gears: (A) shaping; (B) hobbing; (C) milling; and (D) broaching.

used to generate teeth on long rods by pushing the gear material through a die of the required shape and size. The long-extruded material is then eventually cut to get the gears of the desired face width. Materials with excellent drawing properties are generally preferred to manufacture gears by extrusion process. It produces gears with good surface quality and strength. *Gear forging* involves application of localized compressive forces by a die that has the desired shape and size and shapes the metal billet to the desired gear. It can be classified as cold, warm, and hot forging depending upon the working temperature. *Cold drawing* uses hardened dies to draw or form teeth on steel rods. *Gear rolling* involves forming the gear teeth by passing the blank through a series of rolls. It can either be flat or round rolling depending on nature of the rolls used. It is used for producing cylindrical gears having good surface finish and accuracy.

15.1.2.3 Additive manufacturing processes for cylindrical gears

Additive manufacturing processes such as casting, powder metallurgy, and injection molding are also used to manufacture gears. Due to the nature of the processes where gears are built by adding material, these processes

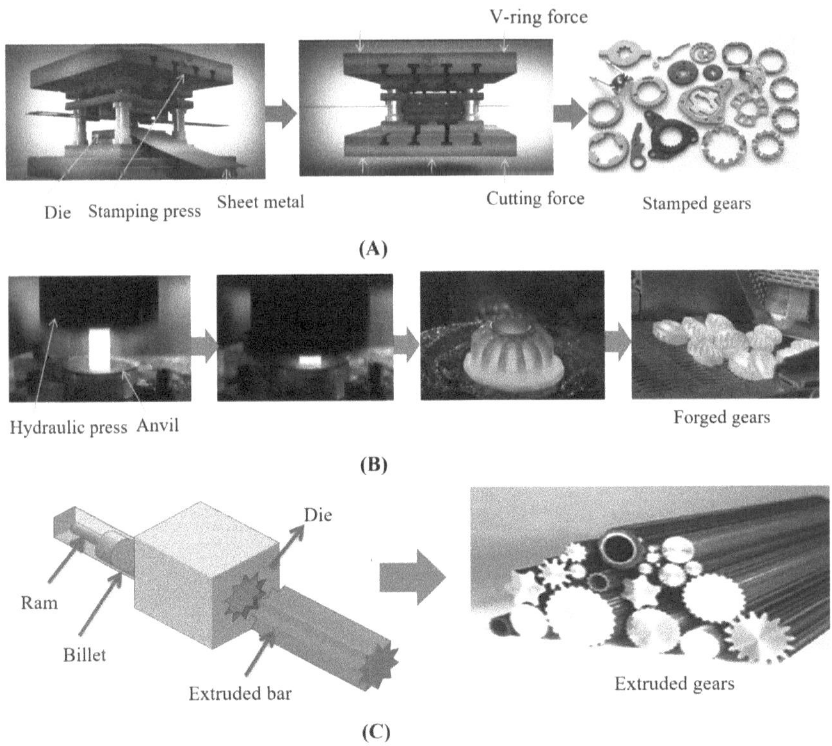

Figure 15.3 Deformative processes for manufacturing cylindrical gears: (A) stamping, (B) forging, and (C) extrusion.

are called additive manufacturing but they completely differ from typical additive layer manufacturing processes in mechanism and characteristics. Gears are manufactured by controlled fusion/injection/forcing of raw materials (powder or liquid form) to a die cavity of required gear shape and size. These processes produce near-net shaped gears in bulk of acceptable quality. In *gear casting*, a liquid material is poured into a mold of required shape and size. It can be used for mass production of gears of various shapes and sizes. Depending on the type of mold material used, gear casting can be subdivided into die, sand, and investment. *Die casting* is used for large volume manufacturing of small gears and is characterized by a very good surface finish and dimensional consistency. *Sand casting* is the most widely used casting process for production of low-quality, large-size gears. *Investment casting* is used to produce small- to medium-sized gears with good surface finish. *Powder metallurgy* involves manufacturing of net shape or near-net shaped gears from metallic powders. It is a precision

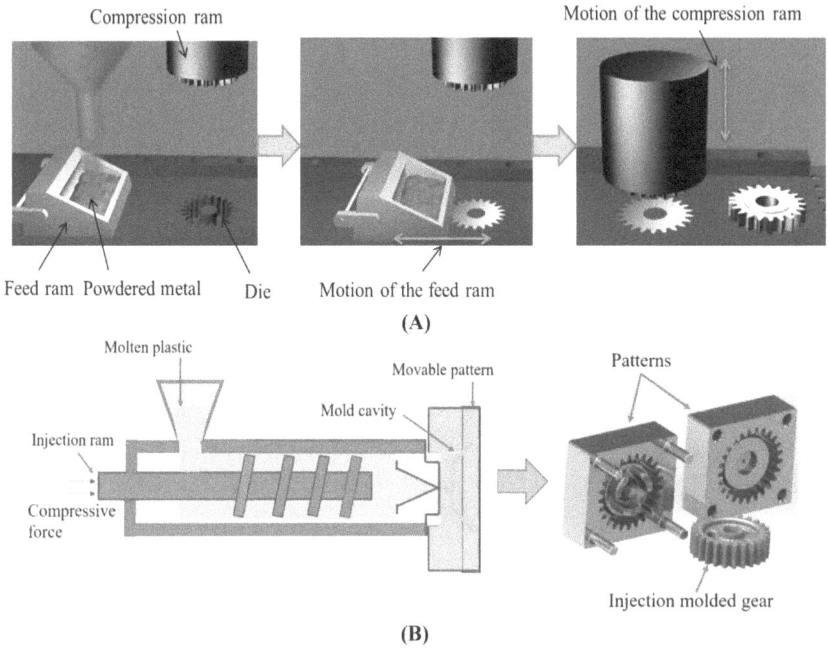

Figure 15.4 Additive processes for gear manufacturing: (A) powder metallurgy, (B) injection molding.

gear manufacturing process used for mass production. *Injection molding* produces nonmetallic or plastic gears by injecting molten material into the mold. It is suitable for high-volume production. Fig. 15.4 illustrates working of powder metallurgy and injection molding.

15.2 Sustainability concerns in conventional gear machining processes

Gear manufacturing industry has been continuously working on improving productivity and product quality while simultaneously minimizing impacts of the process chain on the environment and human health. Gear machining processes such as gear hobbing, gear milling, and gear shaping constitute majority percentage of gear manufacturing processes. These processes use conventional flood lubrication (FL) technique to bring about the cooling and lubrication effects. Chemical composition and physical properties of the lubricating and cooling fluids used in FL induce economic, social, and environmental problems during their handling, use, treatment, and disposal which contribute approximately 12% of the total machining cost (Gupta et al., 2016). Moreover, gears manufactured by these machining processes require subsequent finishing processes to further improve their quality as per the application requirement. Increasing emphasis on improving sustainability in gear manufacturing has led to continuous efforts being made to make these processes more environment-friendly without compromising on the gear quality. Problems associated with conventional gear machining processes can be summarized as:

- These processes necessitate further finishing operations such as shaving, grinding, honing, lapping, skiving, and burnishing to attain the desired gear quality. It requires consumption of additional energy, tools and consumables, and lubricants and coolants thus increasing the overall cost and affecting the sustainability of gear machining processes.
- Large quantities of hydrocarbon-based cutting fluids (CFs) and lubricants supplied to the machining zone in these processes affect the health of the machine operators and also result in higher power consumption.
- Disposal and recycling of wet chips after conventional FL is difficult to deal with and significantly affect the environment.

Various sustainable and efficient manufacturing strategies are being developed to address:

- reduction in the process chain involved in manufacturing of gears;
- minimization of the required amount of the harmful CF;
- substitution of harmful CFs with environment-friendly lubricants and lubrication systems (Gupta et al., 2016);
- minimization of the tool wear thereby reducing the chances of tool failure; and
- manufacturing of net shaped or near-net shaped gears without compromising their quality.

15.2.1 Cutting fluids and their impact on sustainability

Role of CFs can never be neglected in a machining process because they are primarily responsible for increasing its effectiveness and efficiency. CF enhances machining performance by working as an agent by reducing friction between the workpiece–tool and tool–chip interfaces, providing the cooling effect and preventing built-up edge (BUE) formation (Dixit, Sarma, & Davim, 2012a). It also plays a major role in eliminating chips from the machining zone. Consequently, benefits offered by CF in machining operations include (1) better surface finish, (2) longer tool life, (3) lesser energy consumption, and (4) cleaner cutting zone.

CFs are primarily classified into two types: straight fluids and water-mix fluids. *Straight or neat fluids* are derived from petroleum, animal, or vegetable origin. They comprise mineral oils and additives such as fatty materials, chlorinated paraffin, sulfurized oils, and organic phosphorous compounds. They result in enhanced lubrication, but due to their poor cooling effect, they are suitable for low-temperature and low-cutting speed processes. *Water-mix fluids* can be further classified into three categories: (1) soluble or emulsifiable oil, (2) synthetic fluids, and (3) semisynthetic fluids. *Soluble* or *emulsifiable oils* are basically oil droplets comprising mineral or vegetable oil suspended in water at a concentration of 3%–10%. Emulsifiers are then added to help dispersion of oil-in-water thus stabilizing the oil-in-water mixture. Further they comprise extreme pressure (EP) additives, biocides, fungicides, and other performance inhibiting compounds. Presence of EP additives such as sulfur, chlorine, and phosphorous-based chemicals improves lubrication and performance at high temperature operations. Emulsion also helps in reduction of heat generation thereby making it favorable for high cutting speed applications. *Synthetic fluids* have good coolant properties but their performance as lubricants has not been satisfactory. Alkaline inorganic and organic compounds with additives are used to make synthetic fluids. *Semisynthetic fluids* are usually diluted with water and various chemicals with an oil percentage of 2%–30% (Sharma, Singh, & Sorby, 2015).

CFs have been found to pose many hazards to human health and the environment in number of ways. Because of hazardous and toxic substances used in the synthesis of the CFs, they are responsible for some serious health problems such as lung cancer, respiratory diseases, dermatological and genetic diseases (Ozcelik, Kuram, Demirbas, & Şik, 2013). The hydrocarbon part, which is the basis for majority of CFs, is found to be the root cause of cancer. Additives used in *water-mix fluids* are responsible for potential threat to human health and the growth of microorganisms on the shop floor. Biocide additives added to minimize the bacterial growth in the CFs release carcinogenic formaldehyde (Shokrani, Dhokia, Munoz, & Newman, 2013). Moreover, CFs at high temperature are found to release mist, fumes, smoke, and odors containing chemical additives such as sulfur, chlorine, phosphorus, hydrocarbons, and biocides which cause serious respiratory problems among the operators (Kalpakjian & Schmid, 2014).

Contamination of CFs due to metallic chips, fines, and tramp oil during their life cycle deteriorates their efficiency and hinders the possibility of using them repeatedly. Besides, direct discharge of CFs containing biocides adversely pollutes the water bodies, and therefore disposal has been forbidden by the municipalities. Soil and food contaminations can be major consequences. Furthermore, disposal of CFs containing chlorinate is challenging owing to formation of toxic dioxins which can sometimes even result in uncontrolled burning during the disposal procedure. Also, there is a huge amount of cost associated with waste disposal of mineral oils (Kuram, Ozcelik, & Demirbas, 2013). Disposal costs of CFs have been approximately found to be equal with the cost of CF replacement.

In many cases, harmful effects of CFs on the environment and human health remain unknown due to the lack of proper studies. However, CFs have been studied extensively in recent times to characterize their relative benefits and limitations in machining processes.

15.3 Strategies for sustainable machining of gears

Adverse effects of using large amounts of mineral-based CFs in conventional machining processes have already been discussed in previous sections. However, due to necessary requirements of CFs in different machining processes, it is quite difficult to eliminate them completely. An alternate way is to minimize their consumption or replace them with unharmful fluids. A substantial amount of heat is generated as the cutting tool removes the material from the workpiece. It has been reported that 70% of the heat generated in machining is due to the plastic deformation of the workpiece. Remaining 30% of the heat arises as a result of friction at chip–tool and tool–workpiece interfaces. This can lead to significant tool wear and inferior product quality during machining without the use of CFs which increase manufacturing costs and reduce productivity (Canter, 2009). In gear manufacturing too, it is possible to machine gears without the use of lubricants or using the least possible amount of lubricants with a view to achieve sustainability. Alternative cooling and lubrication techniques have been developed to minimize the significant use of conventional CFs which has been found to be the main nonsustainable element of conventional gear machining processes. Environment-friendly machining of gears includes dry machining, and near-dry machining of gears. Their basis and prospects have been discussed in the following sections.

15.3.1 Dry machining

Dry machining or machining without using any CF is becoming increasingly more popular keeping in view of the environmental and health problems related to their use in conventional machining. Advantages of dry machining include (1) elimination of harmful effects of CFs on the environment and human health; (2) elimination of disposal, filtration, and cleaning costs thereby reduction in overall machining costs, and (3) cleaner working environment (Gupta et al., 2017). Though dry machining eliminates the problems related to conventional CFs, it has been a challenge to completely eliminate the use of CFs in machining processes owing to the advantages they offer in terms of productivity and performance in machining processes. Although dry machining can overcome the problems with regards to unsustainable aspects of CFs, but comparatively more friction and adhesion at the cutting tool—workpiece interface results in generation of higher temperature. This results in increased tool wear and hence reduction in tool life. It also leads to the formation of ribbon-like chips which eventually affect the form and dimensional accuracy of the machined gear. Consequently, problems related to dry machining can be summarized as follows (Dixit et al., 2012a):

- Overheating of the cutting tool due to friction.
- Poor surface finish and dimensional tolerances.
- Uses lower values of cutting speeds and production rate so as to increase tool life.
- Formation of BUE.
- Absence of chip removal benefits.
- Dry machining of difficult-to-machine materials is very challenging.

Various grades of carbides and coated carbide tools have been developed for using in dry machining to address the abovementioned problems. They have been helpful but still the productivity rate achieved in dry machining has been significantly smaller than that achieved by using the CF. Though dry machining technology has improved over time, there is a need for developing better technology to further the performance of machining at higher cutting speeds and feed without compromising tool life and product quality. State-of-the-art tool coatings need to be developed that are both economical by ensuring that they do not come off over a long operating time and simultaneously improve the product quality (Canter, 2009).

15.3.2 Minimum quantity lubrication

Minimum quantity lubrication (MQL) is another alternative to conventional FL technique using conventional CFs. It has been identified as an environment-friendly and economically viable method (Sharma, Gurraj,

& Knut, 2014). Zhang, Xia, Chen, and Yin (2012) have described MQL as one of the practical ways to improve sustainability in manufacturing. The MQL system uses the smallest possible amount of fluid supplied directly to the work–tool interface. It is a near-dry machining where green lubricants in microquantities are supplied to the machining zone. Following are the benefits of using MQL system:

- The primary aim of MQL is to significantly reduce quantity of the lubricants used in conventional machining thereby minimizing the adverse effects on environment, society, and health of the operators.
- It eliminates or minimizes the problems associated to preparing, maintenance, treatment, disposal, and recycling of the used CFs used in the conventional machining.
- It significantly reduces harmful impacts of mineral-based CFs on the environment and human health by using environment-friendly and biodegradable lubricants such as vegetable oils, synthetic ester, and fatty alcohols.
- It reduces heat generation by significantly improving the frictional behavior at the workpiece–tool interface thereby improving tool life and workpiece surface integrity (Gupta et al., 2016).
- Reduction in power consumption.
- Reduction in postmachining cleaning work.
- Easy recycling and disposal of the nearly dry chips resulting in cutting down of overall manufacturing cost.
- Significant reduction in manufacturing costs related to CFs and the tooling used in conventional machining by about 10%–17% and 2%–4%, respectively (Kopac, 2009; Weinert, Inasaki, Sutherland, & Wakabayashi, 2004).
- Industrial applications of MQL reduce environmental impact, risk to health of the operators and improve chip recycle value and flexibility (Fratila, 2009).
- Improved performance of machining processes.

15.3.2.1 Principle and mechanism of minimum quantity lubrication

MQL supplies lubricants in very precise quantities ranging from 50 to 150 mL/h directly at the machining zone. It provides a sort of fluid film coating or cushion-like arrangement over tool and workpiece surfaces to cool them down instead of flooding them with the lubricant. It is also known as *mist lubrication* because it makes use of a homogeneous aerosol mixture of cutting oil and compressed air which is supplied to the cutting zone in appropriate amounts such that the entire amount of CF is used during the machining process. Supply of the lubricant in the form of small droplets suspended in the aerosol medium makes its effective use in the zones of heat generation. This results in a higher degree of surface wetting as well as lubricating the inaccessible regions on the workpiece–tool interface (Gupta et al., 2016) as shown schematically in Fig. 15.5 for MQL-assisted hobbing (MQLAH) process.

15.3.2.2 Minimum quantity lubrication systems for machining processes

MQL systems are available with provision of internal and external feed depending on the complexity of the machining process and to have better accessibility of the CF/lubricant to the machining zone (Sharma, Tiwari, & Dixit, 2016).

1. *External feed MQL system*: The lubricant is supplied with the help of external spray nozzles to the machining zone. It distributes the lubricant and air in separate channels, mixing them near the contact point where the lubricant is supplied to the machining zone as an aerosol mixture. Its basic

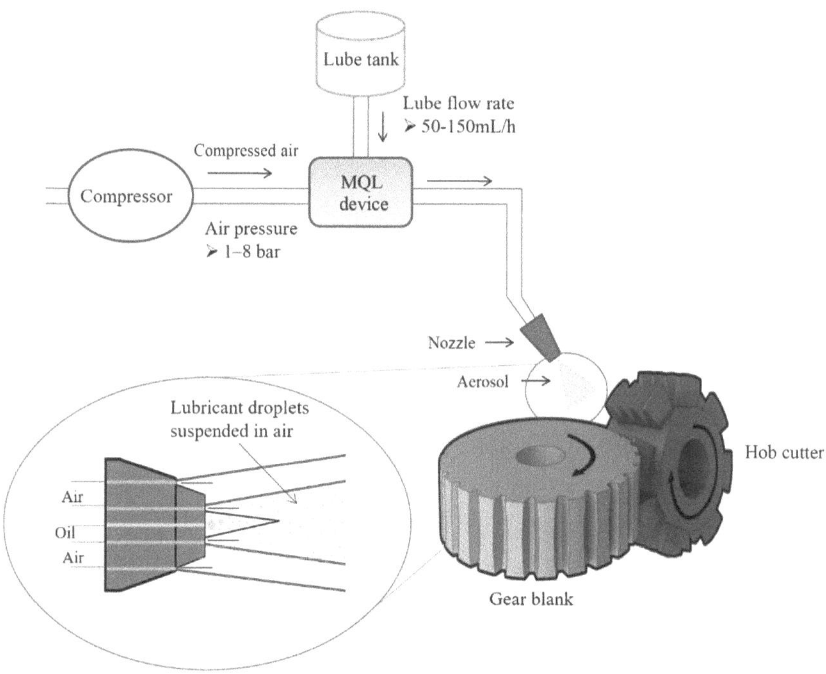

Figure 15.5 Schematic of minimum quantity lubrication system for gear machining.

components include oil reservoir, air compressor, MQL device, and nozzle as shown in Fig. 15.5. The MQL device consists of a micropump and nozzles (depending on the type of operation) along with an aerosol generator system that can produce aerosol with an oil droplet size of about 0.5–5 μm by mixing lubricant oil and compressed air (Gupta et al., 2016). This system offers flexibility to be used in variety of processes because the required spray nozzles can be easily installed on the spindle head.

2. *Internal feed MQL system*: It facilitates aerosol supply directly to the machining zone through microchannels present in the tool, thereby continuously supplying the lubricant to the machining zone throughout the entire machining process. This makes it suitable to use in higher cutting speed operations. But its initial setup cost may be higher because of the requirement of specialized cutting tools. Despite being more complex to implement, internal feed MQL systems exhibit better effectiveness in performing desired machining process as compared to external feed MQL systems. Major challenge with the internal feed MQL system has however been the proper design of tool geometry failing to which can result in dissipation of the mist leading to ineffective lubrication. Hence, there is a need for optimizing the tool design for improving the effectiveness of the technique (Sharma et al., 2015). Internal feed MQL systems can be further divided into following two types based on the method of generating the oil–air mist:

a. *Single-channel or external mixing system:* In this system the aerosol mixture is generated in a separate chamber before entering the spindle. The lubricant and air are mixed externally and transported through the tool to the machining zone. Its main advantage is its relatively simple design void of complex parts inside the tool. It is also comparatively economical and easier to integrate with wide range of machine tools. The main disadvantage is the problem of dispersion and separation of oil mist as it travels through the spindle which eventually affects performance of a machining process.

b. *Two channel or internal mixing system:* In this system the lubricant and air are transported separately through two channels located inside the tool spindle where the required aerosol mixture is generated in an internal mixing device near the tool holder. This system results in less dispersion and can deliver mists with larger droplet sizes as compared to single-channel system. However, this system is expensive and difficult to repair due to complexity of the device and presence of complex parts inside the spindle.

15.3.2.3 Environment-friendly lubricants for minimum quantity lubrication

With continuous increase in awareness about the environmental issues related to usage of CFs, various studies have been carried out to find an alternative to the synthetic CFs without affecting the machining performance. Environment-friendly and biodegradable CFs such as vegetable oils, synthetic esters, and fatty alcohols have been found to be eco-friendly, renewable, less harmful, highly biodegradable, and potential replacements to the hydrocarbon-based CFs (Debnath, Reddy, & Yi, 2014). These toxicologically harmless CFs pose less threat to the workers and offer safe working environment. A comparison of three environmental-friendly CFs/lubricants is given in Table 15.2. *Vegetable oil*—based lubricants are derived from soya bean, castor, rapeseed, and palm oils and have been found to be the most suitable biodegradable CFs. They exhibit superior lubricant properties such as higher flash point, thereby minimizing formation of fumes and fire hazards, higher boiling point, high viscosity index which offer better lubrication at higher temperatures (Gupta et al., 2016). However, they have low thermal and oxidation stability because of which they tend to gum up the machine elements (Sharma et al., 2014). *Synthetic esters* despite having low viscosity possess a high boiling point and flash point because of which much less fumes are emitted in workspace as compared to conventional CFs. They are synthesized by chemical reaction between an alcohol and a fatty acid either thermally or in the presence of a catalyst (Gupta et al., 2016). They exhibit very good biodegradability, and due to their low toxicity, they have been termed as nonhazardous to water. *Fatty alcohols* possess very good biodegradability and are toxically harmless, although they have a lower flash point and offer less lubricity as compared to synthetic esters.

Table 15.2 Environment-friendly cutting fluids along with their advantages and limitations.

Lubricants	Advantages	Limitations
Vegetable oils	High flash point High boiling point High lubricity Low toxicity	Less thermal stability Less oxidation stability
Synthetic esters	High flash point High boiling point Low toxicity	Less lubricity
Fatty alcohols	Good biodegradability Low toxicity	Low flash point Less lubricity

15.3.3 Review of minimum quantity lubrication–assisted gear manufacturing

Sustainability issues in conventional gear manufacturing processes have compelled industries worldwide to implement those means to manufacture gears which minimize the adverse effects on environment, society, operator health, economy, and ecology. Use of MQL in gear machining processes is a major attempt in this direction. Table 15.3 summarizes some relevant research works on MQL-assisted gear machining which is discussed in the following paragraphs.

Fratila (2009) conducted experiments on milling of helical gears using TiN-coated and -uncoated EMo5Co5 HSS tool under different cooling and lubrication techniques, that is, dry cutting (DC), FL, MQL, minimum quantity cooling (MQC), and MQC and lubrication (MQCL) which is a combination of MQC and MQL. A total of 400 helical gears (80 with each technique) were machined on a CNC milling machine, and the output parameters in terms of tool wear, surface quality of the manufactured helical gears, and cooling effects were compared for these different techniques. It was observed that tool wear values were minimum for MQL-assisted gear milling due to effective lubrication action in it. Tool wear was considerably higher for DC and FL as compared to that in MQC and MQCL. Surface quality and surface roughness obtained in MQL were comparable with other techniques. The cooling effect was found to be effective in the case of MQC and MQCL, whereas in MQL the workpiece temperature was higher in comparison to FL. It was concluded that overall performance of MQL equipped gear milling was found to be satisfactory in context of sustainable manufacturing. Subsequently, Fratila and Radu (2010) evaluated thermal stresses and thermo-elastic strains in milling of helical gears using TiN-coated and -uncoated EMo5Co5 HSS tool under DC, FL, MQC, MQL, and MQCL. Finite element analysis revealed that lubrication and cooling techniques significantly influenced thermal stresses and thermo-elastic strains. Their experimental results showed that combination of the cooling and the lubricating effects in near-dry machining reduced the maximum value of thermal stresses by 52% as compared to FL and increased by 55% in comparison to DC. Although DC resulted in lower values of thermal stresses, it was found to be adverse when thermo-elastic displacements were considered. Thermal displacement using MQCL was found to be 20% lower than DC and 32% higher compared to FL. Conventional FL process yielded the worst result in terms of thermal stresses in the gear

Table 15.3 Summary of past research works on minimum quantity lubrication–assisted machining of gears.

References	Machining process	Gear specifications	Tool material	Lubrication techniques	Responses
Fratila (2009)	Milling	Helical gear Material:16MnCr5 m: 2.75 mm N:37; β: 18°50′	TiN-coated EMo5Co5, uncoated EMo5Co5	DC, FL, MQL, MQC, MQCL	TW, SQ, CE
Fratila and Radu (2010)	Milling	Helical gear Material:16MnCr5 m: 2.75 mm N:37; β: 18°50′	TiN-coated and -uncoated EMo5Co5	DC, FL, MQC, MQL, MQCL	T_s, T_{st}
Fratila (2013)	Milling	Helical gear Material:16MnCr5 m: 2.75 mm N:37; β: 18°50′	–	FL, DC, MQL, MQC, MQCL	M_d, $W6$, S, F_α, F_β, f_p
Stachurski (2012)	Hobbing	Spur gear Materials:C45, 42CrMo4	HS6-5-2 HSS	FL, MQL	TW, F_c
Zhang et al. (2012)	Hobbing	Helical gear Material: ZG45, HB 217–235 m: 3.5 mm; N: 85; β: 7°24′	HSS	DC, FL, Cryogenic with MQL	TW
Matsuoka, Ryu, Nakae, Shutou, and Kodera (2013)	Hobbing	Spur gear Material: SCM415 (HB143) chromium molybdenum steel ; m: 3 mm; N: 42	Uncoated HSS, TiN- and (Al, Ti)N-coated HSS	DC, MQL	FW, CW, SR
Sato et al. (2017)	Hobbing, finish hobbing	Spur gear materials: SCM415 (HB143), SCM435 (HB306) m: 2.3 mm; N: 36	TiN- or TiAlN-coated cemented carbide	DC, MQL	FW, CW, SR

CE, Cooling effects; *CW*, crater wear; *DC*, dry cutting; F_c, cutting force; f_p, unitary circular pith deviation; *FL*, flood lubrication; *FW*, flank wear; $F\alpha$, total tooth profile deviation; F_β, total deviations of tooth flank lines (total lead deviation); *HSS*, high speed steel; *MQC*, minimum quantity cooling; *MQCL*, minimum quantity cooling and lubrication; *MQL*, minimum quantity lubrication; *m*, module (mm); M_d, teeth thickness over pins; *N*, number of teeth; *S*, tooth thickness; *SQ*, surface quality; *SR*, surface roughness; *Ts*, thermal stresses; T_{st}, thermo-elastic strains; *TW*, tool wear; *W6*, span measurements over six teeth; α, pressure angle; β, helix angle.

teeth at the end of milling. They concluded that MQCL-assisted gear milling is superior alternatives to DC- and FL-assisted gear milling.

Fratila (2013) conducted experimental research on gear milling of helical gears to study effects of DC, FL, MQC, MQL, and MQCL on deviations in microgeometry of the manufactured helical gears in terms of teeth thickness over pins (M_d), span measurements over six teeth (W_6), tooth thickness (S), total deviations in tooth profile (F_α), total deviations in tooth flank lines or total pitch deviation (F_β), and single pith deviation (f_p). Experimental values of these parameters for gears manufactured using each lubrication technique were compared with the values of gear measurement standard DIN 3961-62. It revealed that MQL-assisted gear milling resulted in stable values for span measurement over six teeth as compared to that in FL, whereas FL gave better results in terms of tooth thickness parameter. Helical gears manufactured by all the lubrication techniques showed similar trends for the values of deviation of total profile deviation F_α, total pitch deviation F_β, and single pitch deviation f_p with respect to standard values implying that the deviation values are independent of type of cooling and lubrication system used in the gear milling process. Therefore it was concluded from the observations that except tooth thickness, other gear accuracy parameters are not affected by type of cooling and lubrication system used in gear milling. Hence, it is recommended to use environment-friendly cooling and lubrication techniques to ensure sustainability in gear manufacturing processes.

Matsuoka and Ryu (2010) investigated influence of TiN- and (Ti, Al) N-coated cemented carbide materials (P10, P20, P30) hob cutter in simulated hobbing of spur gears equipped with MQL system and compared its performance with performance of DC in terms of flank wear, crater wear, and the surface roughness of the hobbed spur gears. Experiments were conducted by simulating hobbing by fly tool cutting on a milling machine. They concluded that hob material P30 showed the longest tool life irrespective of the coatings used. However, tool life of (Ti, Al)N-coated tools was found to be longer than that obtained with fully TiN-coated tools. A significant reduction in flank wear was observed in the case of MQL resulting in prolonged tool life as compared to that in the DC condition. Although a significant reduction was observed even in crater wear, thermal cracks were found to develop on the rake face of fully (Ti, Al)N and TiN-coated P30 tools. Surface roughness values were much lower in the case of MQLAH with all the cemented carbide hob materials and the coatings tested. Thus they summarized that the use of

MQL system for fully (Ti, Al)N-coated P30 tool is the most effective of all the tools and coatings used in terms of flank wear, crater wear, and surface roughness.

Stachurski (2012) conducted experiments on application of MQL in hobbing of spur gears using hob cutter made of HS6-5-2 HSS and compared its performance in terms of tool wear and cutting forces with FL-assisted hobbing (FLAH). Values of tool wear rate and cutting forces in MQLAH were found to be comparable with that obtained in FLAH. Zhang et al. (2012) carried out comparative experiments on cryogenic MQL hobbing, FLAH, and dry hobbing (DH). Their results revealed that cryogenic air with MQL resulted in lesser tool wear as compared to the other two techniques.

Matsuoka et al. (2013) investigated and compared the cutting performances of MQL and DH in terms of flank wear, crater wear, and finished surface roughness. Experiments were carried out by using coated and partially coated tools with MQL and respective performances were compared with that in DH. Their analysis concluded that for uncoated tool, TiN- and (Al, Ti)N-coated tools without coating on the rake face and fully TiN- and (Al, Ti)N-coated tools, use of MQL resulted in decrease of flank wear, crater wear, and surface roughness compared to DH. Localized supply of lubricant at tool–workpiece interface by the MQL system decreases the occurrence of deposited material on the cutting edge of the tool, thereby improving overall cutting performance. Further, a significant decrease in tool wear and surface roughness was observed even at high cutting speeds when MQL was used with the fully (Al, Ti) N-coated tool.

Sato et al. (2017) experimented on MQLAH and finish hobbing using TiN and (Ti, Al)N coatings on hard hob materials P10, P20, and P30. Their results concluded that (1) P30 tool with MQL resulted in lower flank wear as compared to that obtained with P10 and P20 in the case of the coated TiAlN film tool. Although finish hobbing with TiN-coated tool resulted in lesser flank wear values in the case of DH as compared to the values obtained with MQLAH. Flank wear was found to increase in sequence of P10, P20, and P30 hob materials and that the P10 hob material was found to be the most effective. (2) Surface roughness obtained using TiN- and TiAlN-coated tools in MQLAH was smaller than that obtained by DH. However, under the condition of finish hobbing, surface roughness obtained with TiN- and TiAlN-coated P30 was smaller in DH than that in MQLAH.

A review of past literature leads to that further investigations should be done to study the influence of MQL parameters, namely, number and direction of the nozzles, different air pressures, different lubricants, different lube flow rates, etc. on the hob wear. Also, the optimum values of lube flow rates for MQL should be estimated for longer tool life and enhanced product quality. One such research was performed by the authors on a comparative study of MQLAH and FLAH in terms of microgeometry deviations, flank surface roughness, material removal rate (MRR), energy consumption, and flank surface topography, which is described in the following section.

15.4 Comparative study of minimum quantity lubrication and flood lubrication–assisted hobbing

15.4.1 Details of experimentation

Alloy steel 20MnCr5 was chosen as spur gear material due to its excellent machinability and its wide-ranging commercial applications in gear manufacturing industries. Table 15.4 presents specifications of the spur gear and hob cutter. Fig. 15.6 shows photograph of the experimental apparatus for MQLAH process developed by mounting MQL system MT-MQL-V2.2 from *Mecatronik Hong Kong Ltd.* on a manual gear hobbing machine. The MQL system has four micronozzles that can be adjusted at any desired angle with respect to vertical axis of the gear blank

Table 15.4 Specifications of spur gear and hob cutter used in the experiments (Kharka, Jain, & Gupta, 2020a), with kind permission from Elsevier.

Specification of the spur gear	Specifications of the hob cutter
Material: 20MnCr5 alloy steel Chemical composition (by wt.%): 6.40% W; 5% Mo; 4.80% Co; 4.10% Cr; 1.90% V; 0.92% C; and balance Fe Module: 3 mm Pressure angle: 20 degrees Profile: Involute Number of teeth: 16 Pitch circle diameter: 48 mm Face width: 10 mm Tip diameter: 54 mm	Material: Emo5Co5 High speed steel Chemical composition (by wt.%): 0.8%–1.1% Cr; 1%–1.3% Mn; 0.14%–0.19% C; 0.035% P; 0.15%–0.40% S; and balance Fe Module: 3 mm Helix angle: 2.5 degrees Number of starts: 1 Number of gashes: 12 Outer diameter: 80 mm Bore diameter: 32 mm Length: 69 mm

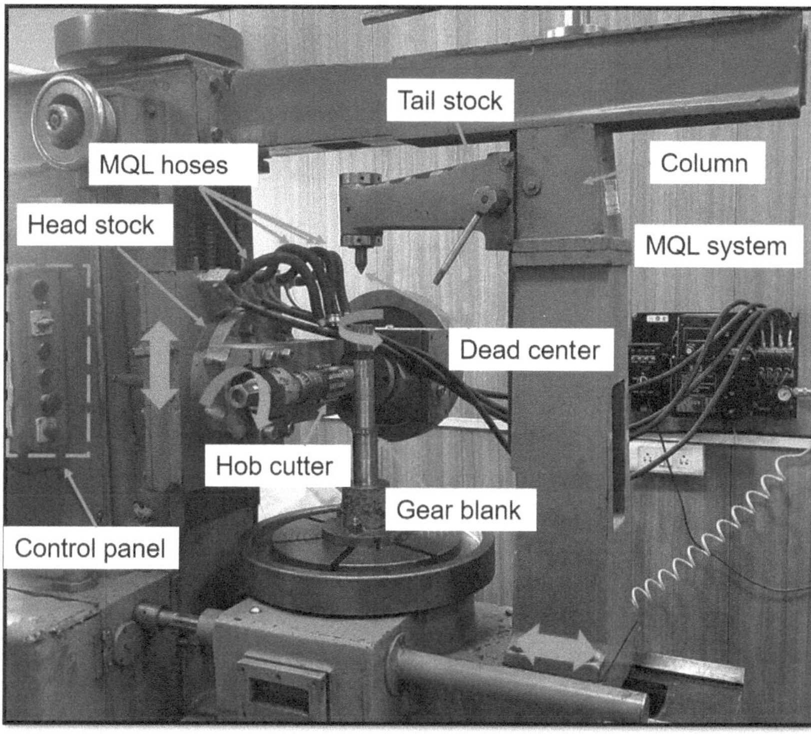

Figure 15.6 Photograph of experimental apparatus for minimum quantity lubrication–assisted hobbing (Kharka, Jain, & Gupta, 2020b). *With kind permission from Springer Nature.*

mounted on the hobbing machine. They supply a very small amount of lubricant in the form of aerosol to the machining zone using compressed air as its carrier. FLAH was carried out at lube flow rate of 2500 mL/h. Hydrocarbon-based oil "Servocut S" mixed with water in a ratio 1:30 and environment-friendly and biodegradable fatty alcohol–based oil "Hyspray A 1536" were used as the lubricants in FLAH and MQLAH processes respectively. Table 15.5 presents the properties of these lubricants.

15.4.2 Procedure of experimentation

Experimental study was conducted in two stages. The first stage of experimentation involved identification of optimum ranges of three parameters related to MQL (i.e., lubricant flow rate, air pressure, and nozzle angle) and three parameters related to gear hobbing (hob cutter speed, axial feed,

Table 15.5 Properties of lubricants used in minimum quantity lubrication–assisted hobbing and flood lubrication–assisted hobbing processes (Kharka et al., 2020b, with permission from Elsevier).

Property (Unit)	Lubricant used in minimum quantity lubrication–assisted hobbing	Lubricant used in flood lubrication–assisted hobbing
	Hyspray A 1536	**Servocut S**
Minimum flash point (°C)	194	150
Kinematic viscosity at 40°C (mm^2/s)	28	20
Kinematic viscosity at 100°C (SUS)	148	NA
Pour point (°C)	NA	0

and depth of cut) in MQLAH process using criteria of minimizing microgeometry deviations and flank surface roughness and maximizing process productivity (Kharka et al., 2020a). It was found during these experiments that the depth of cut does not significantly affect the considered responses and its value of 2.25 mm yielded their optimum values. The second stage of experimentation identified optimum values of the three parameters of MQL, namely, lubricant flow rate Q, air pressure P, and nozzle angle α as 100 mL/h; 4 bar; and 30 degrees, respectively, using minimization of microgeometry deviations and flank roughness as the objectives (Kharka et al., 2020b). These parameters along with the depth of cut were kept constant during the experimentation of the present study in which a total of 18 experiments (9 experiments for MQLAH and 9 for FLAH) were designed and conducted using full factorial design of experiments. In these experiments, hob cutter speed V and axial feed f were varied at three values each within their identified optimum ranges during the first stage of experimentation. Table 15.6 presents details of the fixed and variable parameters.

Deviations in gear microgeometry were evaluated on a computer numerically controlled gear metrology machine *SmartGear 500* from Wenzel GearTec Company of Germany. Deviations in total profile (F_a) and total lead (F_β) were measured on the left flanks (LF) and right flanks (RF) of randomly chosen four teeth of gear manufactured by MQLAH or FLAH process. Evaluation of radial runout (F_r) and deviation in cumulative pitch (F_p) involved their measurement on LF in one go and then on RF on another go for all teeth of an MQLAH or FLAH manufactured

Table 15.6 Details of fixed and variable parameters used in the experiments.

Parameter details	MQLAH	FLAH	
Lubrication technique	Minimum quantity lubrication	Flood lubrication	
Lubricant	Hyspray A 1536	Servocut S	
Lubricant flow rate Q (mL/h)	100	2500	
Air pressure P (bar)	4	NA	
Nozzle angle α (degrees)	30	NA	
Number of micronozzles	4	NA	
Hobbing parameters	Levels for both MQLAH and FLAH		
Hob cutter speed V (m/min)	15	22	29
Axial feed f (mm/rev)	0.32	0.44	0.56
Depth of cut D (mm)	2.25 mm		

FLAH, Flood lubrication–assisted hobbing; *MQLAH*, minimum quantity lubrication–assisted hobbing.

gear. Arithmetic average of all the measured values of F_a, F_β, and F_p and computed value F_r of a gear were used for the analysis purpose. The same process was repeated for all the nine spur gears manufactured by MQLAH and FLAH each. Average and maximum (R_a and R_{max}) values of flank surface roughness of four randomly selected teeth of a manufactured gear were measured using MarSurf LD 130 surface roughness measuring equipment from Mahr Metrology, Germany. The measurements were taken on RF and LF of the selected gear teeth (i.e., total eight values of R_a or R_{max} were measured for a manufactured gear) by tracing 10 μm diameter prob in a direction perpendicular to the hob cutter marks using 0.8 mm cutoff length, 4 mm evaluation length, and Gaussian filter. Mean value of a flank surface roughness parameter (R_a and R_{max}) has been considered for further study. Energy consumed in MQLAH and FLAH processes was measured using three-phase static kWh meter from Powertech Measurement System. Process productivity was measured in terms of volumetric MRR_v, calculated by dividing weight of the material removed by a product of density of the gear material and hobbing time as per the following equation:

$$(MRR)_v = \frac{W_b - W_g}{\rho_g t_g}\left(\frac{\text{mm}^3}{\text{min}}\right) \quad (15.1)$$

where W_b is weight of gear blank (g); W_g is weight of a manufactured gear (g); ρ_g is density of the gear material (g/mm^3); and t_g is total manufacturing time for a gear (minutes). Weights of a gear blank and its corresponding gear were measured on a weighing scale having resolution of 10 mg. Gear manufacturing time was recorded by a stopwatch having the least count of 0.01 s.

15.4.3 Results and discussions

Table 15.7 presents values of microgeometry deviations for MQLAH and FLAH manufactured spur gears corresponding to nine combinations of hob cutter speed and axial feed and Fig. 15.7 depicts bar charts comparing total profile deviation F_a (Fig. 15.7A), total lead deviation F_β (Fig. 15.7B), cumulative pitch deviation F_P (Fig. 15.7C), and radial runout F_r (Fig. 15.7D) for the spur gears manufactured by MQLAH and FLAH processes.

15.4.3.1 Deviations in gear microgeometry

It can be observed from Fig. 15.7 and Table 15.7 that:

- MQLAH gave considerably smaller values of gear microgeometry deviations as compared to FLAH for each combination of hob cutter speed and axial feed. This is due to finer size and better penetration ability of the lubricant particles through the air boundary layer around the rotating hob cutter thus easily lubricating the difficult-to-reach areas of the machining zone in MQLAH process which reduces heat

Table 15.7 Microgeometry deviation parameters given by flood lubrication–assisted hobbing and minimum quantity lubrication–assisted hobbing processes for all the experiments.

			Microgeometry deviation parameters (μm)							
Sl. no.	V (m/min)	f (mm/rev)	Total profile deviation F_a		Total lead deviation F_β		Total pitch deviation F_p		Runout F_r	
			FLAH	MQLAH	FLAH	MQLAH	FLAH	MQLAH	FLAH	MQLAH
1	15	0.32	62.2	51.4	12.9	10.8	111.9	107.9	120.0	114.2
2	15	0.44	66.1	54.8	15.1	11.4	118.0	115.0	130.5	122.9
3	15	0.56	74.4	63.6	19.5	14.1	133.9	121.5	142.9	136.2
4	22	0.32	58.1	47.9	11.1	9.11	104.5	95.7	123.6	110.2
5	22	0.44	61.2	52.3	17.7	13.0	120.2	109.9	134.4	127.4
6	22	0.56	75.6	63.5	21.0	15.1	136.1	125.4	146.6	138.0
7	*29*	*0.32*	*56.4*	*41.9*	*10.6*	*7.83*	*101.6*	*87.4*	*111.9*	*95.3*
8	29	0.44	69.4	45.1	18.5	13.4	124.9	100.7	138.6	112.5
9	29	0.56	82.2	61.2	24.1	16.6	148.0	128.6	162.9	140.9

FLAH, Flood lubrication–assisted hobbing; *MQLAH*, minimum quantity lubrication–assisted hobbing.

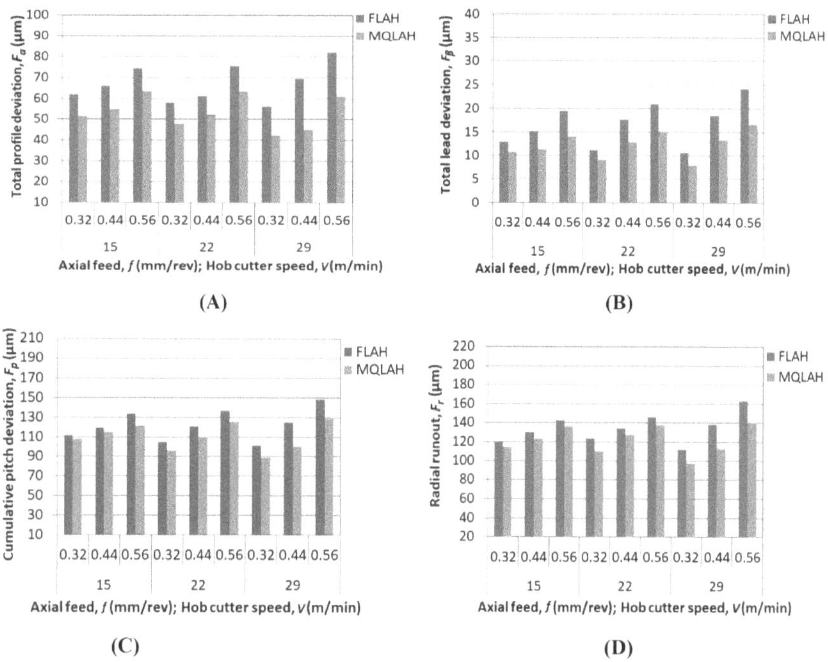

Figure 15.7 Comparison of (A) total profile deviation F_{α}, (B) total lead deviation F_{β}, (C) cumulative pitch deviation F_p, and (D) radial runout F_r of the spur gears manufactured by flood lubrication–assisted hobbing and minimum quantity lubrication–assisted hobbing for different combinations of hob cutter speed and axial feed.

generation and consequently gear microgeometry deviations. Also, the forced convective cooling action by the compressed air also helps in minimizing the heat generation,. whereas larger size of the lubricant particles in FLAH process does not allow them to lubricate the difficult-to-reach areas of the machining zone which causes temperature increase thereby yielding higher gear microgeometry values (Liao, Lin, & Chen, 2007; Tosun & Huseyinoglu 2010).

- Microgeometry deviation parameters of MQLAH and FLAH manufactured spur gears decrease with increase in hob cutter speed and increase with axial feed for a value of the hob cutter speed. This is because of reduction in cutting forces with increase in hob cutter speed which reduces geometrical deviations caused by thrust forces and vibrations (Chuangwen, Jianming, & Yuzhen, 2018). Increase in axial feed increases heat generation at the machining zone which leads to thermal expansion of the gear workpiece, thereby increasing its microgeometry deviations.

- Combination of hob cutter speed of 29 m/min and axial feed as 0.32 mm/rev (corresponding to experiment no. 7 in Table 15.7) gave minimum values of microgeometry deviation parameters for both the MQLAH manufactured spur gear (i.e., 41.86 μm for F_a; 7.83 μm for F_β, 87.42 μm for F_p; and 95.26 μm for F_r) and FLAH manufactured spur gear (i.e., 56.43 μm for F_a; 10.6 μm for F_β; 101.57 μm for F_p; and 111.89 μm for F_r). This suggests reduction by 25.8% in F_a; 26.1% in F_β; 13.9% in F_p; and 14.8% in F_r for the MQLAH manufactured gear than the FLAH manufactured gear. These two gears are referred to as the *better-quality gear* in the subsequent text.

15.4.3.2 Tooth flank surface roughness

Table 15.8 presents average and maximum values of flank surface roughness, energy consumption, and volumetric MRR for the FLAH and MQLAH manufactured spur gears corresponding to nine combinations of hob cutter speed and axial feed. Fig. 15.8 shows bar charts comparing

Table 15.8 Flank surface roughness parameters, energy consumption, and volumetric material removal rate in flood lubrication–assisted hobbing and minimum quantity lubrication–assisted hobbing for different experiments.

			Flank surface roughness						
			R_a (μm)		R_{max} (μm)		Energy consumption (kWh)		
Sl. No	*V* (m/min)	*f* (mm/rev)	FLAH	MQLAH	FLAH	MQLAH	FLAH	MQLAH	Volumetric MRR for FLAH and MQLAH (mm^3/min)
1	15	0.32	0.62	0.56	6.85	5.91	1.15	0.79	87.49
2	15	0.44	0.66	0.59	7.30	6.38	1.24	0.91	120.30
3	15	0.56	0.72	0.66	7.97	7.15	1.55	1.43	153.11
4	22	0.32	0.57	0.49	6.19	5.33	1.26	0.88	131.24
5	22	0.44	0.63	0.60	7.37	6.21	1.82	1.31	180.45
6	22	0.56	0.74	0.65	8.16	7.36	2.31	1.84	229.66
7	*29*	*0.32*	*0.55*	*0.47*	*5.48*	*4.9*	1.18	1.07	174.98
8	29	0.44	0.65	0.61	6.17	5.14	2.26	1.81	240.60
9	29	0.56	0.78	0.67	8.46	7.68	3.19	2.06	306.22

FLAH, Flood lubrication–assisted hobbing; *MQLAH*, minimum quantity lubrication–assisted hobbing; *MRR*, material removal rate.

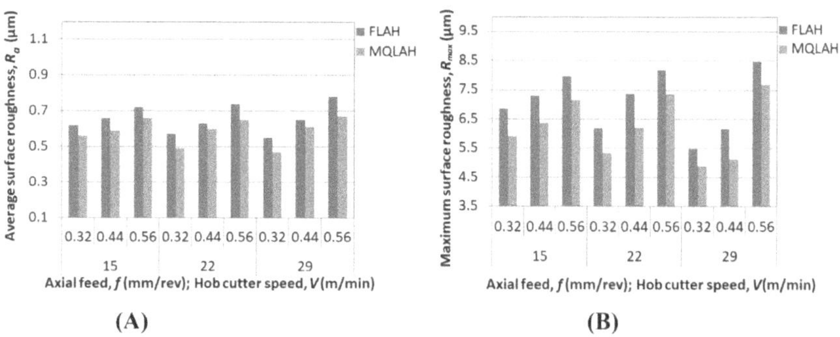

Figure 15.8 Comparison of flank roughness parameters: (A) average roughness and (B) maximum roughness of flood lubrication–assisted hobbing and minimum quantity lubrication–assisted hobbing manufactured spur gears for different combinations of hob cutter speed and axial feed.

values of average roughness (Fig. 15.8A) and maximum roughness (Fig. 15.8B) of flank surfaces of the spur gears manufactured by FLAH and MQLAH processes.

It can be observed from Fig. 15.8 and Table 15.8 that

- MQLAH yielded smaller values of avg. and max. roughness of spur gear flank surfaces than the FLAH manufactured spur gears for all the values of hob cutter speed and axial feed. It is because of the creation of lubricant film coating at the work–tool interface by MQL which improves frictional behavior thereby reducing heat generation thus eliminating formation of BUE which helps to produce smoother flank surfaces. It also facilitates easy sliding of the chips over the hob surface which also aids in smoother flank surfaces.
- Avg. and max. values of flank surface roughness decrease with increase in hob cutter speed and increase with increase in axial feed for both MQLAH and FLAH manufactured spur gears. Their minimum values occur for hob cutter speed of 29 m/min. and axial feed of 0.32 mm/rev. (corresponding to experiment no. 7 in Table 15.8 and corresponding gears are referred to as the *better-finish gears* in the subsequent text). Decrease in flank surface roughness with the increase in hob cutter speed is due to increase in chip flow action thereby minimizing BUE formation, whereas its increase with axial feed is due to the increase in cutting forces acting along the tool–work interface which increases the machine vibrations (Korkut & Donertas, 2007; Morehead, Huang, & Luo, 2007). Moreover, higher axial feed increases MRR which increases temperature at the chip–hob interface

leading to formation of BUE thereby adversely affecting the finish of gear flank surfaces (Das, Dhupal, & Kumar, 2015; Rahman & Kumar, 2001).

- Fig. 15.9 shows surface roughness profiles of the *better-finish* spur gears (corresponding to experiment no. 7 in Table 15.8) manufactured by FLAH (Fig. 15.9A) and MQLAH (Fig. 15.9B). It can be observed from this figure that MQLAH gave 0.47 and 4.9 μm for avg. and max. roughness values of spur gear flank surface, whereas FLAH yielded these values as 0.55 and 5.48 μm, respectively, that is, reductions by 14.5% in R_a and 10.6% in R_{max} values by MQLAH than FLAH.

15.4.3.3 Surface topography

Fig. 15.10 illustrates topography of the flank surfaces of the spur gear manufactured by FLAH (Fig. 15.10A) and MQLAH (Fig. 15.10B) and having *better-finish* (obtained using combination of 29 m/min as hob speed and 0.32 mm/rev as axial feed rate corresponding to experiment no. 7 in Table 15.8). Fig. 15.11 depicts the same for the *worst-finish* (obtained using 29 m/min as hob speed and 0.56 mm/rev as axial feed rate corresponding to experiment no. 9 in Table 15.8) spur gears. Fig. 15.10A reveals that flank surface of the gear manufactured by FLAH consists of distinct adhered BUE fragments and tearing which suggests significant heat accumulation at the machining zone which happened due to the ineffective cooling and lubrication of water-mix lubricants used in FLAH process,

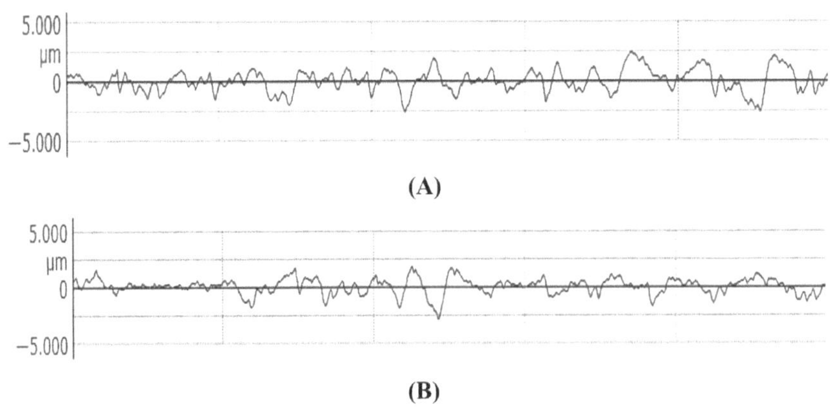

Figure 15.9 Flank surface roughness profiles for the better-finish spur gears manufactured by (A) flood lubrication–assisted hobbing and (B) minimum quantity lubrication–assisted hobbing.

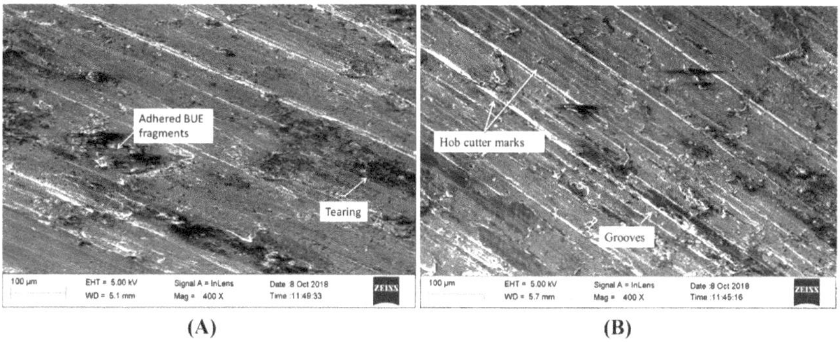

Figure 15.10 Surface topography of the flank surface of the *better-finish* having spur gear manufactured by (A) flood lubrication–assisted hobbing and (B) minimum quantity lubrication–assisted hobbing.

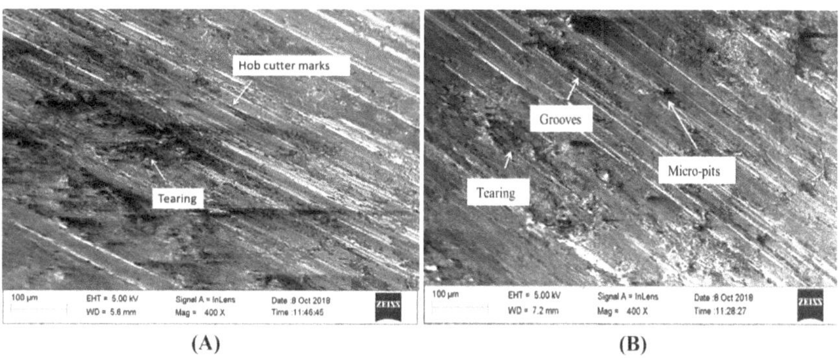

Figure 15.11 Surface topography of the flank surface of the *worst-finish* having spur gear manufactured by (A) flood lubrication–assisted hobbing and (B) minimum quantity lubrication–assisted hobbing.

whereas flank surface of the gear manufactured by MQLAH (Fig. 15.10B) shows no sign of significant adhered BUE fragments and tearing, however hob cutter marks and grooves are visible. Fig. 15.11A shows deteriorated flank surface obtained by FLAH showing formation of the grooves which is caused due to more pronounced vibrations and higher tool wear occurring at higher values of hob cutter speed and axial feed which resulted in the BUE material to get adhered to the workpiece surface as process continued. This happens due to inability of the lubricant particles to reach the tool–work interface owing to its larger particle size and higher values of hob cutter speed and feed leading to ineffective cooling and lubrication. But, MQLAH (Fig. 15.11B) gives improved flank surface due to reduced wear of hob cutter and less formation of BUE caused by effective lubrication facilitated by finer lubricant particle size and ability to reach difficult-to-reach areas of the machining zone.

15.4.3.4 Energy consumption

Fig. 15.12 shows energy consumption by MQLAH and FLAH for different combinations of hob cutter speed and axial feed. It can be observed that:

- Energy consumption increases with hob cutter speed and axial feed for both FLAH and MQLAH processes due to increase in cutting forces with hob cutter speed and axial feed.
- MQLAH consumes significantly less energy than FLAH for all the considered values of hob cutter speed and axial feed. Reduction in energy consumption by MQLAH is around 30.1% for lower speed-feed combination (i.e., 15 m/min and 0.32 mm/rev); it is almost about 35.4% for higher speed-feed combination (i.e., 29 m/min and 0.56 mm/rev). This is due to the formation of a thin film over the tool–work interface owing to the better reachability of environment-friendly fatty alcohol–based lubricant used in MQLAH which enables reduction of frictional forces resulting in lesser energy consumption.

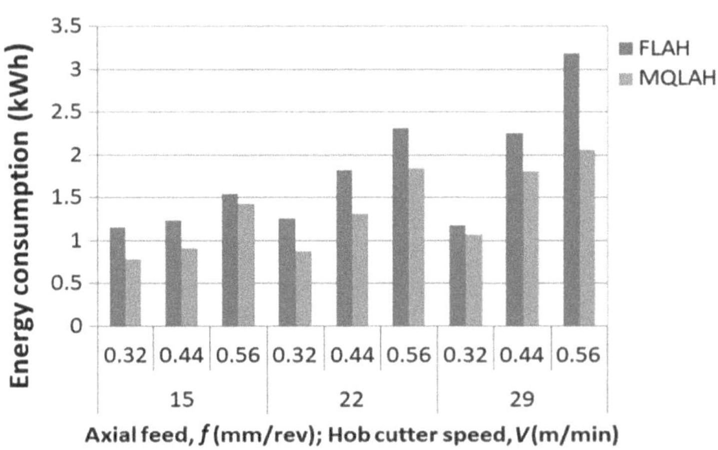

Figure 15.12 Comparison of energy consumption by minimum quantity lubrication–assisted hobbing and flood lubrication–assisted hobbing for different combinations of axial feed and hob cutter speed.

15.4.3.5 Process productivity

It can be observed from Table 15.8 that:

- Volumetric MRR increases with hob cutter speed and axial feed rate for both FLAH and MQLAH processes and attains its maximum value (i.e., 306.2 mm^3/min) at higher values of hob cutter speed and axial feed (i.e., 29 m/min and 0.56 mm/rev).
- For any combination of hob cutter speed and axial feed (i.e., accordingly value of MRR), MQLAH gives smaller values of microgeometry deviations, flank surface roughness, and energy consumption. For example, corresponding to combination of hob cutter speed of 29 m/min and axial feed of 0.56 mm/rev (and MRR of 306.2 mm^3/min), MQLAH gave reductions of 25.5% in total profile error; 31.1% in total lead error; 13.1% in total pitch error; 13.6% in radial runout; 14% in avg. roughness; 9.2% in max. roughness; and 35.4% in energy consumption than those given by FLAH. It implies that MQLAH can use higher values of hob cutter speed and axial feed but still giving smaller values of microgeometry deviations, flank surface roughness values, and energy consumption than FLAH.

15.5 Concluding remarks

This chapter presented a review of conventional gear manufacturing processes, environmental and social issues related to conventional gear machining processes, sustainable alternatives to conventional gear machining, and the role of MQL in achieving sustainability in modern gear manufacturing processes along with research on comparative investigation on environment-friendly lubricant using MQL with water-mixed synthetic lubricant using FLAH for manufacturing of spur gears. Following conclusions can be drawn based on the contents of this chapter:

- Performance of MQL in machining processes is largely dependent on interaction of machining parameters such as cutting speed and feed with number of MQL parameters such as type of lubricant used, lube/oil flow rate, and air pressure. Hence, better research on interaction of these factors is important to effectively implement MQL in gear machining processes.
- MQLAH gives smaller values of microgeometry deviations, flank roughness parameters, and energy consumption than FLAH for all the combinations of hob cutter speed and axial feed. Significant reduction in heat generation caused by finer size and better penetration ability of

the lubricant particles enabling easier lubrication of the difficult-to-reach areas of the machining zone and the forced convective cooling action by the compressed air helps in attaining smaller values of micro-geometry deviations by MQLAH. Creation of lubricant film coating at the workpiece–hob cutter interface by MQL improves frictional behavior and makes sliding of the chips over the hob surface easier thereby reducing heat generation and eliminates formation of BUE which helps to produce smoother flank surfaces by MQLAH. Formation of protective coating by MQL also reduces frictional forces, wear of the hob cutter, and damage to the hob cutting edges, which reduces energy consumption by MQLAH.

- Parameters of microgeometry deviation and flank surface roughness, and energy consumption decrease with increase in hob cutter speed and increase with increase in axial feed rate for both MQLAH and FLAH processes. Combination of hob cutter speed of 29 m/min and axial feed as 0.32 mm/rev gave minimum values of parameters of microgeometry deviation and flank surface for both the MQLAH and FLAH manufactured spur gears, whereas values of 29 m/min as hob cutter speed and 0.56 mm/rev as axial feed gave a maximum value of volumetric MRR.
- Flank surface topography of the FLAH manufactured gear showed adhered BUE fragments and tearing implying significant accumulation of heat at the machining zone caused by ineffective cooling and lubrication by water-mix lubricants used in flood lubrication technique, whereas flank surface topography of the MQLAH manufactured gear showed no adhered BUE fragments and tearing but hob cutter marks and grooves are visible.
- MQLAH can significantly improve productivity by allowing gear hobbing at high speed-feed combinations without compromising on the gear quality as compared to FLAH.
- MQLAH resulted in reduction of energy consumption by around 36% at high speed-feed combination which gave a maximum value of MRR as compared to FLAH which suggests a significant advantage in terms of economic aspect of sustainability.
- MQLAH facilitates the reduction of lubricant usage by about 96% as compared to FLAH thus significantly minimizing the harmful effects of large amount of lubricant along with improving the social aspect of its sustainability.
- Green lubricant–based MQLAH has potential to be established as a sustainable gear machining process.

References

Canter, N. (2009). The possibilities and limitations of dry machining. *Tribology and Lubrication Technology*, 40–44.

Chuangwen, X., Jianming, D., Yuzhen, C., et al. (2018). The relationships between cutting parameters, tool wear, cutting force and vibration. *Advances in Mechanical Engineering*. Available from 10.1177/1687814017750434.

Das, S. R., Dhupal, D., & Kumar, A. (2015). Experimental investigation into machinability of hardened AISI 4140 steel using tin coated ceramic tool. *Measurement, 62*, 108–126. Available from 10.1016/j.measurement.2014.11.008.

Debnath, S., Reddy, M. M., & Yi, Q. S. (2014). Environmentally friendly cutting fluids and cooling techniques in machining: A review. *Journal of Cleaner Production, 83*, 33–47.

Dixit, U. S., Sarma, D. K., & Davim, J. P. (2012a). *Dry machining. Environmentally friendly machining. Springerbriefs in applied sciences and technology*. Boston, MA: Springer.

Fratila D. (2009). Evaluation of near-dry machining effects on gear milling process efficiency, Journal of Cleaner Production, 17, 839–845, Available from https://doi.org/10.1016/j.jclepro.2008.12.010.

Fratila, D., & Radu, A. (2010). Modeling and comparing of steady thermal state at gear milling by conventional and environment-friendly cooling method. *The International Journal of Advanced Manufacturing Technology, 47*, 1003–1012. <https://doi.org/10.1007/s00170-009-2238-x>.

Fratila, D. F. (2013). Research of environment-friendly techniques influence on gear accuracy in context of sustainable manufacturing. *Proceedings of the Romanian Academy – Series A: Mathematics, Physics, Technical Sciences, Information Science, 14*, 56–63.

Gupta, K., Jain, N. K., & Laubscher, R. F. (2017). *Advanced gear manufacturing and finishing*. London: Academic Press ISBN 978-0-12-804460-5. Available from https://doi.org/10.1016/B978-0-12-804460-5.000080.

Gupta, K., Laubscher, R. F., Davim, J. P., & Jain, N. K. (2016). Recent developments in sustainable manufacturing of gears: A review. *Journal of Cleaner Production, 112*, 3320–3330. <https://doi.org/10.1016/j.jclepro.2015.09.133>.

Jawahir, I.S., Dillon, O.W. (2007). Sustainable manufacturing processes: new challenges for developing predictive models and optimization techniques. In: *Proceedings of the first international conference on sustainable manufacturing (SM1)*, (pp. 1–19).

Kalpakjian, S., & Schmid, S. R. (2014). *Manufacturing processes for engineering materials* (5th ed.). Pearson, 12, 1.

Kharka, V., Jain, N. K., & Gupta, K. (2020a). Influence of MQL and hobbing parameters on microgeometry deviations and flank roughness of spur gears manufactured by MQL assisted hobbing. *Journal of Materials Research and Technology, 9*(5), 9646–9656. Available from https://doi.org/10.1016/j.jmrt.2020.06.085.

Kharka, V., Jain, N. K., & Gupta, K. (2020b). Predictive modelling and parametric optimization of minimum quantity lubrication assisted hobbing process. *The International Journal of Advanced Manufacturing Technology, 109*(5), 1681–1694. Available from https://doi.org/10.1007/s00170-020-05757-1.

Kopac, J. (2009). Achievements of sustainable manufacturing by machining. *The Journal of Achievements in Materials and Manufacturing Engineering, 34*, 180–187.

Korkut, I., & Donertas, M. (2007). The influence of feed rate and cutting speed on the cutting forces, surface roughness and tool–chip contact length during face milling. *Materials and Design, 28*, 308–312. Available from https://doi.org/10.1016/j.matdes.2005.06.002.

Kuram, E., Ozcelik., B., Demirbas, E. (2013). Environmentally friendly machining: vegetable based cutting fluids. In: *Green manufacturing processes and systems*, Springer, Berlin, Heidelberg, Available from https://doi.org/10.1007/978-3-642-33792-5_2.

Liao, Y. S., Lin, H. M., & Chen, Y. C. (2007). Feasibility study of the minimum quantity lubrication in high-speed end milling of NAK80 hardened steel by coated carbide tool. *International Journal of Machine Tools and Manufacture*, *47*, 1667–1676. Available from https://doi.org/10.1016/j.ijmachtools.2007.01.005.

Matsuoka, H., Ryu, T., Nakae, T., Shutou, S., & Kodera, T. (2013). Fundamental research on hobbing with minimal quantity lubrication of cutting oil. *JSME International Journal Series C*, *2*, 154–167.

Morehead, M. D., Huang, Y., & Luo, J. (2007). Chip morphology characterization and modeling in machining hardened 52100 steels. *Machining Science and Technology*, *11*, 335–354. Available from https://doi.org/10.1080/10910340701567289.

Ozcelik, B., Kuram, E., Demirbas, E., & Şik, E. (2013). Effects of vegetable-based cutting fluids on the wear in drilling. *Sadhana*, *38*(4), 687–706.

Pusavec, F., Krajnik, P., & Kopac, J. (2010). Transitioning to sustainable production – Part I: Application on machining technologies. *Journal of Cleaner Production*, *18*, 174–184.

Pusavec, F., Kramar, D., Krajnik, P., & Kopac, J. (2010). Transitioning to sustainable production - Part II: Evaluation of sustainable machining technologies. *Journal of Cleaner Production*, *18*, 1211–1221.

Rahman, M., & Kumar, A. S. (2001). Evaluation of minimal of lubricant in end milling. *The International Journal of Advanced Manufacturing Technology*, *18*, 235–241. Available from https://doi.org/10.1007/s001700170063.

Sato, Y., Matsuoka, H., Ryu, T., Nakae, T., Kubo, A., Qiu, H., ... Anan, R. (2017). Fundamental research on hobbing and finish-hobbing in dry and with MQL system. *Key Engineering Materials*, *740*, 139–144. <https://doi.org/10.4028/www.scientific.net/KEM.740.139>.

Sharma, A. K., Tiwari, A. K., & Dixit, A. R. (2016). Effects of minimum quantity lubrication (MQL) in machining processes using conventional and nanofluid based cutting fluids: A comprehensive review. *Journal of Cleaner Production*, *127*, 1–18.

Sharma, V. S., Singh, G., & Sorby, K. (2015). A review on minimum quantity lubrication for machining processes. *Materials and Manufacturing Processes*, *30*, 935–953.

Shokrani, A., Dhokia, V., Munoz, E. P., & Newman, S. (2013). State-of-the-art cryogenic machining and processing. *International Journal of Computer Integrated Manufacturing*, *26*, 616–648.

Stachurski, W. (2012). Application of minimal quantity lubrication in gear hobbing. *Mechanics and Mechanical Engineering*, *16*, 133–140.

Tosun, N., & Huseyinoglu, M. (2010). Effect of MQL on surface roughness in milling of AA7075-T6. *Materials and Manufacturing Processes*, *25*, 793–798. Available from https://doi.org/10.1080/10426910903496821.

Weinert, K., Inasaki, I., Sutherland, J. W., & Wakabayashi, T. (2004). Dry machining and minimum quantity lubrication. *CIRP Annals – Manufacturing Technology*, *53*, 511–537.

Zhang, X. H., Xia, C., Chen, P., & Yin, G. F. (2012). Comparative experimental research on cryogenic gear hobbing with MQL. *Advanced Materials Research*, 2259–2264. <https://doi.org/10.4028/www.scientific.net/amr.479-481.2259>.

Further reading

Adler, D. P., Hii, W. S., Michalek, D. J., & Sutherland, J. W. (2006). Examining the role of cutting fluids in machining and efforts to address associated environmental/health concerns. *Machining Science and Technology*, *10*(1), 23–58. <https://doi.org/10.1080/10910340500534282>.

Ali, S. M., Dhar, N. R., & Dey, S. K. (2011). Effect of minimum quantity lubrication (MQL) on cutting performance in turning medium carbon steel by uncoated carbide insert at different speed-feed combinations. *Advances in Production Engineering and Management, 6*, 185–196.

Attanasio, A., Gelfi, M., Giardini, C., & Remino, C. (2006). Minimal quantity lubrication in turning: effect on tool wear. *Wear, 260*, 333–338.

Boubekri, N., & Shaikh, V. (2012). Machining using minimum quantity lubrication: A technology for sustainability. *International Journal of Applied Science and Technology, 2*, 111–115.

Boubekri, N., Shaikh, V., & Foster, P. R. (2010). A technology enabler for green machining: minimum quantity lubrication (MQL). *Journal of Manufacturing Technology Management, 21*(5), 556–566.

Chetan., Behera, B. C., Ghosh, S., & Rao, P. V. (2016). Application of nanofluids during minimum quantity lubrication: A case study in turning process. *Tribology International, 101*, 234–246.

Dhar, N. R., Kamruzzaman, M., & Ahmed, M. (2006). Effect of minimum quantity lubrication (MQL) on tool wear and surface roughness in turning AISI-4340 steel. *Journal of Materials Processing Technology, 172*, 299–304.

Diaz-Elsayed, N., Dornfeld, D., & Horvath, A. (2015). A comparative analysis of the environmental impacts of machine tool manufacturing facilities. *Journal of Cleaner Production, 95*, 223–231.

Dixit, U.S., Sarma, D.K., Davim, J.P., (2012b). Environmentally friendly machining. Applied Sciences and Technology.

Günay, M., & Yücel, E. (2013). An evaluation on machining processes for sustainable manufacturing. *Gazi University Journal of Science, 26*, 241–252.

Haapala, K.R., Zhao, F., Camelio, J., Sutherland, J.W. (2014). MSEC2011-50300.

Kovar, E.P. (n.d.). Dry gear hobbing – July/Aug 1995 gear technology.

Lawal, S. A., Choudhury, I. A., & Nukman, Y. (2012). Application of vegetable oil-based metalworking fluids in machining ferrous metals – A review. *International Journal of Machine Tools and Manufacture, 52*, 1–12.

Lodhia, P. (2007). Dry machining process vs wet machining process.

Matsuoka, H., Ono, H., Ryu, T., Nakae, T., & Shutou, S. (2012). Research on minimal quantity lubrication of cutting oil in turning (effect of quantity of oil supply). *Transactions of the Japan Society of Mechanical Engineers, Series C, 78*, 3630–3641.

Matsuoka, H., Tsuda, Y., Suda, S., & Yokota, H. (2006). Fundamental research on hobbing with minimal quantity lubrication of cutting oil. *JSME International Journal Series C Mechanical Systems, Machine Elements and Manufacturing, 49*(4), 1140–1150. Available from https://doi.org/10.1299/jsmec.49.1140.

Najiha, M. S., Rahman, M. M., & Yusoff, A. R. (2016). Environmental impacts and hazards associated with cutting fluids and recent advances in the sustainable systems: A review. *Renewable and Sustainable Energy Reviews, 60*, 1008–1031.

Neugebauer, R., Putz, M., & Hellfritzsch, U. (2007). Improved process design and quality for gear manufacturing with flat and round rolling. *CIRP Annals. Manufacturing Technology, 56*, 307–312.

Sharma, J., & Sidhu, B. S. (2014). Investigation of effects of dry and near dry machining on AISI D2 steel using vegetable oil. *Journal of Cleaner Production, 66*, 619–623. <https://doi.org/10.1016/j.jclepro.2013.11.042>.

Sharma, V. S., Dogra, M., & Suri, N. M. (2009). Cooling techniques for improved productivity in turning. *International Journal of Machine Tools and Manufacture, 49*, 435–453.

Silva, L. R., Bianchi, E. C., Catai, R. E., Fusse, R. Y., França, T. V., & Aguiar, P. R. (2005). Study on the behavior of the minimum quantity lubricant – MQL technique under different lubricating and cooling conditions when grinding ABNT 4340 steel. *Brazilian Society of Mechanical Sciences and Engineering*, *27*, 192–199.

Sreejith, P. S. (2008). Machining of 6061 aluminum alloy with MQL, dry and flooded lubricant conditions. *Materials Letters*, *62*, 276–278.

Tai, B. L., Stephenson, D. A., Furness, R. J., & Shih, A. J. (2014). Minimum quantity lubrication (MQL) in automotive powertrain machining. *Procedia CIRP*, *14*, 523–528.

Taylor, P., Barczak, L. M., & Batako, A. D. (2012). Application of minimum quantity lubrication in grinding. *Materials and Manufacturing Processes*, *27*, 37–41, 6914.

Taylor, P., Filipovic, A., & Stephenson, D. A. (2007). Minimum quantity lubrication (MQL) applications in automotive power- train machining. *Machining Science and Technology*, *0344*, 37–41.

Upadhyay, V., Jain, P. K., & Mehta, N. K. (2013). Machining with minimum quantity lubrication: a step towards green manufacturing. *International Journal of Machining and Machinability of Materials*, *13*, 349.

Yamada, N., Takeno, T., & Adachi, K. (2014). Lubricity of carbon nitride coatings in humid nitrogen gas environment. *Transactions of the JSME (Japanese)*, *80*, 1–16.

Index

Note: Page numbers followed by "*f*" and "*t*" refer to figures and tables, respectively.

J

K

L

M

T

Kolibri 360 GmbH
Ferdinand-Jühlke-Straße 7
99095 Erfurt, Deutschland
produktsicherheit@kolibri360.de